# Applied Mathematical Sciences



Volume 223

The mathematization of all sciences, the fading of traditional scientific boundaries, the impact of computer technology, the growing importance of computer modeling and the necessity of scientific planning all create the need both in education and research for books that are introductory to and abreast of these developments. The purpose of this series is to provide such books, suitable for the user of mathematics, the mathematician interested in applications, and the student scientist. In particular, this series will provide an outlet for topics of immediate interest because of the novelty of its treatment of an application or of mathematics being applied or lying close to applications. These books should be accessible to readers versed in mathematics or science and engineering, and will feature a lively tutorial style, a focus on topics of current interest, and present clear exposition of broad appeal. A compliment to the Applied Mathematical Sciences series is the Texts in Applied Mathematics series, which publishes textbooks suitable for advanced undergraduate and beginning graduate courses.

Ferdinand Verhulst

# Averaging for Nonlinear Dynamics with Applications and Numerical Bifurcations

## Parametric and autoparametric systems, Hamiltonian systems, FPU systems, coupled oscillators and chaos

Springer

Ferdinand Verhulst
Mathematisch Instituut
University of Utrecht
Utrecht, The Netherlands

ISSN 0066-5452 ISSN 2196-968X (electronic)
Applied Mathematical Sciences
ISBN 978-3-032-12744-0 ISBN 978-3-032-12745-7 (eBook)
https://doi.org/10.1007/978-3-032-12745-7

Mathematics Subject Classification: 34-E10, 34-C29, 35-B20, 70-G60, 70K65



This Springer imprint is published by the registered company Springer Nature Switzerland AG
The registered company address is: Gewerbestrasse 11, 6330 Cham, Switzerland

# Preface

The basic idea of the Toolbox of Averaging Theorems [158] that appeared in 2023 is to present a complete survey of averaging for differential equations including definitions and theorem formulations. The references that were added can be helpful to study the proofs that are available for the theorems. The Toolbox can be used for studying or for a seminar on approximation theory, and at the same time, it can serve as a reference text for scientists from the applied fields like engineering, physics, biology, etc.

The present monograph has an important aspect added. We know that the outlook of the mathematician and the scientist in an applied field is very different. The mathematician is content when the problem formulation admits solutions and any result, just existence of a solution, a general expression describing the solution, or a purely qualitative indication of behaviour may suffice. But at such a point the work of the applied scientist is only beginning.

This provides the motivation to add a Part II to the Toolbox that discusses the use of averaging for differential equations in a number of application fields. The applications are both aimed at physics and at engineering. Also, Part II has a more hybrid character as it has been very stimulating and inspiring to include in our research numerical bifurcation analysis by using tools as Auto, Mathematica, and Matcont. For the applications, more theory is needed than presented in Part I, so the text is interspersed with theoretical intermezzo's.

Chapter 9 on Hamiltonian systems is a typical example of problems where questions in theoretical physics and mechanical engineering regarding resonance play a part. Chapter 10 on Fermi-Pasta-Ulam chains answers basic questions about these chains and extends the theory. Parametric excitation in Chap. 11 and instability induced by dissipation in Chap. 13 discuss classical but not widely known results in engineering. A large amount of space is taken by Chap. 12 on coupled oscillators and chaos. For this topic, as for the chapter on diffusion and waves Chap. 14, we restrict ourselves as much more material is available.

Utrecht, The Netherlands | Ferdinand Verhulst

# Declarations

**Competing Interests** The author has no competing interests to declare that are relevant to the content of this manuscript.

# Contents

# Chapter 1
# Introduction

**Abstract** To study perturbation problems of differential equations it is essential to use our knowledge of the 'unperturbed' problem. An example using the Mathieu-equation shows erroneous results arising from crude averaging. Variational equations suitable for averaging can be obtained from amplitude-phase variables, comoving variables, amplitude-angle transformation and forced linear equations. Contraction and iteration approximations prepare the way for use of the Poincaré-Lindstedt method. Various perturbation methods are discussed showing the advantages of averaging.

The purpose of this book is to present in Part I a concise survey of averaging theorems for applied mathematicians, and especially also for physicists and engineers. Most of the theoretical background in Part I can be found in various parts of the scientific literature; we aim at a summary for practical use. Part II contains advanced problems and results of various application fields; for the applications we add more theoretical background. Also we will add results of numerical bifurcation techniques.

As the emphasis is on the use of averaging as an approximation tool we generally leave out proofs but refer to the references in the literature where proofs and validation can be found. Proofs are necessary, but including them would strongly increase the size of the book; however, if we want to establish the validity of averaging results we need proofs, also if we find it necessary to extend the theory. That is why we add some discussion on the theorems.

In addition to the theorems we give in Part I elementary and slightly more complicated examples to illustrate the theoretical results.

We suppose that the reader is acquainted with basic (bachelor) knowledge of calculus, ordinary differential equations (ODEs) and partial differential equations (PDEs). Apart from being a toolbox the book can also be useful for a mathematics seminar on averaging as the references to proofs of the theorems can be used to analyse the theoretical background.

In this chapter we discuss possible approaches to perturbation problems. The formulation of variational equations for averaging is useful, but the detailed form



F. Verhulst, *Averaging for Nonlinear Dynamics with Applications and Numerical Bifurcations*, Applied Mathematical Sciences 223,
https://doi.org/10.1007/978-3-032-12745-7_1

depends on the type of problem. The chapter ends with a comparison of various perturbation theories and points out a hybrid strategy using both numerical and analytical approaches.

## 1.1 Perturbation Problems

Suppose we are interested in the eigenvalues of a real, $n \times n$ matrix $A$. If $n = 2$, the calculation is easy; if $n = 3$, it can still be done explicitly but with a little bit more work. Considering a perturbation of the matrix $A + \varepsilon A_1$ with $\varepsilon$ a small parameter and $A_1$ again a $n \times n$ matrix, are the eigenvalues of the perturbed matrix close to the eigenvalues of $A$? Such questions are classical and, as it turns out, highly non-trivial; see appendix 15.3 of [148] for examples and references. Even simple problems pose non-trivial questions. Consider the eigenvalues of a two-dimensional problem with characteristic equation

$$\lambda^2 + b\lambda + 1 = -\varepsilon\lambda.$$

One question is: can we represent or approximate the perturbed problem in the form of a series expansion? If this is the case, the second question is what is the form of such an expansion? In this problem we can write down the solution

$$\lambda_{1,2} = \frac{1}{2}(-b - \varepsilon \pm \sqrt{b^2 - 4 + \varepsilon 2b + \varepsilon^2}.$$

If $b \neq \pm 2$, the expansion is $\lambda_{1,2} = \frac{1}{2}(-b - \varepsilon \pm \sqrt{b^2 - 4} \pm \frac{2b}{\sqrt{b^2-4}}\varepsilon) + O(\varepsilon^2)$.

If $b = \pm 2$, the expansion is $\lambda_{1,2} = -\frac{b}{2} \pm \frac{1}{2}\sqrt{2b\varepsilon} - \frac{1}{2}\varepsilon + O(\varepsilon^{3/2})$.

It is clear in this simple problem that expansion with respect to $\varepsilon$ is possible but, depending on $b$, the form of the expansion is different. Such a result could have been predicted by the implicit function theorem.

In the sequel we will consider similar questions for differential equations, for instance, with perturbations of the harmonic equation

$$\ddot{x} + x = \varepsilon f(x).$$

Can we use the known harmonic solutions of the linear equation if $\varepsilon = 0$ to obtain insight in the solutions if $\varepsilon \neq 0$? We start the problem formulation quite general. Suppose we have a perturbation problem of the form

$$\dot{x} = f_0(t, x) + \varepsilon f_1(t, x) + \varepsilon^2 \ldots \tag{1.1}$$

with $x$, $f_0$, $f_1$ $n$-vectors and suitable initial conditions. The parameter $\varepsilon$ is small and positive. A dot like in $\dot{x}$ will always describe differentiation with respect to time.

System (1.1) describes a perturbation of the ODE

$$\dot{y} = f_0(t, y). \tag{1.2}$$

Suppose now that we have obtained knowledge about the solutions $y(t)$ of the 'unperturbed' problem. Does this enable us to obtain information about the perturbed equation (1.1)? Such are the questions formulated and often answered in this book. We replaced $x$ by $y$ to indicate the quantitative and possibly qualitative difference between the perturbed (1.1) and unperturbed problem (1.2).

In later chapters we will see other perturbation formulations. The analysis of perturbation problems is subtle and full of surprises. To motivate our approach to develop variational equations in the subsequent sections we give two examples of a natural but deficient approach.

**Example 1.1** A simple example that we can solve explicitly is damped harmonic oscillation described by the scalar initial value problem

$$\ddot{x} + \varepsilon\dot{x} + x = 0, x(0) = 1, \dot{x}(0) = 0. \tag{1.3}$$

Equations with higher-order derivatives can be put in the form (1.1), for instance, in this case by putting $x = x_1, \dot{x} = x_2$ and $x = (x_1, x_2)$. The unperturbed equation is the harmonic equation with goniometric functions as solutions. The perturbed equation produces quantitative and qualitative different solutions as the oscillations are slowly damped and tend to zero.

Ignoring the known exact solution one could quite naturally try to find an approximation by proposing the expansion

$$x(t) = y_0(t) + \varepsilon y_1(t) + \varepsilon^2 \ldots$$

Substitution into equation (1.3) produces

$$\ddot{y}_0 + \varepsilon\ddot{y}_1 + \varepsilon(\dot{y}_0 + \varepsilon\dot{y}_1) + y_0 + \varepsilon y_1 + \cdots = 0.$$

Collecting terms with equal powers of $\varepsilon$ we find to first order

$$\ddot{y}_0 + y_0 = 0, \ddot{y}_1 + \dot{y}_0 + y_1 = 0.$$

Applying the initial conditions to $y_0$ we find

$$y_0(t) = \cos t, \ddot{y}_1 + y_1 = \sin t.$$

As we know, the equation for $y_1$ yields resonant solutions that increase with time and become unbounded, such terms are called *secular* terms. The terminology derives from eighteenth-century astronomy. The solutions of Eq. (1.3) tend to zero, so we

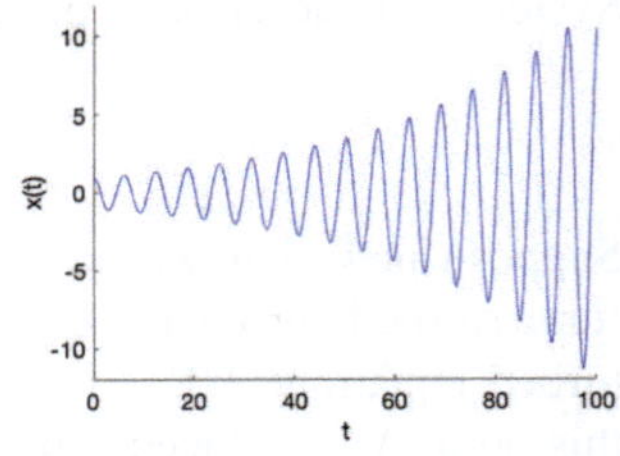


**Fig. 1.1** Numerical approximation of the initial value problem (1.4)

have to look for a different perturbation approach instead of a straightforward expansion. This will lead us in the next subsection to variational equations associated with system (1.1).

The next example deals with a classical equation of mathematical physics and engineering, the Mathieu-equation. We try another deficient approach.

**Example 1.2** If there are periodic terms in an equation, can we not immediately average without using complicated transformations? It would be attractive to apply such a procedure to ODEs with periodic terms. We consider the Mathieu-equation, a simple looking equation, but this is deceptive. Consider an oscillator with a small modulation of the frequency

$$\ddot{x} + (1 + \varepsilon \cos 2t)x = 0, x(0) = 1, \dot{x}(0) = 0. \tag{1.4}$$

We will use this equation several times in the applications. Solving the equation requires special functions and is not easy, but if $\varepsilon = 0$, the equation (1.4) is the harmonic equation. Can we use the harmonic solutions as a first approximation? An additional argument to use these solutions as an approximation for $\varepsilon$ small is that averaging over time of the coefficient $\cos 2t$ gives zero. Applying the initial conditions we find for $\varepsilon = 0$

$$x_0(t) = \cos t.$$

We expect $x_0(t)$ to be an approximation of the initial value problem (1.4). However, a numerical approximation of the initial value problem shows in Fig. 1.1 that the solution $x(t)$ is rather quickly increasing in size. The solution is in fact unbounded; we will discuss the treatment of Mathieu-equations again later on and in Part II, Chap. 11.

Our approach in Example 1.2 seemed natural, and it is surprising that this 'crude averaging' gives erroneous results. Already in the eighteenth century mathematicians and astronomers realised this. It will turn out that the Lagrange 'method of variation of constants' leads to slowly varying variational equations that are suitable for averaging.

In physics and engineering it is quite natural to consider differential equations of the form (1.1) (or other equation types) that contain terms that are periodic in time $t$ or periodic in a spatial variable. Such studies started when one considered gravitational two-body systems as the Sun and the Earth while one wanted to describe the motion of a smaller object in this system like the Moon or a comet. The idea is then to use the known solutions of the gravitational problem of two masses to study by variation of constants the problem with the perturbing small mass. Variation of constants is here in general a non-linear transformation leading to *variational equations* that contain periodic terms that can be averaged. We will discuss various formulations.

## 1.2 Amplitude-Phase Transformation

Consider a second-order equation of the form

$$\ddot{x} + \omega^2 x = \varepsilon f(t, x, \dot{x}). \tag{1.5}$$

The frequency $\omega$ is a positive constant. We can solve the equation if $\varepsilon = 0$ with solutions of the form

$$x(t) = r_0 \cos(\omega t + \phi_0).$$

The constant amplitude $r_0$ and phase $\phi_0$ are determined by the initial conditions. We propose to apply the Lagrange method of *variation of constants* by transforming $x, \dot{x} \mapsto r, \phi$:

$$x = r(t)\cos(\omega t + \phi(t)),\ \dot{x} = -r(t)\omega \sin(\omega t + \phi(t)). \tag{1.6}$$

Requiring that the derivative of $x$ equals $\dot{x}$ and substituting these expressions into Eq. (1.5) we find

$$\begin{cases} \dot{r}\cos(\omega t + \phi) - r\sin(\omega t + \phi)\dot{\phi} = 0, \\ -\dot{r}\omega \sin(\omega t + \phi) - r\omega\cos(\omega t + \phi)\dot{\phi})) = \varepsilon f(.), \end{cases} \tag{1.7}$$

where we abbreviated $f(.) = f(t, r(t)\cos(\omega t + \phi(t)), -r(t)\omega\sin(t + \phi(t))$. Solving algebraically for $\dot{r}, \dot{\phi}$ we find the so-called variational equations:

$$\begin{cases} \dot{r} &= -\frac{\varepsilon}{\omega}\sin(t + \phi) f(.), \\ \dot{\phi} &= -\frac{\varepsilon}{\omega r}\cos(t + \phi) f(.). \end{cases} \tag{1.8}$$

System (1.8) suggests that the variables amplitude and phase $r(t), \phi(t)$ slowly change with time. The solutions of system (1.8) give us with transformation (1.6) the exact solutions of Eq. (1.5). In general we cannot solve system (1.8) analytically, but we can average the system over time keeping $r, \phi$ constant during the averaging process. In Chap. 2 we will give a precise formulation of the process and refer to proofs.

The variation of constants method leaves us the freedom to choose other independent solutions of the unperturbed system instead of the amplitude-phase representation, for instance, a combination of $\cos t$ and $\sin t$. It turns out that for autonomous equations so if $f(t, x, \dot{x})$ does not explicitly depends on $t$, the amplitude-phase transformation (1.6) is quite effective.

We demonstrate the use of variational equations by some examples.

**Example 1.3** We consider an oscillator with small damping $\varepsilon\dot{x}$ and small anharmonic force $\varepsilon x^3$:

$$\ddot{x} + \varepsilon\dot{x} + x = \varepsilon x^3, \tag{1.9}$$

and initial values $x(0), \dot{x}(0)$ given. The equation is autonomous. Comparing with Eq. (1.5) we have $\omega = 1$ and $f = -\varepsilon\dot{x} + \varepsilon x^3$.

Introducing amplitude-phase transformation (1.6) we find with (1.9) the variational system:

$$\begin{cases} \dot{r} &= -\varepsilon \sin(t+\phi)\left(r\sin(t+\phi) + r^3\cos^3(t+\phi)\right), \\ \dot{\phi} &= -\frac{\varepsilon}{r}\cos(t+\phi)\left(r\sin(t+\phi) + r^3\cos^3(t+\phi)\right). \end{cases} \tag{1.10}$$

The right-hand side of the system is $2\pi$-periodic in t. Anticipating the formulation in Chap. 2 we find the averages

$$\frac{1}{2\pi}\int_0^{2\pi} \sin(t+\phi)\left(r\sin(t+\phi) + r^3\cos^3(t+\phi)\right)dt = \frac{1}{2}r,$$

$$\frac{1}{2\pi}\int_0^{2\pi} \cos(t+\phi)\frac{1}{r}\left(r\sin(t+\phi) + r^3\cos^3(t+\phi)\right)dt = \frac{3}{8}r^2.$$

The averaged variational system becomes

$$\frac{dr}{dt} = -\frac{\varepsilon}{2}r, \ \frac{d\phi}{dt} = -\varepsilon\frac{3}{8}r^2. \tag{1.11}$$

We kept the notation $r, \phi$ although the averaged system is supposed to produce an approximation. The approximations are with initial values $r(0), \phi(0) = 0$:

$$r(t) = r(0)e^{-\frac{\varepsilon}{2}t}, \phi(t) = \frac{3}{8}e^{-\varepsilon t} - \frac{3}{8}.$$

As a second example we consider a more general anharmonic equation without damping.

**Example 1.4** Consider the anharmonic oscillator described by the non-linear equation

$$\ddot{x} + x - \varepsilon x^n = 0, \tag{1.12}$$

with initial values $x(0), \dot{x}(0)$ given, and $n$ is a natural number ($n = 2, 3, 4, \ldots$). As in the preceding example we transform $x, \dot{x} \mapsto r, \phi$ by Eq. (1.6) and apply again the Lagrange method of *variation of constants*.

For $\dot{r}, \dot{\phi}$ we find from system (1.8) the variational equations

$$\begin{cases} \dot{r} &= -\varepsilon r^n \sin(t+\phi)\cos^n(t+\phi), \\ \dot{\phi} &= -\varepsilon r^{n-1}\cos^{n+1}(t+\phi). \end{cases} \tag{1.13}$$

Averaging the right-hand sides over $t$ keeping $r$ and $\phi$ constant we find

$$\frac{d\tilde{r}}{dt} = 0, \quad \frac{d\tilde{\phi}}{dt} = 0 \text{ if n is even,} \tag{1.14}$$

whereas $d\tilde{\phi}/dt \neq 0$ if $n =$ odd. For instance if $n = 3$, we find

$$\frac{d\tilde{\phi}}{dt} = -\varepsilon\,\frac{3}{8} r(0)^2.$$

The result for the approximate amplitude $\tilde{r}$ should not surprise us. As we can verify by differentiation Eq. (1.12) has the first integral

$$\frac{1}{2}\dot{x}^2 + \frac{1}{2}x^2 - \varepsilon\frac{1}{n+1}x^{n+1} = \text{constant}, \tag{1.15}$$

corresponding with closed curves, cycles, in the phase-plane. The constant is determined by the initial conditions. The cycles are $\varepsilon$-close to the circles $\dot{x}^2 + x^2 =$ constant. A question is then whether a more accurate approximation might produce non-trivial results for the amplitude; such questions are studied in Chap. 4. On the other hand, the phase changes non-trivially if $n$ is odd, corresponding with an $O(\varepsilon)$ increase of the period $2\pi$ depending on the initial amplitude $r(0)$.

## 1.3 Comoving Variables

Consider again the perturbed harmonic equation (1.5). Using amplitude-phase transformation (1.6) may give problems when considering small values of the amplitude as polar coordinates become singular near $r = 0$. We can see this in the variational system (1.8) where $r$ is found in the denominator of the phase equation.

However, if we can write after transformation to amplitude-phase coordinates

$$\varepsilon f(.) = \varepsilon r(t)\bar{f}(t, r(t)\cos(\omega t + \phi)(t), -r(t)\omega\sin(t + \phi(t))$$

with $\bar{f}$ non-singular in $r$ we have a removable singularity at $r = 0$ and we can still use amplitude-phase variables. This is, for instance, the case in Example 1.4.

If we have a singularity in the variational system for $r = 0$ we choose other independent solutions of the unperturbed system, in this case $x, \dot{x} \mapsto y_1, y_2$ with:

$$x = y_1 \cos\omega t + \frac{y_2}{\omega}\sin\omega t,\ \dot{x} = -y_1\omega\sin\omega t + y_2\cos\omega t. \tag{1.16}$$

The variational equations become with the same procedure as before

$$\begin{cases} \dot{y}_1 &= -\frac{\varepsilon}{\omega}\sin(\omega t) f(t, y_1\cos\omega t + \frac{y_2}{\omega}\sin\omega t, -y_1\omega\sin\omega t + y_2\cos\omega t), \\ \dot{y}_2 &= \varepsilon\cos(\omega t) f(t, y_1\cos\omega t + \frac{y_2}{\omega}\sin\omega t, -y_1\omega\sin\omega t + y_2\cos\omega t). \end{cases} \tag{1.17}$$

It turns out that for non-autonomous perturbations transformation (1.16) has advantages; for autonomous problems transformation (1.6) is useful as it gives directly equations for the change of amplitude and phase. We will see many examples in later chapters, but we discuss an iconic example already here.

**The Mathieu-Equation with Prime Resonance**

**Example 1.5** Consider again the so-called Mathieu-equation from Example 1.2:

$$\ddot{x} + (1 + 2\varepsilon\cos 2t)x = 0.$$

Because of the presence of the term $2\varepsilon\cos 2t\, x$ this linear equation is called *parametrically excited*. Such equations are considered in more detail in Chap. 11.

In slowly varying phase-amplitude form, using the transformation (1.6) we find the variational system

$$\begin{cases} \dot{r} &= 2\varepsilon\sin(t + \phi)\cos 2t\, r\cos(t + \phi), \\ \dot{\psi} &= 2\varepsilon\cos^2(t + \phi)\cos 2t. \end{cases} \tag{1.18}$$

Averaging over the $2\pi$-periodic right-hand side keeping $r, \phi$ constant produces

$$\dot{r} = \frac{1}{2}\varepsilon r\sin 2\phi,\ \dot{\phi} = \frac{1}{2}\varepsilon\cos 2\phi.$$

This system is not convenient to analyse; it is even non-linear. Also we would like to study the behaviour near the origin where $r = 0$.

Trying the second transformation (1.16) we obtain the variational system

$$\begin{cases} \dot{y}_1 &= 2\varepsilon \sin t \cos 2t (y_1 \cos t + y_2 \sin t), \\ \dot{y}_2 &= -2\varepsilon \cos t \cos 2t (y_1 \cos t + y_2 \sin t). \end{cases} \tag{1.19}$$

Averaging the $2\pi$-periodic system keeping $y_1, y_2$ constant produces

$$\dot{y}_1 = -\frac{1}{2}\varepsilon y_2, \ \dot{y}_2 = -\frac{1}{2}\varepsilon y_1. \tag{1.20}$$

These are linear equations with constant coefficients and are easier to analyse; the eigenvalues near (0, 0) are $\pm \frac{1}{2}\varepsilon$ so that the approximations of $y_1, y_2$ are a linear combination of $\exp(\pm\frac{1}{2}\varepsilon t)$. If for instance $x(0) = x_0, \dot{x}(0) = 0$, we have $y_1(0) = x_0, y_2(0) = 0$ and

$$y_1(t) = \frac{1}{2}x_0 e^{-\frac{1}{2}\varepsilon t} + \frac{1}{2}x_0 e^{\frac{1}{2}\varepsilon t}, \ y_2(t) = \frac{1}{2}x_0 e^{-\frac{1}{2}\varepsilon t} - \frac{1}{2}x_0 e^{\frac{1}{2}\varepsilon t},$$

and for $x(t)$

$$x(t) = \frac{1}{2}x_0 e^{-\frac{1}{2}\varepsilon t}(\cos t + \sin t) + \frac{1}{2}x_0 e^{\frac{1}{2}\varepsilon t}(\cos t - \sin t).$$

These approximate solutions of the Mathieu-equation show the instability of the trivial solution at, what is usually called, the prime resonance of the Mathieu-equation.

## 1.4 Amplitude-Angle Transformation

Consider again the perturbed harmonic equation (3.3). Instead of amplitude-phase transformation we can also transform to amplitude and angle $x, \dot{x} \mapsto r, \alpha$:

$$x = r \sin \alpha, \ \dot{x} = \omega r \cos \alpha \tag{1.21}$$

The transformation (1.21) leads by variation of constants to

$$\begin{cases} \dot{r} &= \frac{\varepsilon}{\omega} \cos \alpha f(t, r \sin \alpha, \omega r \cos \alpha), \\ \dot{\alpha} &= \omega - \frac{\varepsilon}{\omega r} \sin \alpha f(t, r \sin \alpha, \omega r \cos \alpha). \end{cases} \tag{1.22}$$

The angle variable $\alpha$ replaces $t + \phi$ of the amplitude-phase transformation, $\alpha$ will be *timelike* if the equation for $\alpha$ has no zeroes; the phase $\phi$ of the oscillations is also a type of angle but not necessarily timelike. If we have more, say $n$ coupled oscillators, we can introduce in the same way $n$ amplitudes and $n$ angles. We shall see theory and exercises with amplitude-angle formulations in Chap. 7.

For other special equations from mathematical physics, for instance, Hamiltonian systems, other variational equations may be appropriate; see Chap. 9.

## 1.5 Perturbed Linear Systems and Forcing

When studying perturbation problems it is natural to use our knowledge of the unperturbed ($\varepsilon = 0$) problem, but as we have seen there are various ways to formulate the solutions of the 'unperturbed' problems. To obtain variational (slowly varying) equations is easiest in the case that the unperturbed problem is linear, and especially if the coefficients are constant. Consider in $\mathbb{R}^n$ the quasi-linear system with continuous $n \times n$ matrix $A(t)$ and perturbation term $\varepsilon g(t,x) + \varepsilon^2 \ldots$:

$$\frac{dx}{dt} = A(t)x + \varepsilon g(t,x) + \varepsilon^2 \ldots$$

The vector function $g \in \mathbb{R}^n$ and the dots represent functions that are smooth (continuously differentiable) in $x$ and continuous in $t$ on suitable domains (to be specified later). The unperturbed problem is linear and has $n$ independent solutions that we group together in a fundamental matrix $\Phi(t)$. In many problems $A(t)$ is constant and this is the easiest case.

The variation of constants method of Lagrange suggests the transformation $x \mapsto z$

$$x = \Phi(t)z$$

producing the *variational equation*

$$\frac{dz}{dt} = \varepsilon \Phi^{-1}(t) g(t, \Phi(t)z) + \ldots$$

We expect $z(t)$ to be a slowly varying function of time. As we have seen, the choice of independent solutions may change the appearance of the variational equation, of course without changing the resulting solutions of the perturbed ODE. However, a good choice of independent solutions helps us to understand the calculations of the perturbation process.

If the unperturbed ($\varepsilon = 0$) system consists of a linear system with constant coefficients plus a forcing term we can exploit our knowledge of such problems as we often know the solutions of such forced linear systems explicitly.

Consider the system in vector form

$$\dot{x} = A(t)x + \phi(t) + \varepsilon g(t,x), \tag{1.23}$$

with $x \in \mathbb{R}^n$, $A(t)$ a $n \times n$ continuous matrix, $\phi(t)$ a continuous $n$-vector and $g(t, x)$ a sufficiently smooth $n$-vector. Suppose that $\Phi(t)$ is a fundamental matrix of the homogeneous equation $\dot{x} = A(t)x$, and the $n$-vector $\eta(t)$ is a particular solution of the inhomogeneous equation if $\varepsilon = 0$. In general we will know $\Phi(t)$ and $\eta(t)$ explicitly only if $A(t)$ is constant. A general solution $x_l(t)$ of the unperturbed system is

$$x_l(t) = \Phi(t)c + \eta(t),$$

with $c$ a constant $n$-vector. We transform $x \to y$ by

$$x = \Phi(t)y + \eta(t). \tag{1.24}$$

Substitution into system (1.23) produces

$$\Phi(t)\dot{y} = \varepsilon g(t, \Phi(t)y + \eta(t)).$$

By definition, a fundamental matrix is non-singular, and its inverse exists. It follows that we have

$$\dot{y} = \varepsilon \Phi(t)^{-1} g(t, \Phi(t)y + \eta(t)). \tag{1.25}$$

An initial condition $x(t_0)$ applied to system (1.23) will transform to

$$y(t_0) = \Phi^{-1}(t_0)(x(t_0) - \eta(t_0)).$$

**Example 1.6** In the case of the forced harmonic equation

$$\ddot{x} + \omega^2 x = \psi(t) + \varepsilon g(t, x, \dot{x}), \tag{1.26}$$

with non-trivial scalar solution $x(t) = \eta(t)$ if $\varepsilon = 0$, we transform

$$x = r\cos(\omega t + \phi) + \eta(t),\ \dot{x} = -r\omega \sin(\omega t + \phi) + \dot{\eta}(t). \tag{1.27}$$

This leads to the slowly varying system obtained by transformation (1.27) in the form

$$\begin{cases} \dot{r} &= -\frac{\varepsilon}{\omega}\sin(t+\phi) g(t, r\cos(\omega t+\phi)+\eta(t), -r\omega\sin(\omega t+\phi)+\dot{\eta}(t)), \\ \dot{\phi} &= -\frac{\varepsilon}{\omega r}\cos(t+\phi) g(t, r\cos(\omega t+\phi)+\eta(t), -r\omega\sin(\omega t+\phi)+\dot{\eta}(t)). \end{cases} \tag{1.28}$$

In Chap. 6 this technique is applied to the forced Duffing equation. The forcing complicates the dynamics considerably.

## 1.6 Unperturbed Problems That Are Not Elementary

When studying a perturbation problem it is natural that one has sufficient knowledge of the unperturbed problem. In practice one can run into difficulties if the unperturbed (or reduced) problem $\dot{x} = f_0(t, x)$ of system (1.1) is "understood" but has no elementary solutions like the trigonometric ones or if the solutions are known only in implicit form. In such cases one has to improvise. We consider two of such problem fields.

### 1.6.1 *Elliptic Functions for the Mathematical Pendulum*

To show briefly what kind of technical complications arise we discuss the perturbed mathematical pendulum which involves using elliptic functions. We sketch the derivation of variational equations. These remarks are based on [123], but the first edition (1985) on section 8.3. The equation to be considered is

$$\ddot{\phi} + \sin\phi = \varepsilon F(\ldots).$$

We omit details about the function $F(\ldots)$, and we will use the classic treatise on elliptic functions [35]. As before we have the first integral of the unperturbed system

$$\frac{1}{2}\dot{\phi}^2 - \cos\phi = E.$$

Putting

$$\phi = 2\theta,\ k = \left(\frac{2}{1+E}\right)^{1/2},$$

we have

$$\dot{\theta}^2 = \frac{1}{k^2}(1 - k^2\sin^2\theta)^{1/2}.$$

It follows that we have the relation

$$t = k\int_0^{\theta} \frac{ds}{1 - k^2\sin^2 s)^{1/2}} = ku.$$

Using the definitions of Jacobian elliptic functions $am, sn, dn$ in [35] we have

$$\sin\theta = sn(ku, k),\ \theta = am(ku, k),\ \dot{\theta} = \pm\frac{1}{k}dn(ku, k).$$

We will use again variation of constants to obtain variational equations. Here we transform $(\theta, \dot{\theta}) \to (u, k)$ by

$$\theta = am(ku, k), \; \dot{\theta} = \frac{1}{k} dn(ku, k).$$

We have some restrictions on the sign of $\dot{\theta}$. Note that for the perturbation problem we also have to transform $F(\ldots)$. A rather long calculation in [123] (first edition, section 8.3) leads to equations for $(\dot{k}, \dot{u})$ with three elliptic functions on the right-hand sides. Averaging the expressions is possible to some extent; it produces certain estimates, but it is a non-trivial affair.

A more elementary but also in the phase-plane more local analysis of the perturbed mathematical pendulum is found in Sect. 12.5.

### 1.6.2 The Perturbed Gravitational Two-Body Problem

Consider the motion of two point masses in 3-space acting upon each other by a Newtonian force field with a perturbation of the force field. In relative vector coordinates $\mathbf{r}$ we have for the unperturbed motion the equation

$$\ddot{\mathbf{r}} = -\frac{\mu}{r^3}\mathbf{r}, \tag{1.29}$$

with $r$ the Euclidean norm (distance between the masses) of $\mathbf{r}$; the constant $\mu = Gm$ depends on the gravitational constant $G$ and the sum $m = m_1 + m_2$ of the two masses. The solutions of Eq. (1.29) are 'understood', the equation of motion is called super-integrable and the solutions are for a large range of initial values planar, elliptic orbits. We owe the reduction to the motion in a plane to the super-integrability; together the scalar energy and the three-dimensional angular momentum $\mathbf{h} = \mathbf{r} \times \dot{\mathbf{r}}$ represent four independent first integrals. However, we have no elementary description of the dynamics in time.

In appendix D of [123] a number of perturbed gravitational two-body problems is considered. We mention 'motion around an oblate planet' and 'atmospheric drag'. Here we discuss briefly a third problem that plays a part in astrophysics as it involves the evolution of stars and galaxies; see also [143]. All stars are losing mass, either slowly or explosively. As stars are usually formed in gravitationally bound small groups, loss of mass is changing their orbital motion. It has been shown that if the loss of mass from the two-body system is by isotropic ejection, angular momentum is conserved, and the motion is planar. The equations of motion are

$$\ddot{r} = -\frac{\mu(t)}{r^2} + \frac{h^2}{r^3}, \; r^2\dot{\phi} = h, \tag{1.30}$$

with $h = |\mathbf{h}|$, $\mu(t)$ consisting now of the sum of the two variable masses multiplied by the gravitational constant. We introduce the timelike variable $\tau$ by $\dot{\tau} = hu^2$ and following Laplace $u = 1/r$. Note that with $h > 0$ and $r$ bounded $\tau$ is 'like' time. Putting $u'$ for $du/d\tau$ the equation for $r$ becomes

$$u'' + u = \frac{\mu(t(\tau))}{h^2}.$$

To illustrate the treatment we choose $\mu(t) = \mu_0 + (\mu_r - \mu_0)\exp(-\varepsilon t)$ with the assumption that for $t \to \infty$, a rest mass $m_r$ producing $\mu_r$ results (the analysis in [123] and [143] is much more general). As $\dot{\mu} = -\varepsilon(\mu - \mu_r)$, the system becomes in $u$, $\tau$ and $\mu$

$$u'' + u = \frac{\mu}{h^2},\ \mu' = -\varepsilon\frac{\mu - \mu_r}{hu^2}. \tag{1.31}$$

If $\varepsilon = 0$, we have $u = \mu_0/h^2 + \rho_0\cos(\tau + \theta_0)$, $du/d\tau = -\rho_0\sin(\tau + \theta_0)$. So we propose the variation of constant transformation

$$u = \frac{\mu}{h^2} + \rho\cos(\tau + \theta),\ \frac{du}{d\tau} = -\rho\sin(\tau + \theta)$$

producing the variational equations

$$\frac{d\rho}{d\tau} = \varepsilon\cos(\tau + \theta)\frac{\mu}{h^3u^2},\ \frac{d\theta}{d\tau} = -\frac{\varepsilon}{\rho}\sin(\tau + \theta)\frac{\mu}{h^3u^2},\ \frac{d\mu}{d\tau} = -\varepsilon\frac{\mu - \mu_r}{hu^2}, \tag{1.32}$$

where $u$ still has to be replaced following the transformation. The averages of the right-hand sides can be found in tables of definite integrals. Explicit expressions can be found in [143].

A more classical but equivalent approach was given in [69]. In such a case one formulates perturbation equations for the orbital elements eccentricity, semi-major axis, an angle describing the motion, etc. This is again a suitable approach for this specific type of problem.

If the mass ejection is explosive, we have to replace $\varepsilon$ by a large parameter. In such a case the method with variables $u$, $du/d\tau$, $\mu$ is to be preferred.

## 1.7 Order Functions and Timescales

To estimate accuracy and validity of approximations we have to introduce some concepts that may look boring and tedious but that are absolutely necessary to obtain precise statements. Consider, for instance, the function $f(t) = \sin t + \sqrt{\varepsilon}\cos t + \varepsilon^2 t$. The function $\sin t$ can be considered an approximation of $f(t)$ but in what sense?

A guess is that it is an approximation with accuracy $\sqrt{\varepsilon}$, but this is not true for all time. We will introduce the concepts for statements on accuracy to make sense.

We will give a short description of norms and estimates for ODEs. The extension to PDEs is to a large extent analogous; see Chap. 8. In the sequel, for ODEs, the norm of a constant $n$-vector $u$ is $\|u\| = \sum_{i=1}^{n} |u_i|$, the norm of an $n \times n$ matrix $A$ is $\|A\| = \sum_{i,j=1}^{n} |a_{ij}|$. When considering finite-dimensional vector functions depending on variables and defined on a certain domain $D$, we will use the supremum norm on $D$.

Estimating a function $f(t, x, \varepsilon)$ dependent on time, $n$-dimensional variable $x \in D$ and parameter $\varepsilon$ means that we take the supremum norm of $f(t, x, \varepsilon)$ on the chosen time interval and $D$ resulting in an expression in $\varepsilon$. We have some abuse of language here as the norm of a vector function is supposed to be a real number instead of a function of parameter $\varepsilon$.

Suppose that we want to approximate a (vector) function $x_\varepsilon(t)$ for $t \geq 0$ by another function $y_\varepsilon(t)$ in terms of the small parameter $\varepsilon$. For the concept of 'approximation' we need so-called order functions and timescales.

**Definition 1.1** A real, continuous function $\delta(\varepsilon)$ on $(0, 1]$ for which $\lim_{\varepsilon \to 0}$ exists will be called an order function.

Examples of order functions are $1 + \varepsilon$, $\varepsilon$, $\varepsilon^2$, $\varepsilon \ln \varepsilon$, $\varepsilon^{2/5}$, etc.

An important concept is the timescale, where an approximation is valid. A timescale indicates the length of a time interval measured in $\varepsilon$ or better $1/\varepsilon$.

**Definition 1.2** Consider the vector functions $x_\varepsilon(t)$ and $y_\varepsilon(t)$. If we have for $t \geq 0$ the estimate

$$x_\varepsilon(t) - y_\varepsilon(t) = O(\varepsilon^m), \quad 0 \leq \varepsilon^n t \leq C,$$

with $m, n, C$ positive constants independent of $\varepsilon$, we shall say that $y_\varepsilon(t)$ is an *$O(\varepsilon^m)$-approximation of $x_\varepsilon(t)$ on the timescale $1/\varepsilon^n$*.

Note that if necessary we can generalise the formulation somewhat by replacing $\varepsilon^m$ by the order function $\delta_1(\varepsilon)$ and $\varepsilon^n$ by the order function $\delta_2(\varepsilon)$.

**Example 1.7** Consider the function $x_\varepsilon(t) = (1 + \varepsilon + \varepsilon^2 t) \sin t$ with $O(\varepsilon)$-approximation $\sin t$ on the timescale $1/\varepsilon$. However, $\sin t$ is also an $O(\varepsilon^{\frac{1}{2}})$-approximation on the timescale $1/\varepsilon^{\frac{3}{2}}$. It is also easy to see that $(1 + \varepsilon) \sin t$ is an $O(\varepsilon^2)$-approximation on the timescale 1.

We shall often use this abbreviation of 'an approximation on a timescale'. Usually we will be happy to obtain an approximation on a timescale such as $1/\varepsilon$ or $1/\varepsilon^2$. As the estimates are valid for $\varepsilon \to 0$, these are already long timescales.

We started this section with the example $f(t) = \sin t + \sqrt{\varepsilon} \cos t + \varepsilon^2 t$ that can be handled in a similar way.

## 1.8 On Contraction and Iteration Approximations

A basic assumption for the existence and uniqueness of ODE initial value problems is to assume that the ODE ($\dot{x} = f(t, x)$) has a right-hand side that is Lipschitz-continuous in $x$ and continuous in $t$. A more precise formulation runs as follows: Consider the equation

$$\dot{x} = f(t, x),\ x(t_0) = x_0,$$

with $x, x_0 \in D \subset \mathbb{R}^n$, $t, t_0 \in \mathbb{R}$, $f(t, x)$ continuous in $[t_0, \infty)$. For $f(t, x)$ we have with positive constant $k$ and $x_1, x_2$ in a neighbourhood of $x_0$ and for $t_0 \le t \le t_0 + L$ ($L$ a positive constant) that in the Euclidean norm holds the estimate

$$\|f(t, x_1) - f(t, x_2)\| \le k\|x_1 - x_2\|.$$

One can interpret the estimate as stating that locally the change of $f(t, x)$ is bounded by linearisation. Note that if the vector field $f(t, x)$ is continuously differentiable in a neighbourhood of $x_0$, it is Lipschitz-continuous. We have now a remarkable result; see [46] chs. 1 and 2.

For the initial value problem we formulate an equivalent Volterra integral equation

$$x(t) = x_0 + \int_{t_0}^{t} f(s, x(s))ds. \tag{1.33}$$

If we have no idea of the form of the solution, we may use as a very crude first approximation $x^0(t) = x_0$. The next approximations will be

$$x^{(1)}(t) = x_0 + \int_{t_0}^{t} f(s, x_0)ds,\ x^{(2)}(t) = x_0 + \int_{t_0}^{t} f(s, x^{(1)}(s))ds,\ x^{(3)}(t) = \ldots,$$

etc. The surprise is that even with this constant start function we have that after an infinite number of iteration steps the sequence of functions $x_0, x^{(1)}(t), x^{(2)}(t), \ldots$ converges to the solution of the initial value problem. The iteration process is called *Picard-Lindelöf iteration* and in a more abstract setting *contraction.*

The process with this simple start function is important to obtain existence and uniqueness of solutions but not for actual constructions or approximations of the solution. It is instructive to apply the iteration process to the harmonic oscillator with initial conditions, for instance

$$\ddot{x} + x = 0,\ x(0) = 0, \dot{x}(0) = 1,$$

with solution $\sin t$. We change it into a first-order vector ODE by putting $x = x_1, \dot{x} = \dot{x}_1 = x_2, \dot{x}_2 = -x_1, x_1(0) = 0, x_2(0) = 1$. The integral equations become

$$x_1(t) = \int_0^t x_2(s)ds, x_2(t) = 1 - \int_0^t x_1(s)ds.$$

Using as start function the initial conditions 0, 1 we find at the second step $t, 1$, next $t, 1 - \frac{1}{2}t^2$, etc. The iteration process produces a series of polynomials in $t$ that we recognise as the Taylor expansion of $\sin t$. But in general it will not be possible to identify the nature of the iteration series. Also, truncation of the series restricts the timescale of validity of the Taylor approximation severely.

However, the iteration process works equally well and sometimes better with a good choice of starting function. We will use this idea in Chap. 3 and in the applications for periodic solutions and quasi-periodic solutions. The general approach is as follows. Consider again Eq. (1.1), but for simplicity leave out the $O(\varepsilon^2)$ terms:

$$\dot{x} = f_0(t, x) + \varepsilon f_1(t, x).$$

Assume that we can solve the unperturbed problem with initial condition $x_0$; call this solution $x^{(0)}(t)$. Use this solution as a start of the iteration process, so

$$\begin{aligned} x^{(1)}(t) &= x^{(0)}(t) + \varepsilon \int_{t_0}^t f_1(s, x^{(0)}(s))ds, \\ x^{(2)}(t) &= x^{(1)}(t) + \varepsilon \int_{t_0}^t f_1(s, x^{(1)}(s))ds, \\ \ldots & \quad \ldots \\ x^{(n)}(t) &= x^{(n-1)}(t) + \varepsilon \int_{t_0}^t f_1(s, x^{(n-1)}(s))ds. \end{aligned}$$

The process will converge for $n \to \infty$ by contraction, but there is the problem stated above. The iteration process is in general convergent for $t_0 \le t \le t_0 + L$ and so on a timescale $O(1)$ with respect to $\varepsilon$ in a neighbourhood of the initial value in $D$. This makes the iteration process useful only for bounded solutions on an $O(1)$ interval of time, for instance, for periodic solutions and quasi-periodic solutions if they exist.

## 1.9 Comparison of Methods

This section contains general information on approximation methods and is independent of the sequel. The reader will notice that in this book other analytic approximation methods get scarce attention, but as perturbation theory is an old subject, we should realise that many different methods have been formulated in the literature. See [152] for a more extensive comparison with other methods and references.

**Multiple Timing**

In particular the method of multiple timescales or multiple timing is intuitively attractive; it has become a popular method in engineering. The idea of multiple timing is as follows: Suppose we have a perturbation problem like

$$\ddot{x} + x = \varepsilon f(t, x)$$

with $x$ and $f$ scalar functions, or more general a problem like (1.1). The solutions and its approximations are expected to be functions of $t, \varepsilon t, \varepsilon^2 t$ or even $\varepsilon^n t$ with $n = 3, 4, \ldots$ Assume then that the approximation of $x(t)$ is a function of different variables $t, \tau_1 = \varepsilon t, \tau_2 = \varepsilon^2 t$, etc., so $x = x(t, \tau_1.\tau_2, \ldots)$. It is possible in this way to derive equations that produce a formal approximation of the solutions with independent variables $t, \tau_1, \tau_2, \ldots$ However, the formal approximations obtained by the procedure are *completely dependent on the a priori assumptions* of certain timelike variables.

For elementary problems that can be treated by *first-order averaging*, see Chap. 2; the averaging approximations are dependent on $t$ and $\varepsilon t$, and they are equivalent to multiple timing calculations. For second-order calculations and more advanced research problems multiple timing gives sometimes erroneous results; the method depends on luck and correct guesses. This is caused by the *a priori choice* of relevant timelike variables, whereas in other methods, like averaging, the timelike variables emerge naturally during the calculations; for more discussion see [148] and [152] where this is explicitly shown for the Mathieu-equation.

The emergence of unexpected timelike variables is often connected with bifurcations. We present a simple example

$$\dot{x} = \varepsilon^2 y,\ \dot{y} = -\varepsilon x,\ \ddot{z} + z = x, \tag{1.34}$$

with initial values $x(0) = 0, y(0) = 1, z(0) = \dot{z}(0) = 0$. Solving the system to $O(\varepsilon)$ and indicating the approximation by $\tilde{x}$, etc., we find

$$\tilde{x} = 0,\ \tilde{y} = 1, \tilde{z} = 0.$$

We can easily solve the system with the initial values exactly to find the solutions

$$x(t) = \sqrt{\varepsilon} \sin \varepsilon^{\frac{3}{2}} t,\ y(t) = \cos \varepsilon^{\frac{3}{2}} t, z(t) = \frac{\sqrt{\varepsilon}}{1 - \varepsilon^3} \sin \varepsilon^{\frac{3}{2}} t - \frac{\varepsilon^2}{1 - \varepsilon^3} \sin t.$$

So the (exact) solutions show unexpected timelike variables. See Sect. 4.2.3 for more discussion on timelike variables.

**Normalisation**

The basic approach for normalisation is to transform the differential equation to a more tractable form. The resulting equations after transformation have usually to be truncated at the higher-order terms; the truncated result is called a normal form.

Consider for instance the system

$$\dot{x} = Ax + f(x),$$

with $x$ a $n$-vector, $A$ a $n \times n$ constant matrix, $f(x) = f_2(x) + f_3(x) + \ldots + f_n(x)$, where $f_n(x)$ is a homogeneous polynomial of degree $n$. We introduce a certain resonance characteristic of the eigenvalues of matrix $A$; for details see [146] ch. 13 or [66]. If the eigenvalues of the matrix $A$ are non-resonant, the equation can be put into linear form by a near-identity transformation $x = y+h_2(y)+h_3(y)+\ldots$ where the terms $h_j,\ j = 2, 3, \ldots$ are homogeneous polynomials in $y$. Poincaré showed that if the eigenvalues are non-resonant the equation for $x$ can be transformed by a series to the linear equation $\dot{y} = Ay$. However, in general the series is not convergent.

In [66], section 3.3, an explicit calculation is given for the perturbed harmonic oscillator like the Van der Pol equation. In this case $n = 2$ and the eigenvalues of $A$ are $\lambda_{1,2} = \pm i$. There is resonance, but all quadratic terms can still be removed.

The names of Poincaré and Dulac are connected with the formulation of normalisation. Near-identity transformations are related to the analysis that takes place to produce averaging approximations. Details about this relation are given in [123]; see also Remark 2.1 in Chap. 2.

**The Poincaré-Lindstedt Method**

This method focuses on periodic solutions of ODEs. Lindstedt developed the method for the equation (1.5) and two coupled equations of this type; Lindstedt's analysis is ingenious but formal in the sense that it produces no existence results and no approximation estimates. Poincaré extended the method for general ODEs and introduced as basic principle the use of continuation based on the implicit function theorem and contraction. He concluded to existence of the periodic solutions and convergence of the expansions. See also Sect. 12.1 and for examples Chap. 3 and references [146]. There is a relation with averaging as one needs expansion of amplitudes and phases (or other variables) and one eliminates secular terms by imposing conditions. We will discuss the method again in Chap. 3.

**Renormalisation**

This is a formal method (no existence and no approximation estimates) that has some elements of the Poincaré-Lindstedt method but considers more general problems and solutions. The method avoids making assumptions on timelike variables and so obtains better results than multiple timing. For references and more details see [152].

**The Method of Harmonic Balance**

We follow the description in [130]. Looking for periodic solutions of an ODE in $x$ one can substitute a Fourier expansion of the form

$$x(t) = \sum_{n=0}^{\infty} (a_n \cos n\Omega t + b_n \sin n\Omega t).$$

The parameter $\Omega$ is still unknown. The idea is then to equate after substitution the coefficients of equal harmonics. A simple example is to apply this for the equation

$$\ddot{x} + x = cx^3, \tag{1.35}$$

with parameter $c$. Usually one chooses a finite expansion, for instance, $n = 0, 1, a_0 = 0$; we find after substitution of the Fourier expression $a_1 \cos \Omega t + b_1 \sin \Omega t$

$$-a_1\Omega^2 \cos \Omega t - b_1\Omega^2 \sin \Omega t + a_1 \cos \Omega t + b_1 \sin \Omega t = c(a_1 \cos \Omega t + b_1 \sin \Omega t)^3.$$

We have the conditions

$$-a_1\Omega^2 + a_1 = ca_1^3 \frac{3}{4}, \; -b_1\Omega^2 + b_1 = cb_1^3 \frac{3}{4},$$

where we have neglected higher-order harmonics. The parameters $\Omega$ and $a_1, b_1$ depend on each other. This makes sense as we know that Eq. (1.35) is conservative and has a family of periodic solutions with period depending on the amplitude. However, we have in general no clear insight in the existence and approximation character of our results.

**Conclusion**
Apart from explicit error estimates, an important aspect of approximation results is the question whether a formally approximate result like a periodic solution or a quasi-periodic family of solutions corresponds with the same phenomena in the original unperturbed system. Only when using averaging methods and the Poincaré-Lindstedt method we will have both qualitative and approximation statements; see for instance Chap. 3 on periodic solutions and later applications on quasi- and almost-periodic solutions.

## 1.10 Analytic and Numerical Approximations

An interesting question is why, with the advance of numerical computing, one would bother with calculating analytical approximations. The answer is that one should realise that in real-life problems of engineering, physics, biology and other sciences we have many parameters, much more than in the examples of this chapter, and also more variables leading to spatial dimensions much higher than two. With so many choices to make without advance knowledge of the solutions, numerical calculations look arbitrary. How do we know to make a choice? Numerically one obtains illustration and confirmation but not validation of global results.

Considering differential equations with several parameters and also initial or boundary conditions a sensible research strategy is combining analytic approximation theory with high-accuracy numerics. Useful numerical routines can be found, for instance, in the programs AUTO, MATCONT and MATHEMATICA; we will show more details and examples later. The analytic approximation computations point at dependence on parameters and side conditions; it also supplies more general qualitative insight. The numerics adds experimentation with inspiring phenomena and adds necessary special illustrations and confirmation.

Examples of such a hybrid approach can be found in [7], [9] and part II.

# Part I
# Averaging Theorems and Examples

# Chapter 2
# First Order Periodic Averaging

**Abstract** We consider a slowly varying ODE $\dot{x} = \varepsilon f(t, x)$ with terms periodic in time. A theorem is formulated to approximate the solutions by averaging over time with error $O(\varepsilon)$ on a long timescale of order $1/\varepsilon$. The results can be extended to a sum of quasi-periodic vector fields as long as the sum is finite. The examples discuss the Van der Pol-equation and generalisations, the forced Duffing-equation, and problems with damping.

We have seen averaging results in the Introduction. We will explain now the theory for $O(\varepsilon)$ approximations on a long timescale. Consider the variational equation with initial value:

$$\dot{x} = \varepsilon f(t, x) + \varepsilon^2 g(t, x, \varepsilon),\ x(t_0) = x_0. \tag{2.1}$$

Suppose that $x \in \mathbb{R}^n$ and $f(t, x)$ is $T$-periodic with respect to time $t$, $T$ is a positive constant independent of $\varepsilon$; we average the vector function $f$ with respect to time keeping $x$ fixed:

$$f^0(x) = \frac{1}{T}\int_{t_0}^{t_0+T} f(t, x)dt.$$

We will use the notation $Df(t, x)$ or for short $Df$ for the Jacobian. If we have the $n$-dimensional vector function $f(t, x) = (f_1, f_2, \ldots, f_n)^t$, then $Df(t, x)$ is the $n \times n$ matrix where we differentiate the elements only to the spatial variable $x = (x_1, x_2, \ldots, x_n)$:

$$Df = \begin{pmatrix} \partial f_1/\partial x_1 & \partial f_1/\partial x_2 & \ldots & \partial f_1/\partial x_n \\ \partial f_2/\partial x_1 & \partial f_2/\partial x_2 & \ldots & \partial f_2/\partial x_n \\ \ldots & \ldots & \ldots & \ldots \\ \partial f_n/\partial x_1 & \partial f_n/\partial x_2 & \ldots & \partial f_n/\partial x_n \end{pmatrix}. \tag{2.2}$$




F. Verhulst, *Averaging for Nonlinear Dynamics with Applications and Numerical Bifurcations*, Applied Mathematical Sciences 223,
https://doi.org/10.1007/978-3-032-12745-7_2

## 2.1 The Basic Averaging Theorem

We have the following basic theorem for periodic averaging:

**Theorem 2.1** *Consider the initial value problem* (2.1) *and the initial value problem*

$$\dot{y} = \varepsilon f^0(y), \; y(t_0) = x_0. \tag{2.3}$$

*Assume for* $x, y, x_0 \in D \subset \mathbb{R}^n$, $t_0 \leq t \leq \infty, 0 \leq \varepsilon \ll 1$*:*

1. $f, g, Df$ *are defined, continuous, and bounded by a constant* $M$ *independent of* $\varepsilon$ *in* $[t_0, \infty] \times D$.
2. $g$ *is Lipschitz-continuous with respect to* $x \in D$.
3. $f$ *is* $T$*-periodic in* $T$ *with* $T$ *a constant independent of* $\varepsilon$.
4. $y(t)$ *belongs to a bounded interior subset of* $D$ *on the timescale* $1/\varepsilon$.

*Then we have for the solutions of the initial value problem* (2.3) *the estimate:*

$$x(t) - y(t) = O(\varepsilon) \text{ as } \varepsilon \to 0 \text{ on the timescale } 1/\varepsilon.$$

**Remark 2.1** The proof of Theorem 2.1 is rather complicated. Although averaging has been in use for a very long time, the first proof had to wait till 1928 and was given by Fatou [55]. Later proofs were given in [20] and [117]. The complications in the proof are caused by the fact that we have to use an approximating equation which contains terms of the same order in the small parameter $\varepsilon$ as the original variational equation. See [123] ch. 2 or [146] ch. 11.3.

For better understanding second order averaging in Chap. 4, we discuss transformations that are used in modern proofs. We start with a near-identity transformation $x \mapsto z$ for Eq. (2.1):

$$x(t) = z(t) + \varepsilon u(t, z(t)) \text{ with } \mathrm{u(t, x)} = \int_{\mathrm{t_0}}^{\mathrm{t}} \left(\mathrm{f(s, x) - f^0(x)}\right) \mathrm{ds}. \tag{2.4}$$

The variable $z$ is implicitly defined by Eq. (2.4). As we subtracted $f(s, x) - f^0(x)$, the integral will be uniformly bounded (think of the $T$-periodic $f(t, x)$ as a Fourier series in $t$ where we subtract the "constant" part $f^0(x)$). Substitution of $x(t)$ into Eq. (2.1) leads after some manipulations to the equation

$$\dot{z} + \varepsilon \frac{\partial}{\partial z} u(t, z) \dot{z} = f^0(z) + O(\varepsilon^2).$$

Details of the calculations can, for instance, be found in [146]. We have given the expressions to show that $\frac{\partial}{\partial z} u(t, z)$ is expected to play a part in second order approximations (Chap. 4). The near-identity transformation (2.4) is in [123] related to normalisation theory.

**Remark 2.2** Condition 4 in Theorem 2.1 uses the notion of "interior subset". Such a set is a subset of domain $D$ as used in Theorem 2.1 that consists of the points that are at $O(1)$ distance with respect to $\varepsilon$ from the boundary of $D$. The reason for this condition is that it may happen that the initial conditions of the solution are close to a manifold that separates oscillating and escaping solutions. A small perturbation may push the solution into the domain where solutions have undesirable behaviour.

In addition we have the problem of the interval of time where the solution exists. For linear equations with continuous coefficients the domain of existence extends to all time, $t > t_0$, and for non-linear equations the interval of time remains a point of discussion. We give a simple example.

**Example 2.1** Consider the first order equation with initial value $x(0) = 1$:

$$\dot{x} = \varepsilon 2 \cos^2(t) x^2. \tag{2.5}$$

Averaging over $t$ produces the averaged equation

$$\dot{y} = \varepsilon y^2,$$

with solution

$$y(t) = \frac{1}{1 - \varepsilon t}.$$

The function $y(t)$ approximates $x(t)$ with error $O(\varepsilon)$ on intervals of time of size $q/\varepsilon$ with for $q$ the condition $0 < q < 1$. The domain $D$ remains of finite size $O(1)$ on these intervals.

**Remark 2.3** When introducing and using constants, we assume always that they are independent of $\varepsilon$. This also holds for the constant involving the boundedness of the interior subset of domain $D$. It is important to note that the finite dimension $n$ of the configuration space plays no part in the estimate, and the condition is only that the solution starting in $x_0$ does not leave the subset. *This opens the possibility of averaging very large systems as long as the solutions are bounded.*

**Remark 2.4** In the sequel we will often put $t_0 = 0$. Keeping $t_0$ as a parameter can be useful when connecting different asymptotic approximations in neighbouring domains.

## 2.2 Quasi-periodic Averaging

A simple extension of Theorem 2.1 is the case that the vector field $f(t, x)$ in system (2.1) is of the quasi-periodic form:

$$f(t, x) = \sum_{i=1}^{N} f_i(t, x), \tag{2.6}$$

with the vector fields $f_i(t, x), i = 1, \ldots, N$ periodic with periods $T_i, i = 1, \ldots, N$ independent of each other (incommensurable). We introduce the averaged vector field:

$$f^0(x) = \sum_{i=1}^{N} \frac{1}{T_i} \int_{t_0}^{t_0+T_i} f_i(t, x)dt. \tag{2.7}$$

This leads to the theorem:

**Theorem 2.2** *Consider the initial value problem* (2.1) *and the initial value problem with quasi-periodic* $f(t, x)$ *as in* (2.6)*. We have for the solutions of the initial value problems for* $x$ *and* $y$ *the estimate:*

$$x(t) - y(t) = O(\varepsilon) \text{ as } \varepsilon \to 0 \text{ on the timescale } 1/\varepsilon.$$

## 2.3 Exercises

It follows from theorem (2.1) that the results in Chap. 1 for Eqs. (1.9), (1.12) and Example 1.5 (Mathieu) represent $O(\varepsilon)$ approximations on the timescale $1/\varepsilon$. We will study now a few other problems suitable for periodic averaging, in particular some iconic examples. Other applications of first order averaging will appear in subsequent chapters. Visualisations of the results are often difficult, and we can use time series, projections or some geometric features if applicable.

### *2.3.1 A Linear Example with Damping and Forcing*

A well-known problem is to add forcing to the damped harmonic equation. This enables us to show how averaging handles linear resonance.

**Example 2.2** The equation is externally excited with $O(\varepsilon h)$ forcing in the form:

$$\ddot{x} + \mu\varepsilon\dot{x} + x = \varepsilon h \sin(\omega t), \tag{2.8}$$

with initial values $x(0), \dot{x}(0)$ given; the constants $\mu, h, \omega$ are positive. The term $\varepsilon h \sin(\omega t)$ represents a periodic external force on a vibrating mass with size 1. We write this system as

$$\ddot{x} + x = -\varepsilon\mu\dot{x} + \varepsilon h \sin(\omega t).$$

As we have seen in Chap. 1, in the case of non-autonomous perturbations a different form of variational equations than in amplitude-phase is more effective. We transform $x, \dot{x} \to y_1, y_2$ following Eq. (1.16):

$$x = y_1 \cos t + y_2 \sin t, \ \dot{x} = -y_1 \sin t + y_2 \cos t. \tag{2.9}$$

We find the variational system:

$$\begin{aligned}
\dot{y}_1 &= -\varepsilon \sin t(\mu y_1 \sin t - \mu y_2 \cos t + h \sin(\omega t)),\\
\dot{y}_2 &= \varepsilon \cos t(\mu y_1 \sin t - \mu y_2 \cos t + h \sin(\omega t)).
\end{aligned}$$

Averaging the variational system, we will use, to avoid too much different notation, again $y_1, y_2$ to indicate the approximate variables. If $\omega \neq 1$, we find by averaging over $t$:

$$\dot{y}_1 = -\frac{\varepsilon}{2}\mu y_1, \ \dot{y}_2 = -\frac{\varepsilon}{2}\mu y_2,$$

showing damping of the oscillations. The forcing is not effective within error $O(\varepsilon)$. Of course we know this from the exact solution. If $\omega \neq 1$, the result to this order of approximation is the damped solution without forcing:

$$x_h(t) = e^{-\frac{\varepsilon}{2}\mu t}(c_1 \cos t + c_2 \sin t),$$

with $c_1, c_2$ determined by the initial conditions.

The resonant case is $\omega = 1$, which changes the averaged variational system to:

$$\dot{y}_1 = -\frac{\varepsilon}{2}[\mu y_1 + h], \ \dot{y}_2 = -\frac{\varepsilon}{2}\mu y_2,$$

leading to the $O(\varepsilon)$ approximation valid on the timescale $1/\varepsilon$:

$$x(t) = x_h(t) - \frac{\varepsilon}{2} h t \cos t + O(\varepsilon).$$

$O(1)$ **Forcing**

Consider now a change of the problem where the forcing is not $O(\varepsilon)$. The equation becomes

$$\ddot{x} + \mu\varepsilon\dot{x} + x = h \sin(\omega t). \tag{2.10}$$

We know that the solutions for $\varepsilon = 0$ are unbounded if $\omega = 1$ or very large if $\omega$ is $\varepsilon$-close to 1. So assume $\omega \neq 1$. A particular solution of the unperturbed equation is with this assumption:

$$\eta(t) = \frac{h}{1-\omega^2} \sin \omega t.$$

Transforming

$$x = r\cos(t+\phi) + \frac{h}{1-\omega^2} sin\, \omega t, \; \dot{x} = -r\sin(t+\phi) + \frac{\omega h}{1-\omega^2} \cos\, \omega t,$$

we find with (1.28) the variational equations:

$$\begin{cases} \dot{r} &= \varepsilon \sin(t+\phi)\left(-r\sin(t+\phi) + \frac{\omega h}{1-\omega^2}\cos\, \omega t\right), \\ \dot{\phi} &= \varepsilon \cos(t+\phi)\left(-r\sin(t+\phi) + \frac{\omega h}{1-\omega^2}\cos\, \omega t\right). \end{cases} \tag{2.11}$$

If $\omega \neq 1$ is rational, the right-hand side of the variational system (2.11) is periodic, and if $\omega$ is irrational, the system is quasi-periodic. In both cases averaging produces

$$\dot{r} = -\frac{\varepsilon}{2} r, \dot{\phi} = 0,$$

where we kept $r, \phi$ for the approximate quantities. The approximate $r(t)$ tends to zero so we conclude

$$x(t) \to \frac{h}{1-\omega^2} sin\, \omega t, \; \omega \neq 1,$$

a result known for the exact solution.

### 2.3.2 *The Van der Pol-Equation*

As a classical example of non-linear oscillations, we present the *Van der Pol-equation*. The equation is basically analysed by first order averaging, multiple timing with $t$, and $\varepsilon t$ produces the same results in this case.

**Example 2.3** The Van der Pol-equation was developed to model electronic oscillations, and it was also used to mimic heart-beat oscillations. It is a standard example of the so-called self-excited oscillations. The equation is

$$\ddot{x} + x = \varepsilon \dot{x}(1 - x^2). \tag{2.12}$$

The term $\varepsilon \dot{x}(1 - x^2)$ is called the self-excitation term. We obtain variational equations by using the transformation (1.6) $x, \dot{x} \to r, \phi$ for a perturbed harmonic equation (1.5). We find the slowly varying variational system:

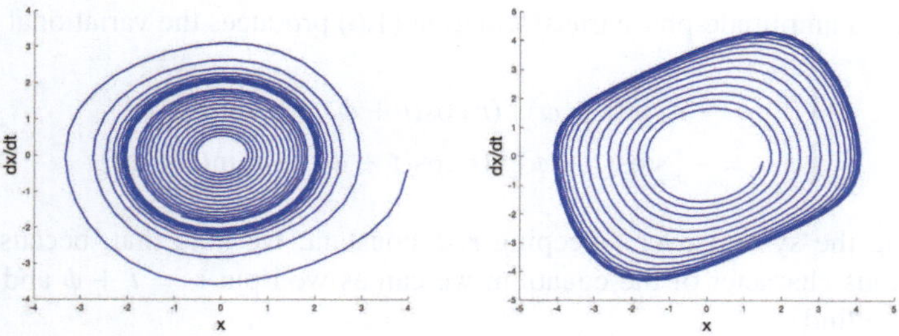


**Fig. 2.1** The $x, \dot{x}$ phase-plane of the Van der Pol-equation (left) with $\varepsilon = 0.05, x(0) = 0.1, x(0) = 4, \dot{x}(0) = 0$ (2 orbits). Right the phase-plane of the generalised Van der Pol-equation (2.17) with $\varepsilon = 0.05, a = 2, b = -0.6, x(0) = 1.5, \dot{x}(0) = 0$

$$\begin{cases} \dot{r} &= \varepsilon r \sin^2(t+\phi)(1 - r^2\cos^2(t+\phi)), \\ \dot{\phi} &= \varepsilon \sin(t+\phi)\cos(t+\phi)(1 - r^2\cos^2(t+\phi)). \end{cases} \tag{2.13}$$

The right-hand side is $2\pi$-periodic in $t$; averaging over time $t$ produces the system:

$$\dot{r} = \frac{\varepsilon}{2} r(1 - \frac{1}{4}r^2),\ \dot{\phi} = 0.$$

Solving the equations with initial values $r(0) = r_0, \phi(0) = \phi_0$ yields the approximations:

$$r(t) = r_0 \frac{e^{\frac{\varepsilon t}{2}}}{\left(1 + \frac{1}{4}r_o^2(e^{\varepsilon t} - 1)\right)^{\frac{1}{2}}} + O(\varepsilon),\ \phi(t) = \phi_0 + O(\varepsilon),$$

valid on the timescale $1/\varepsilon$. Taking the limit of $r(t)$ for $t \to \infty$, we find $r(t) \to 2 + O(\varepsilon)$.

As we can see from the averaged equation and also from the approximate $r(t)$, putting $r_0 = 2$ produces as approximation the periodic solution $2\cos(t + \phi_0)$. In Chap. 3 we consider periodic solutions by averaging. See Fig. 2.1 (left).

### 2.3.3 Averaging Autonomous Equations

Apart from the interesting presence of a periodic solution, the Van der Pol-equation is an example of an autonomous equation, and time does not occur explicitly in Eq. (2.12). A more general form of such an autonomous second order ODE is

$$\ddot{x} + x = \varepsilon f(x, \dot{x}). \tag{2.14}$$

Using again amplitude-phase transformation (1.6) produces the variational system:

$$\begin{cases} \dot{r} &= -\varepsilon \sin(t+\phi) f(r\cos(t+\phi), -r\sin(t+\phi)), \\ \dot{\phi} &= -\frac{\varepsilon}{r}\cos(t+\phi) f(r\cos(t+\phi), -r\sin(t+\phi)). \end{cases} \tag{2.15}$$

Averaging the system over $t$ keeping $r, \phi$ constant, we note that, because of the autonomous character of the equation, we can as well put $s = t + \phi$ and average over $s$. We find

$$-\frac{1}{2\pi}\int_0^{2\pi} \sin(t+\phi) f(r\cos(t+\phi), -r\sin(t+\phi))dt = F_1(r),$$

$$-\frac{1}{2\pi}\int_0^{2\pi} \cos(t+\phi) f(r\cos(t+\phi), -r\sin(t+\phi))dt = F_2(r),$$

and so for the averaging approximations:

$$\dot{r} = \varepsilon F_1(r), \ \dot{\phi} = \varepsilon F_2(r). \tag{2.16}$$

The implication is that averaging of an autonomous second order equation reduces the system to one dimension as solving the equation for $r$ leads to direct integration of the equation for $\phi$. We will meet such reductions also for higher-dimensional equations.

### *2.3.4 A Generalised Van der Pol-Equation*

In the Van der Pol-equation the self-excitation term $(1-x^2)\dot{x}$ produces alternating positive and negative damping if $x^2(t)$ becomes alternating larger and smaller than 1. For the value near $r = 2$, this results in a periodic solution. We consider a straightforward generalisation to see what happens for more general self-excitation. Consider the equation:

$$\ddot{x} + x = \varepsilon \dot{x}(1 - ax^2 - b\dot{x}^2). \tag{2.17}$$

The variational equations obtained by the transformation (1.6) become

$$\dot{r} = \varepsilon r \sin^2(t+\phi)(1 - ar^2\cos^2(t+\phi) - br^2\sin^2(t+\phi)),$$
$$\dot{\phi} = \varepsilon \sin(t+\phi)\cos(t+\phi)(1 - ar^2\cos^2(t+\phi) - br^2\sin^2(t+\phi)).$$

The right-hand side is $2\pi$-periodic in $t$; averaging over time $t$ produces the system:

$$\dot{r} = \frac{\varepsilon}{2}r(1 - \frac{a}{4}r^2 - \frac{3}{4}br^2),\ \dot{\phi} = 0.$$

If $a+3b > 0$, we have an equilibrium at $r = 2/\sqrt{a+3b}$. In Chap. 3 it will be shown that the equilibrium corresponds with a periodic solution. See Fig. 2.1 (right).

### 2.3.5 *The Forced Duffing-Equation Without Damping*

The harmonic equation has a family of $2\pi$-periodic solutions, and the equation is called isochronous. Forcing the harmonic equation with a $2\pi$-periodic function leads to unbounded solutions (linear resonance), and forcing with a different period leads to oscillation with bounded amplitude. What happens if we consider this equation with a cubic term added, the Duffing-equation?

So consider the forced Duffing-equation:

$$\ddot{x} + x + \varepsilon\gamma x^3 = \varepsilon h \cos t, \tag{2.18}$$

with $\gamma, h \neq 0$ constants. Transforming with (1.6), we find in amplitude-phase variables $r, \phi$ the variational system and from this the averaged system:

$$\begin{aligned}\dot{r} &= -\frac{1}{2}\varepsilon h \sin\phi,\\ \dot{\phi} &= -\frac{1}{2}\varepsilon\left(-\frac{3}{4}\gamma r^2 + h\frac{\cos\phi}{r}\right).\end{aligned}$$

Stationary solutions $(r_0, \phi_0)$, $r_0 > 0$ satisfy the transcendental equations

$$h \sin\phi_0 = 0,\ -\frac{3}{4}\gamma r_0^2 + h\frac{\cos\phi_0}{r_0} = 0.$$

If $\sin\phi_0 = 0$, we have $\cos\phi_0 = \pm 1$. This produces for the stationary solution the condition:

$$r_0^3 = \mp\frac{4h}{3\gamma}. \tag{2.19}$$

We are looking for a positive solution of $r_0$. The corresponding approximation for $x(t)$ is valid with error $O(\varepsilon)$ on the timescale $1/\varepsilon$. We will return later to the question whether the approximation describes a periodic solution. In general we will discuss later on the interesting question whether a perturbation result is qualitatively persistent under higher order perturbations.

### 2.3.6 Non-linear Damping

If the damping is progressive, for instance in the equation:

$$\ddot{x} + \varepsilon(\dot{x} + \dot{x}^3) + x = 0,$$

we can see that damping to zero equilibrium is stronger than in the linear case. This can be shown by averaging but also by differentiation of the so-called action $I = \frac{1}{2}(\dot{x}^2 + x^2)$. We find

$$\frac{dI}{dt} = \dot{x}\ddot{x} + x\dot{x} = -\varepsilon(\dot{x}^2 + \dot{x}^4).$$

The action decreases stronger than for the linear case.

What happens if the damping weakens?

**Example 2.4** Consider the equation:

$$\ddot{x} + \varepsilon(\dot{x} - \dot{x}^3) + x = 0. \tag{2.20}$$

Transformation (1.6) produces the variational equations:

$$\begin{cases} \dot{r} &= -\varepsilon \sin(t+\phi)(r\sin(t+\phi) - r^3\sin^3(t+\phi)), \\ \dot{\phi} &= -\frac{\varepsilon}{r}\cos(t+\phi)(r\sin(t+\phi) - r^3\sin^3(t+\phi)). \end{cases} \tag{2.21}$$

Averaging over $t$ produces

$$\dot{r} = -\frac{\varepsilon}{2}r(1 - \frac{3}{4}r^2),\ \dot{\phi} = 0.$$

We find two equilibria, $r = 0$ and $r = 2/\sqrt{3}$. This is qualitatively a new phenomenon, but by linearisation near the equilibria we find that the trivial solution is stable, and the second equilibrium is unstable. For the stability by linearisation in more general theory, one can consult [146].

# Chapter 3
# Periodic Solutions

**Abstract** Averaging an ODE of the form $\dot{x} = \varepsilon f(t, x)$, the averaged equation will, in many cases, contain equilibria (stationary solutions). Under rather general conditions, these stationary solutions will have in an $\varepsilon$-neighbourhood a periodic solution of the original equation. Examples are a generalised Van der Pol-equation and the Duffing-equation with small forcing. The Poincaré-Lindstedt method is introduced to obtain a convergent series for these periodic solutions.

A mathematical theory of approximation should, because of its expected precision, also produce *qualitative* statements. In other words: if we find a phenomenon by approximation, does it persist for the original problem with the original equations? We will discuss a few basic cases in various chapters of Part I and also in the applications of Part II. The next theorem on the existence of periodic solutions by averaging is sometimes called "Bogoliubov's second theorem". We will see in Chap. 12 that the extension to the presence of tori in dynamical systems has a relation with the Bogoliubov theory of periodic solutions.

In addition we will consider the Poincaré-Lindstedt method that we met already in the introductory Chap. 1. This useful result is, like the second Bogoliubov theorem, based on the implicit function theorem but of a different type as it leads to *convergent* approximations. As we shall see, the averaging aspects are hidden in a periodicity condition.

## 3.1 Periodic Solutions

Equilibria of averaged variational systems.

**Theorem 3.1** *Consider the variational n-dimensional system* (2.1) $\dot{x} = \varepsilon f(t, x)$ *with* $f(t, x)$ *T-periodic in t and the averaged system* $\dot{y} = \varepsilon f^0(y)$. *Suppose that the conditions of Theorem 2.1 are satisfied. Assume in addition that* $y_0$ *is a stationary solution (critical point) of the averaged equation,* $f^0(y_0) = 0$. *If:*



F. Verhulst, *Averaging for Nonlinear Dynamics with Applications and Numerical Bifurcations*, Applied Mathematical Sciences 223,
https://doi.org/10.1007/978-3-032-12745-7_3

1. *$f(t, x)$ is a smooth vector field*
2. *For the determinant of the Jacobian $J$ in $y_0$, we have*

$$\left| \frac{\partial f^0(y)}{\partial y}|_{y=y_0} \right| \neq 0$$

*then a $T$-periodic solution of the equation $\dot{x} = \varepsilon f(t, x)$ exists in an $\varepsilon$-neighbourhood of $x = y_0$.*

**Remark 3.1** A *proof* can be found in [20] and in ch. 11.8 of [146] based on the implicit function theorem. The use of this theorem means that the periodic solutions obtained in this way have a period close to $T$ and amplitudes close to the equilibrium value. Note that apart from the periodic solutions obtained from the equilibria, the variational system may obtain other periodic solutions.

We can even establish the stability of the periodic solution, as it matches exactly the stability of the stationary solution of the averaged equation if the real parts of the eigenvalues are non-zero. This reduces the stability problem of the periodic solution to determining the eigenvalues of a matrix.

Later we will meet the notion of hyperbolic and hyperbolicity. A $n \times n$ matrix is called structurally stable if the $n$ eigenvalues have all of them a non-zero real part. We call this also *hyperbolicity* of the critical point. However, see Remark 3.3 for autonomous systems.

**Remark 3.2** Following the error estimate of theorem (2.1), the error of the approximation is $O(\varepsilon)$ on the timescale $1/\varepsilon$. However, as we have existence of the periodic solution, we have for the amplitude validity of the error estimate for all time.

**Remark 3.3** We have seen before that when starting with an *autonomous* equation or system and averaging the corresponding variational equations, one eliminates one angle or, more generally, one reduces the system with one dimension. The implication is that a stationary solution obtained from an autonomous equation always has one eigenvalue zero and so the determinant of $J$, $det(J) = 0$. This is, for instance, the case for the Van der Pol-equation. In accordance with this observation, we have an important modification of the existence requirement for periodic solutions in the case of $n$-dimensional autonomous equations: The Jacobian in the equlibrium $y_0$ in Theorem 3.1 must have rank $n - 1$.

The eigenvalue zero is a general feature of solutions of autonomous equations, see [146].

## 3.2 Exercises

### 3.2.1 A First Order Equation

Consider the variational equation:

$$\dot{x} = \varepsilon(x - 2x^2 \sin^2 t). \tag{3.1}$$

Averaging over $t$ produces

$$\dot{y} = \varepsilon(y - y^2).$$

The two equilibria of the averaged equation are $y_0 = 0, 1$. We have from Theorem 3.1:

$$\frac{\partial f^0(y)}{\partial y}|_{y=y_0} = 1 - 2y_0.$$

For both equilibria the Jacobian does not vanish, and we conclude that both equilibria have an $\varepsilon$-close $2\pi$-periodic solution. It is easy to see that for $y_0 = 0$ the periodic solution is unstable, near $y_0 = 1$ stable.

In this case we can compute the stable periodic solution $x_s(t)$ by variation of constants exactly:

$$x_s(t) = \frac{4 + \varepsilon^2}{4 + \varepsilon^2 - 2\varepsilon \sin 2t - \varepsilon^2 \cos 2t}. \tag{3.2}$$

### 3.2.2 A Generalised Van der Pol-Equation

Consider an example of the equations formulated in Chap. 2:

$$\ddot{x} + x = \varepsilon \dot{x}(1 - 3x^2 - 5\dot{x}^2). \tag{3.3}$$

Introducing amplitude-phase coordinates by transformation (1.6) and averaging produces

$$\dot{r} = \frac{\varepsilon}{2} r(1 - \frac{9}{2} r^2), \; \dot{\phi} = 0. \tag{3.4}$$

We can conclude that $r = \sqrt{2}/3$ is a stable critical point of the averaged equation for $r$. Applying Theorem 3.1 and Remark 3.3, we conclude that equation (3.3) contains an asymptotically stable periodic solution with approximation

$$x(t) = \frac{\sqrt{2}}{3}\cos(t + \phi(0)) + O(\varepsilon).$$

### 3.2.3 The Duffing-Equation with Small Forcing

Consider again the Duffing-equation with small forcing and damping, $\omega$ near 1:

$$\ddot{x} + \varepsilon\mu\dot{x} + \varepsilon\gamma x^3 + x = \varepsilon h \cos \omega t,$$

parameters $\mu \geq 0, h \neq 0, \omega^{-2} = 1 - \beta\varepsilon$. This is a typical mechanical example representing non-linear oscillations with damping and forcing close to resonance. Transforming $\tau = \omega t$, we have in amplitude-phase variables the averaged system

$$r' = -\frac{1}{2}\varepsilon(\mu r + h \sin \psi),$$
$$\psi' = -\frac{1}{2}\varepsilon\left(\beta - \frac{3}{4}\gamma r^2 + h\frac{\cos\psi}{r}\right).$$

Stationary solutions (equilibria) $r_0 > 0$, $\psi_0$ satisfy the transcendental equations

$$\mu r_0 + h \sin \psi_0 = 0,\ \beta - \frac{3}{4}\gamma r_0^2 + h\frac{\cos\psi_0}{r_0} = 0.$$

For the derivative of the right-hand side of the averaged system, leaving out the scale factor $-\frac{1}{2}\varepsilon$, we find the Jacobian at $r_0$, $\psi_0$ ($r_0 > 0$):

$$J = \begin{pmatrix} \mu & h\cos\psi_0 \\ -\frac{3}{2}\gamma r_0 - h\frac{\cos\psi_0}{r_0^2} & -h\frac{\sin\psi_0}{r_0} \end{pmatrix}.$$

Leaving out the scale factor $\frac{1}{4}\varepsilon^2$, the determinant of the Jacobian $J$ becomes at equilibrium:

$$det(J) = \left(-\mu\sin\psi_0 + \frac{3}{2}\gamma r_0^2\cos\psi_0 + h\frac{\cos^2\psi_0}{r_0}\right)\frac{h}{r_0}.$$

For periodic solutions to exist, we have from Theorem 3.1 the requirement $|J| \neq 0$ if the stationary solutions are substituted.

To be more explicit, we will look at the case of exact resonance $\beta = 0$. For the stationary solutions (equilibria), we have

$$h\sin\psi_0 = -\mu r_0,\ h\cos\psi_0 = \frac{3}{4}\gamma r_0^3.$$

Using that $\cos^2 \psi_0 + \sin^2 \psi_0 = 1$, we find

$$\frac{9}{16}\gamma^2 r_0^6 + \mu^2 r_0^2 - h^2 = 0,$$

which always admits at least one positive solution for $r_0$. For the Jacobian, we have, eliminating $\psi_0$,

$$det(J) = r_0 \left(\mu + \frac{27}{16}\gamma^2 r_0^4\right).$$

If $\mu$ and $\gamma$ do not vanish simultaneously, we have $det(J) > 0$, so a stationary solution $r_0 > 0, \psi_0$ corresponds with a periodic solution of the original forced Duffing-equation. The stability of the periodic solution follows from the eigenvalues of the equilibrium and depends on the values of the parameters.

## 3.3 The Poincaré-Lindstedt Method

We discussed the Poincaré-Lindstedt method briefly in Section 1.9. This method was developed for periodic solutions and uses a small parameter and a periodicity condition that makes it related to averaging. At the same time this focus on periodic solutions excludes results on more general orbits and stability. A strong point is that the Poincaré-Lindstedt method leads to a convergent series of approximations whereas averaging results are asymptotic.

A basic ingredient of the method is the Poincaré-expansion theorem:

**Theorem 3.2** *Consider the initial value problem*

$$\dot{x} = F(t, x, \varepsilon), x(t_0) = \mu, \tag{3.5}$$

*with* $|t - t_0| < h$, $x, \mu \in D \subset \mathbb{R}^n, 0 \leq \varepsilon \leq \varepsilon_0$. *If* $F(t, x, \varepsilon)$ *is continuous with respect to* $t$, $x$ *and* $\varepsilon$ *and can be expanded in a convergent power series with respect to* $x$ *and* $\varepsilon$ *for* $||x|| \leq \rho, 0 \leq \varepsilon \leq \varepsilon_0$ *(*$\rho$ *a positive constant), then* $x(t)$ *can be expanded in a convergent power series with respect to* $\varepsilon$ *and* $\mu$ *in a neighbourhood of* $\varepsilon = 0$,

$$x(t) = x_0(t) + \varepsilon x_1(t) + \ldots + \varepsilon^n x_n(t) + \ldots$$

*convergent on the timescale 1.*

Although the timescale of validity for the expansion is $O(1)$ and so relatively short, it is sufficient for periodic solutions with $O(1)$ period. The proof of theorem (3.2) was given by Poincaré in the 1st volume of [110], and a modern proof is given in [146]. The title of the three volumes [110] is deceptive, although a large number of applications in the books are on celestial mechanics and the books can be considered as the first systematic text on non-linear dynamics. See also the discussion in [150].

The procedure of Poincaré-Lindstedt is based on theorem (3.2) and the Picard-Lindelöf method of iteration; this iteration is actually based on contraction of maps applied to ODEs that satisfy a Lipschitz condition as formulated in the Introduction. See for more details of the ODE background [46] chs. 1 and 2. It consists of writing the ODE with initial value as an integral equation. Substituting a suitable guess for the solution (as we saw in the Introduction even using the initial value is sufficient), the iteration process converges to the exact solution. However the convergence is slow, and many integrations over polynomials and other functions are involved. So the process of iteration is correct but without additional assumptions impractical. The contraction is directly related to the implicit function theorem, and the iteration process can be drastically improved by using Poincaré-expansion and imposing special conditions. For periodic solutions we expand and apply periodicity conditions.

To demonstrate the procedure, we start with the variational equation (3.1) of the exercise that we studied by averaging.

**Example 3.1** We write the equation (3.1) as an integral equation:

$$x(t) = x_0 + \varepsilon \int_0^t (x(s) - 2x(s)^2 \sin^2(s))ds, \tag{3.6}$$

where we have chosen $t_0 = 0$; as for $\varepsilon = 0$ we have $\dot{x} = 0$, we choose for the unknown initial value $x(0)$ the expansion:

$$x(0) = \rho_0 + \varepsilon\rho_1 + \varepsilon^2\rho_2 + \varepsilon^3 \ldots$$

As the equation is $2\pi$-periodic, we will look for $2\pi$-periodic solutions. Substituting the expansion into the integral equation (3.6), we find with higher order terms omitted:

$$x(0) + \varepsilon x_1(t) + \varepsilon^2 x_2(t) + \varepsilon^3 \ldots =$$
$$x(0) + \varepsilon \textstyle\int_0^t [(x(0) + \varepsilon x_1(s) + \varepsilon^2 x_2(s) - 2(x(0) + \varepsilon x_1(s) + \varepsilon^2 x_2(s))^2 \sin^2 s]ds.$$

Using the expansion for $x(0)$ and arranging by order of $\varepsilon$, we find to first order:

$$x_1(t) = \int_0^t [\rho_0 - 2\rho_0^2 \sin^2 s]ds = (\rho_0 - \rho_0^2)t + \frac{1}{2}\rho_0^2 \sin 2t.$$

For the periodic solution, one has the *periodicity condition*:

$$\int_0^{2\pi} x_1(s)ds = 0 \Rightarrow \rho_0 - \rho_0^2 = 0.$$

We conclude that $\rho_0 = 0$ or $\rho_0 = 1$. Choose $\rho_0 = 1$ as earlier for Eq. (3.1).

Keeping the $O(\varepsilon)$ terms in the integral, we have $x_2(t) =$

$$\int_0^t [\rho_1 + x_1(s) - 2(2\rho_0\rho_1 + 2\rho_0 x_1(s)) \sin 2s]ds$$
$$= \int_0^t [\rho_1 + \frac{1}{2}\sin 2s - 4(\rho_1 + \frac{1}{2}\sin 2s)^2 \sin 2s]ds.$$

Integration leads to the secular term for $x_2(t)$: $-\rho_1 t$ so $\rho_1 = 0$. We conclude for the periodic solution:

$$x(t) = 1 + \frac{1}{2}\varepsilon \sin 2t + O(\varepsilon^2). \tag{3.7}$$

We compare with the exact solution (3.2) by expanding:

$$x_s(t) = \frac{4+\varepsilon^2}{4+\varepsilon^2 - 2\varepsilon \sin 2t - \varepsilon^2 \cos 2t} = 1 + \frac{1}{2}\varepsilon \sin 2t + O(\varepsilon^2),\ x(0) = 1 + \frac{1}{4}\varepsilon^2.$$

### 3.3.1 General Approach for Variational Systems

Generalising to $n$-dimensional variational systems runs in the same way. Consider a system as derived from a perturbation problem as in the introductory Chap. 1

$$\dot{x} = \varepsilon f(t, x, \varepsilon), \tag{3.8}$$

that satisfies the conditions of theorem (14.35) and with the vector field $f(t, x)$ $T$-periodic in $t$. We write down the corresponding integral equation

$$x(t) = x(0) + \int_0^t f(s, x(s), \varepsilon)ds,$$

and we expand the vector field $f(t, x)$ with respect to $\varepsilon$, substitute the Poincaré-expansion for $x(t)$, and expand the unknown initial condition with respect to $\varepsilon$ in constant $n$-vectors $x(0) = \rho_0 + \varepsilon\rho_1 + \ldots$ Equating powers of $\varepsilon$, we find integral equations for the terms $x_n(t), n = 0, 1, 2, \ldots$ like (3.6). If we find by the periodicity condition a unique, isolated solution $\rho_0$ for $x_1(t)$ to be periodic, we have pinpointed the periodic solution to first order. If we have no uniqueness, we can try to find this at next order. This is tied in with the implicit function theorem as will be made clear in the next subsection.

Using the Poincaré-Lindstedt method for autonomous systems, we will, in general, have a perturbation of the period $T$ suggested by the original perturbation problem. Using as a start the corresponding variational system in amplitude-phase coordinates, this has as advantage a time-dependent phase that may perturb the unperturbed period $T$.

In some cases, as for Hamiltonian systems, we have families of periodic solutions, and we find no uniqueness at any order of $\varepsilon$. In such cases we have to modify the Poincaré-Lindstedt method to prove existence.

### 3.3.2 Poincaré-Lindstedt for Autonomous Systems

As we discussed, the iteration process using the integral equation form of an ODE with initial value is correct and produces results on a timescale of $O(1)$. However, the effectiveness of the iteration depends on the type of solution we are looking for and on the ODE. For instance in the case of Eq. (3.8), we are looking for $T$-periodic solutions. Consider now the second order autonomous equation

$$\ddot{x} + x = \varepsilon f_1(x, \dot{x}) + \varepsilon^2 f_2(x, \dot{x}) + O(\varepsilon^3) \tag{3.9}$$

and assume that the assumptions of the Poincaré-expansion theorem (3.2) are satisfied. A periodic solution, if it exists, may have a period close to the $2\pi$-periodic solutions of the unperturbed equation, but its actual period is still unknown. We will start with ignoring the presence of variational equations, and in the applications we return to this. Following Poincaré, we will look for $T$-periodic solutions with $T = T(\varepsilon) = 2\pi + \varepsilon \ldots$. We introduce a new time-like variable $\theta$ by the near-identity transformation:

$$\omega t = \theta, \omega^{-2} = 1 - \varepsilon\eta(\varepsilon) = 1 - \eta_0\varepsilon - \eta_1\varepsilon^2 + O(\varepsilon^3),$$

with the purpose of looking for $2\pi$-periodic solutions in $\theta$. The initial values are

$$x(0) = \rho_0 + \varepsilon\rho_1 + O(\varepsilon^2), \dot{x}(0) = 0.$$

The assumption $\dot{x}(0) = 0$ is no restriction for second order autonomous ODEs. Eq. (3.9) becomes with these transformations

$$\frac{d^2x}{d\theta^2} + x = \varepsilon\eta x + \varepsilon(1 - \varepsilon\eta) f(x, (1 - \varepsilon\eta)^{-1/2}\frac{dx}{d\theta}) + \varepsilon^2 \ldots \tag{3.10}$$

Consider the integral equation that is equivalent to the initial value problem on $\theta$:

$$x(\theta) = x(0)\cos\theta + \varepsilon \int_0^\theta \sin(\theta - s)[\eta x(s) + (1 - \varepsilon\eta) f_1(x(s), (1 - \varepsilon\eta)^{-1/2}\frac{dx}{d\theta})]ds + \varepsilon^2 \ldots, \tag{3.11}$$

where the $O(\varepsilon^2)$ terms are determined by $f_1, f_2, \ldots$ For $\varepsilon = 0$, Eq. (3.9) has an infinite set of periodic, synchronous (harmonic) functions. We will look for periodic solutions in $\theta$ that branch off of one or more harmonic functions in this set. The periodicity condition for the first term $f_1$ becomes

$$\int_0^{2\pi} sin(\theta - s)[\eta x(s) + (1 - \varepsilon\eta) f_1(x(s), (1 - \varepsilon\eta)^{-1/2}\frac{dx}{d\theta})]ds = 0. \tag{3.12}$$

Writing $\sin(\theta - s) = \sin\theta \cos s - \cos\theta \sin s$, we find from the integral 2 conditions:

$$\begin{cases} F_1 &= \int_0^{2\pi} \cos s[\eta x(s) + (1 - \varepsilon\eta) f_1(x(s), (1 - \varepsilon\eta)^{-1/2}\frac{dx}{d\theta})]ds = 0, \\ F_2 &= \int_0^{2\pi} \sin s[\eta x(s) + (1 - \varepsilon\eta) f_1(x(s), (1 - \varepsilon\eta)^{-1/2}\frac{dx}{d\theta})]ds = 0. \end{cases} \tag{3.13}$$

To lowest order, we have two unknowns, $\rho_0$ and $\eta_0$, at each order of $\varepsilon$ again two unknowns. At each step the system (3.13) is uniquely solvable if for the Jacobian $\Delta = det(J) \neq 0$:

$$J = \frac{\partial(F_1, F_2)}{\partial(\rho_n, \eta_n)}, n = 0, 1, 2 \ldots \tag{3.14}$$

If $det(J)$ does not vanish, the periodicity condition satisfies the implicit function theorem.

To lowest order we have the conditions:

$$\begin{cases} F_1 &= \pi\eta_0\rho_0 + \int_0^{2\pi} \cos s\, f_1(\rho_0 \cos s, -\rho_0 \sin s)ds = 0, \\ F_2 &= \int_0^{2\pi} \sin s\, f_1(\rho_0 \cos s, -\rho_0 \sin s)ds = 0. \end{cases} \tag{3.15}$$

If the lowest order system (3.15) satisfies already the uniqueness condition, we have existence of the periodic solution corresponding with $\rho_o, \eta_0$.

**Remark 3.4** In the spirit of Lindstedt we have started with one perturbed harmonic equation. The extension to systems of autonomous equations is straightforward.

### 3.3.3 *The Poincaré-Lindstedt Approach for Non-autonomous Systems*

A second order equation like (3.9) with $T$-periodic solutions for $\varepsilon = 0$ requires an expansion of time to describe periodic solutions with period close to $T$. Also, the initial condition will, in general, depend on $\varepsilon$. We have two parameter expansions.

This changes if we have an equation with $T$-periodic terms like

$$\ddot{x} + x = \varepsilon f(t, x, \dot{x}, \varepsilon) \tag{3.16}$$

that satisfies the Poincaré-expansion theorem (3.2) and with $f(t, x, \dot{x}, \varepsilon)$ $T$-periodic. A periodic solution, if it exists, will have period $T$. For autonomous equations the transformation for time and phase $\phi$: $t + \phi = \tau$ leaves the equations invariant, but this is not the case for non-autonomous equations. So we introduce the phase $\phi$ as a parameter to be determined. The initial value provides the second parameter for equation (3.16).

An alternative is to use the transformation (1.16) that introduces implicitly a phase. The set-up of the Poincaré-Lindstedt method, the use of the integral equation, periodicity conditions, and application of the implicit function theorem run as before. In [146] ch. 10.3 this is shown for the forced Duffing equation with damping:

$$\ddot{x} + \varepsilon\mu\dot{x} + x - \varepsilon x^3 = \varepsilon h \cos\omega t,$$

see also subsection 3.2.3 for the treatment by averaging. We will see another application, the Mathieu-equation with damping, in the sequel.

## 3.4 Applying the Poincaré-Lindstedt Method

Although the Van der Pol-equation has been extensively discussed in various forms and generalisation, also in many books, we will discuss the equation again because of its iconic character. The equation was also used often in papers demonstrating formula-manipulating computer routines; see for an early example [3]. Other typical examples illustrating the method will follow.

### *3.4.1 The Periodic Solution of the Van der Pol-Equation*

We can directly copy from the treatment of autonomous equations in the case:

$$\dot{x} + x = \varepsilon\dot{x}(1 - x^2).$$

Eq. (3.10) becomes

$$\frac{d^2x}{d\theta^2} + x = \varepsilon\eta x + \varepsilon(1 - \varepsilon\eta)^{1/2}(1 - x^2)\frac{dx}{d\theta}.$$

The corresponding integral equation is

$$x(\theta) = x(0)\cos\theta + \varepsilon\int_0^\theta \sin(\theta - s)[\eta x(s) + (1 - \varepsilon\eta)^{1/2}(1 - x(s)^2)\frac{dx(s)}{ds}]ds.$$

Comparing with (3.15), we have the periodicity conditions:

$$F_1 = \pi \eta_0 \rho_0 = 0,$$

$$F_2 = \frac{1}{2}\rho_0(1 - \frac{1}{4}\rho_0^2) = 0.$$

The periodicity conditions yield $\rho_0 = 2$, $\eta_0 = 0$.

It is not difficult to implement the iteration process for automatic computation. In [3] this was done for 164 steps.

**Using the variational equations**

As noted in subsection 3.3.2, we can use the variational equations (2.13):

$$\dot{r} = \varepsilon r \sin^2(t + \phi)(1 - r^2 \cos^2(t + \phi)),$$
$$\dot{\phi} = \varepsilon \sin(t + \phi)\cos(t + \phi)(1 - r^2 \cos^2(t + \phi)).$$

The integral equations are

$$\begin{cases} r(t) &= r(0) + \varepsilon \int_0^t r(s) \sin^2(s + \phi(s)) \left(1 - r^2(s) \cos^2(s + \phi(s))\right) ds, \\ \phi(t) &= \phi(0) + \varepsilon \int_0^t \sin(s+\phi(s)) \cos(s+\phi(s)) \left(1 - r^2(s) \cos^2(s+\phi(s))\right) ds. \end{cases} \tag{3.17}$$

As stated earlier putting $\phi_0 = 0$ is no restriction for autonomous second order equations. To obtain only periodic terms, we require the secular terms for the amplitude to vanish. This demonstrates that the Poincaré-Lindstedt method is complementary to averaging as instead of removing the terms with average zero, we remove the secular terms.

Using the expansion theorem, we have $\phi(t) = 0 + \varepsilon\phi_1 + \ldots$, $r(t) = r_0 + \varepsilon r_1 + \ldots$ with the dots standing for higher order terms dependent on time, so to $O(\varepsilon)$ we start with $r_0 \cos t$. The first integral equation produces

$$\int_0^t \frac{1}{2} r_0 \left(1 - \frac{1}{4} r_0^2 - \cos 2s + \frac{1}{4} r_0^2 \cos 4s\right) ds.$$

As expected we find the periodic secularity condition $r_0 = 2$ and to first order in $\varepsilon$:

$$r(t) = 2 + \varepsilon(-\frac{1}{2}\sin 2t + \frac{1}{4}\sin 4t) + O(\varepsilon^2). \tag{3.18}$$

For $\phi(t)$ we find to first order in $\varepsilon$ the $2\pi$-periodic expression:

$$\phi(t) = \varepsilon(\frac{1}{4}\cos 2t + \frac{1}{8}\cos 4t) + O(\varepsilon^2). \tag{3.19}$$

Including the $O(\varepsilon)$ terms, the solution is still $2\pi$-periodic, and a period change will be at most $O(\varepsilon^2)$.

### 3.4.2 A Case of Non-uniqueness

In some cases where the Jacobian vanishes, this is caused by the presence of a continuous family of periodic solutions. Additional information may help to identify such cases. In Example 1.4 we discussed a simple Hamiltonian system with a continuous family of periodic solutions. Consider the system again.

**Example 3.2** The equation from Example 1.4 for $n = 3$ is

$$\ddot{x} + x - \varepsilon x^3 = 0.$$

In the formulation of Eq. (3.11), we have the integral equation:

$$x(\theta) = x(0)\cos\theta + \varepsilon \int_0^\theta \sin(\theta - s)[\eta x(s) + (1 - \varepsilon\eta)x^3(s)]ds.$$

To first order this leads to the periodicity conditions:

$$F_1 = \eta_0\rho_0 + \int_0^{2\pi} \cos s\, \rho_0^3 \cos^3 s ds = \rho_0(\eta_0 + \frac{3}{8}\rho_0^2) = 0,\ F_2 = \int_0^{2\pi} \sin s\, \rho_0^3 \cos^3 s ds = 0.$$

The second periodicity condition holds for any $\rho$ (as expected from Example 1.4), and for $\eta_0$ we find

$$\eta_0 = -\frac{3}{8}\rho_0^2.$$

# Chapter 4
# Second Order Periodic Averaging

**Abstract** An approximation with error $O(\varepsilon^2)$ may be useful as qualitatively different phenomena may arise at second order. An example with mixed $O(\varepsilon)$ and $O(\varepsilon^2)$ terms illustrates the theorem. Different timescales play a part and are discussed.

To improve the averaging-approximation to second order in $\varepsilon$, we can be motivated by the necessity of more precision but, more importantly, also by the fact that qualitative new phenomena arise at higher order. The following theorem gives such an improvement in the case of periodic vector fields.

## 4.1 Second Order Precision

**Theorem 4.1** *Consider the initial value problems*

$$\dot{x} = \varepsilon f(t, x) + \varepsilon^2 g(t, x) + \varepsilon^3 R(t, x, \varepsilon), x(t_0) = x_0, \tag{4.1}$$

*and*

$$\dot{y} = \varepsilon f^0(y) + \varepsilon^2 f_1^0(y) + \varepsilon^2 g^0(y), y(t_0) = x_0, \tag{4.2}$$

*with*

$$f_1(t, x) = Df(t, x)u^1(t, x) - Du^1(t, x)f^0(x),$$

*and*

$$u^1(t, x) = \int_{t_0}^{t_0+t} (f(s, x) - f^0(x))ds.$$




F. Verhulst, *Averaging for Nonlinear Dynamics with Applications and Numerical Bifurcations*, Applied Mathematical Sciences 223,
https://doi.org/10.1007/978-3-032-12745-7_4

*Assume for $x, y, x_0 \in D \subset \mathbb{R}^n$, $t_0 \leq t \leq \infty, 0 \leq \varepsilon \ll 1$:*

1. *$f, g, R, Df$ are defined, continuous, and bounded by a constant $M$ independent of $\varepsilon$ in $[t_0, \infty] \times D$ (the square matrix $Df$ refers to the Jacobian derivative with respect to the spatial variable $x$ only).*
2. *$R$ is Lipschitz-continuous with respect to $x \in D$.*
3. *$f$ and $g$ are smooth and $T$-periodic in $t$ with $T$ a constant independent of $\varepsilon$, and $f^0, f_1^0, g^0$ are the averages of $f, f_1, g$.*
4. *$y(t)$ belongs to a bounded interior subset of $D$ on the timescale $1/\varepsilon$.*

*Then we have for the solutions of the initial value problems the estimate:*

$$x(t) - (y(t) + \varepsilon u^1(t, y)) = O(\varepsilon^2) \text{ as } \varepsilon \to 0 \text{ on the timescale } 1/\varepsilon.$$

**Remark 4.1** For a proof see [123] ch. 2.9. As stated before a second order approximation is only required when the first order approximation produces trivial results or if we expect new phenomena at second order. This may happen more often than the literature on perturbation theory suggests. Interesting examples are the instability tongues of the Mathieu-equation and higher order resonances in Hamiltonian systems, see Chap. 9.

**Remark 4.2** As $u^1(t, x)$ has average zero, in the periodic case $Du^1(t, x)f^0(x)$, will have average zero.

**Remark 4.3** As in the case of first order averaging, the case of quasi-periodic averaging carries over to second order. In the case of almost-periodic vector fields where we have an infinite number of independent periods, the analysis is more complicated, see also Chap. 5.

## 4.2 Exercises

The periodic solution of the Van der Pol equation is calculated to second order in ch. 2 of [123]. The important qualitative change is that in the expression $x(t) = r(t)\cos((t + \phi(t))$ we have in second order approximation dependence on $(t - \frac{1}{16}\varepsilon^2 t)$.

An interesting question is if the periodic solution of the Van der Pol equation (2.12) persists if the self-excitation takes place at $O(\varepsilon^2)$, and at order $O(\varepsilon)$ the oscillations are not harmonic. More applications can be found in Chap. 6 and in part II. For Hamiltonian and dissipative systems with dimension larger than 2, qualitative phenomena will often shift to higher order. We start with some straightforward questions about two-dimensional conservative systems.

### 4.2.1 Simple Conservative Systems

Consider the equation:

$$\ddot{x} + x = \varepsilon(ax^2 + bx^3). \tag{4.3}$$

The equation (4.3) can be derived from the Hamiltonian function:

$$H = \frac{1}{2}\dot{x}^2 + \frac{1}{2}x^2 - \varepsilon\frac{a}{3}x^3 - \varepsilon\frac{b}{4}x^4. \tag{4.4}$$

We are interested in the influence of the non-linear terms in Eq. (4.3) on the radius $r$ and phase $\phi$ that affects the period. Suppose $x(0) = r_0 > 0, \dot{x}(0) = 0$. From the Hamiltonian we have the first integral:

$$\frac{1}{2}\dot{x}^2 + \frac{1}{2}x^2 - \varepsilon\frac{a}{3}x^3 - \varepsilon\frac{b}{4}x^4 = C,\ C = \frac{1}{2}r_0^2 - \varepsilon\frac{a}{3}r_0^3 - \varepsilon\frac{b}{4}r_0^4.$$

With transformation (1.6) we have $\dot{x}^2(t) + x^2(t) = r^2(t)$, so it is clear that near the origin of phase space $r(t) = r_0 + O(\varepsilon)$ for $t \geq 0$. We will investigate with more precision what happens to radius and phase (period) in the quadratic and cubic case.

**Example 4.1** Consider the **quadratic** case $a \neq 0, b = 0$. Applying transformation (1.6) with $r(0) = r_0, \phi(0) = 0$, we find

$$f(t, r, \phi) = \begin{pmatrix} -ar^2 \sin(t + \phi)\cos^2(t + \phi) \\ -ar\cos^3(t + \phi) \end{pmatrix}$$

with

$$f^0(r, \phi) = \begin{pmatrix} 0 \\ 0 \end{pmatrix}.$$

So $r(t) = r_0 + O(\varepsilon), \phi(t) = 0 + O(\varepsilon)$. To obtain a second order approximation following Theorem 4.1, we have to compute and average the expression $f_1 = Df.u^1$. Abbreviating $t + \phi = \alpha$, we find $Df(t, r, \phi) =$

$$\begin{pmatrix} -2ar\sin\alpha\cos^2\alpha & -ar^2\cos^3\alpha + 2ar^2\sin^2\alpha\cos\alpha \\ -a\cos^3\alpha & 3ar\cos^2\alpha\sin\alpha \end{pmatrix}.$$

We find

$$u^1(t, r, \phi) = \begin{pmatrix} \frac{1}{3}ar^2\cos^3\alpha \\ -ar\sin\alpha + \frac{1}{3}ar\sin^3\alpha \end{pmatrix}.$$

We find from $Df.u^1$ after averaging $(0, -\frac{5}{8}a^2r^2)^T$.
In the notation of Theorem 4.1, we have

$$\dot{y}_1 = 0,\ \dot{y}_2 = -\varepsilon^2 \frac{5}{8} a^2 r^2,$$

and so the $O(\varepsilon^2)$ approximation for $r(t)$ becomes

$$r(t) = r_0 + \varepsilon \frac{1}{3} a r_0^2 \cos^3(t - \varepsilon^2 \frac{5}{8} a^2 r_0^2 t). \tag{4.5}$$

The approximation for the phase can be obtained in the same way.

In the case of a cubic term, the first order approximation is not trivial.

**Example 4.2** Consider the **cubic** case $a = 0, b \neq 0$, the Duffing equation. Applying transformation (1.6) with $r(0) = r_0, \phi(0) = 0$, we find

$$f(t, r, \phi) = \begin{pmatrix} -br^3 \sin(t+\phi)\cos^3(t+\phi) \\ -br^2 \cos^4(t+\phi) \end{pmatrix},\ f^0(r, \phi) = \begin{pmatrix} 0 \\ -\frac{3}{8}br^2 \end{pmatrix}.$$

Using initial values $r(0) = r_0, \phi(0) = 0$, we have a first order approximation valid on the timescale $1/\varepsilon$:

$$x(t) = r_0 \cos(1 - \varepsilon \frac{3}{8} b r_0^2)t + O(\varepsilon),\ \dot{x}(t) = -r_0 \sin(1 - \varepsilon \frac{3}{8} b r_0^2)t + O(\varepsilon),$$

corresponding with a set of periodic solutions with an $O(\varepsilon)$ shifted period that depends on the initial $r_0$. Following Theorem 4.1, we have to compute and average again $f_1 = Df.u^1$. Abbreviating again $t + \phi = \alpha$, we find

$$Df(t, r, \phi) = \begin{pmatrix} -3br^2 \sin\alpha \cos^3\alpha & -br^3 \cos^4\alpha + 3br^3 \sin^2\alpha \cos^2\alpha \\ -2br\cos^4\alpha & 4br^2 \cos^3\alpha \sin\alpha \end{pmatrix},$$

and, using $\cos^4\alpha = \frac{3}{8} + \frac{1}{2}\cos 2\alpha + \frac{1}{8}\cos 4\alpha$,

$$u^1(t, r, \phi) = \begin{pmatrix} \frac{1}{8}br^3 \cos 2\alpha + \frac{1}{32}br^3 \cos 4\alpha \\ -\frac{1}{4}br^2 \sin 2\alpha - \frac{1}{32}br^2 \sin 4\alpha \end{pmatrix}.$$

Multiplying $Df$ and $u^1$, we find $f_1$, and after averaging

$$f_1^0(r, \phi) = \begin{pmatrix} 0 \\ -\frac{51}{256}b^2r^4 \end{pmatrix}.$$

In the notation of Theorem 4.2, we obtain

$$\dot{y}_1 = 0,\ \dot{y}_2 = -\varepsilon\frac{3}{8}by_1^2 - \varepsilon^2\frac{51}{256}b^2y_1^4. \tag{4.6}$$

We can use this result to improve the approximation of $r, \phi$, but, more interesting, we will use this second order approximation in the next subsection.

### 4.2.2 Van der Pol Excitation at Second Order

We summarise a result of [148] ch. 12.5. Consider the equation:

$$\ddot{x} + x - \varepsilon bx^3 = \varepsilon^2\dot{x}(1 - x^2)$$

with $b$ a suitable constant independent of $\varepsilon$. Introducing amplitude-phase variables with transformation (1.6), we find (see preceding example)

$$f(t, r, \phi) = \begin{pmatrix} -br^3\sin(t+\psi)\cos^3(t+\phi) \\ -br^2\cos^4(t+\phi) \end{pmatrix},\ f^0(r, \phi) = \begin{pmatrix} 0 \\ -\frac{3}{8}br^2 \end{pmatrix}.$$

Using initial values $r(0) = r_0, \phi(0) = 0$, we have a first order approximation valid on the timescale $1/\varepsilon$:

$$x(t) = r_0\cos(1 - \varepsilon\frac{3}{8}br_0^2)t + O(\varepsilon),\ \dot{x}(t) = -r_0\sin(1 - \varepsilon\frac{3}{8}br_0^2)t + O(\varepsilon),$$

corresponding with a set of periodic solutions with an $O(\varepsilon)$ shifted period that also depends on the initial $r_0$. This family of periodic solutions does not persist as a periodic solution branches off from one of the first order approximate solutions when considering second order approximations. We show this. Following Theorem 4.1, we use the result from the preceding example for $f_1 = Df.u^1$ after averaging

$$f_1^0(r, \phi) = \begin{pmatrix} 0 \\ -\frac{51}{256}b^2r^4 \end{pmatrix}.$$

Now we can write down the equation for $y$ as formulated in the procedure for second order calculations. Note that $g^0(y)$ (notation Theorem 4.1) was calculated already. We obtain

$$\dot{y}_1 = \varepsilon^2\frac{y_1}{2}\left(1 - \frac{1}{4}y_1^2\right)$$
$$\dot{y}_2 = -\varepsilon\frac{3}{8}by_1^2 - \varepsilon^2\frac{51}{256}b^2y_1^4.$$

If $y_1(0) = r_0 = 2$, we have a stationary solution $y_1(t) = 2$, and with $y_2(0) = \phi(0) = 0$

$$y_2(t) = -\varepsilon\frac{3}{2}bt - \varepsilon^2\frac{51}{16}b^2t.$$

The equilibrium is obtained at $O(\varepsilon^2)$ of the averaged system and is asymptotically stable. An $O(\varepsilon^2)$-approximation of the asymptotically stable periodic solution is obtained by inserting $y_1, y_2$ into $u^1$ so that

$$r(t) = 2 + \varepsilon b\cos 2(t + y_2(t)) + \varepsilon\frac{b}{4}\cos 4(t + y_2(t)) + O(\varepsilon^2),$$

$$\phi(t) = y_2(t) - \varepsilon b\sin 2(t + y_2(t)) - \varepsilon\frac{b}{8}\sin 4(t + y_2(t)) + O(\varepsilon^2),$$

valid on the timescale $1/\varepsilon$. At this order of approximation, the timelike variables for the periodic solution are $t, \varepsilon t, \varepsilon^2 t$.

### 4.2.3 Intermezzo on the Nature of Timescales

As discussed earlier, an assumption is often made in research that the natural timelike variables in expansions with respect to a small parameter $\varepsilon$ will be of the form $t, \varepsilon t, \varepsilon^2 t, \ldots, \varepsilon^n t$, etc. As we discussed in the Introduction, this assumption is usually based on simple examples, and it is sometimes correct but in general not. For instance in Chaps. 7 and 9, we will meet systems where local domains play a part with completely different timelike variables.

In [148] (Sect. 12.5) and [152], a practical problem, the Mathieu-equation, is studied in the form:

$$\ddot{x} + (1 + \varepsilon a + \varepsilon^2 b + \varepsilon\cos 2t)x = 0. \tag{4.7}$$

In [148] a first order approximation produces the boundaries of the instability tongue or Floquet tongue sketched in Fig. 11.1 (without damping) with $a = \pm 0.5$. The $O(\varepsilon)$ approximation is valid on the interval of time $1/\varepsilon$. A second order calculation shows that the boundary of the tongue is pinpointed with $b = -1/32$; the timelike variables that play a part at second order are $t, \varepsilon t, \varepsilon^{3/2}t, \varepsilon^2 t$. This is quite unexpected, and this result would not have been obtained with simple a priori assumptions on the timelike variables.

The unexpected timelike variables arise in a neighbourhood of transitions like the curves of periodic solutions of the Mathieu-equation that separate domains of stability and instability and are in general associated with bifurcations; see for details Chap. 11. *Such phenomena are often of special interest in non-linear analysis as such transitions correspond with qualitative changes in the system.* As an illustration we discuss a few examples involving low-dimensional bifurcations (involving few variables and parameters); we can find such bifurcations as subsystems after first or second order averaging of a larger system.

A simple example is the saddle-node bifurcation described by

$$\dot{x} = a - bx^2. \tag{4.8}$$

If $ab < 0$, there is no critical point; suppose $ab > 0$, then we have the critical points $x_0 = \pm\sqrt{a/b}$. Their stability and local behaviour with time are described by the coefficient of linearisation $-2bx_0 = \mp 2\sqrt{ab}$ near the critical points. If for instance $a = \varepsilon^2, b = \varepsilon$, the leading timelike variable is $\varepsilon^{3/2}t$.

Consider the system inspired by the pitchfork bifurcation:

$$\dot{x} = \varepsilon^2 y - \varepsilon y^3,\ \dot{y} = \varepsilon x. \tag{4.9}$$

Three critical points (equilibria) are $(x_0, y_0) = (0, 0), (0, \pm\sqrt{\varepsilon})$. Linearisation near the critical points $(0, y_0)$ produces

$$\dot{x} = \varepsilon^2 y - 3\varepsilon y_0^2 y + \ldots,\ \dot{y} = \varepsilon x,$$

where the dots represent the neglected non-linear terms. We have the characteristic eigenvalue equations and timelike variables near the critical points:

$(0, 0)$: $\lambda^2 - \varepsilon^3 = 0$, timelike variable $\varepsilon^{3/2}t$.

$(0, \pm\sqrt{\varepsilon})$: $\lambda^2 - 2\varepsilon^3 = 0$, timelike variable $\varepsilon^{3/2}t$.

See for extensive discussions of bifurcations and timelike variables also [92]. In general we expect in regions where bifurcations occur the presence of timelike variables of the form $\varepsilon^q t$ with $q$ a positive rational number.

As we will see, an example of the pitchfork bifurcation is found in the Van der Pol equation.

**Example 4.3 (Emergence of the Van der Pol Periodic Solution)**
Consider the Van der Pol equation in slightly more general form than Eq. (2.12):

$$\ddot{x} + x = \varepsilon\dot{x}(a - x^2). \tag{4.10}$$

If parameter $a < 0$, the oscillations will be damped, and there is no periodic solution. If $a > 0$, we have nearly the case of Eq. (2.12), and first order averaging produces

$$\dot{r} = \frac{\varepsilon}{2}r(a - \frac{1}{4}r^2),\ \dot{\phi} = 0. \tag{4.11}$$

If parameter $a$ starts at a negative value and we let it increase, it will pass through zero, and for $a > 0$ a periodic solution emerges with amplitude $2\sqrt{a}$ by a pitchfork bifurcation. It starts with $a$ being small, say $a = \varepsilon$. This is the situation sketched above where we have the timelike variable $\varepsilon^{3/2}t$. We improve the reasoning by writing Eq. (4.10) as

$$\ddot{x} + x = -\varepsilon \dot{x} x^2 + \varepsilon^2 \dot{x}.$$

In amplitude-phase variables $r, \phi$ the variational system to $O(\varepsilon)$ becomes

$$\dot{r} = -\varepsilon r^3 \sin^2(t+\phi)\cos^2(t+\phi),$$
$$\dot{\phi} = -\varepsilon r^2 \sin(t+\phi)\cos^2(t+\phi).$$

First order averaging produces

$$\dot{r} = -\varepsilon \frac{1}{8} r^3, \ \dot{\phi} = 0.$$

Computing the quantities $Df$ and $u^1$ in the notation of Theorem 4.1, we find no $O(\varepsilon^2)$ contribution for the amplitude and a small contribution to the phase. Averaging the term $\varepsilon^2 \dot{x}$, we find

$$\dot{r} = -\varepsilon \frac{1}{8} r^3 + \varepsilon^2 \frac{r}{2}. \tag{4.12}$$

As predicted above we find for the amplitude of the periodic solution $2\sqrt{\varepsilon}$.

# Chapter 5
# First Order General Averaging

**Abstract** If the ODE $\dot{x} = \varepsilon f(t, x)$ is not time-periodic or quasi-periodic, a more general way of averaging can be used sometimes. This is the case for systems with almost-periodic time dependence or for systems with slowly varying terms like decreasing or increasing friction.

We can apply averaging in a more general sense with as important examples equations with terms that have a limit as $t \to \infty$ and if almost-periodic vector fields are present. Consider again Eq. (2.1) and assume that the following "general" average exists:

$$f^0(x) = \lim_{T \to \infty} \frac{1}{T} \int_{t_0}^{t_0+T} f(t, x) dt. \tag{5.1}$$

As we shall see, general averages of the form (5.1) play a part in certain applications. It may also generate a different order function that arises in the estimates. We have with $x \in D$ an error $O(\delta(\varepsilon))$ on the timescale $1/\varepsilon$ with

$$\delta(\varepsilon) = \sup_{x \in D} \sup_{0 \leq \varepsilon t \leq L} \varepsilon \left| \int_{t_0}^{t_o+t} (f(s, x) - f^0(x)) ds \right|, \tag{5.2}$$

with $L$ a positive constant.

## 5.1 The Basic Theorem for General Averaging

**Theorem 5.1** *Consider the initial value problem* (2.1) *(brief version* $\dot{x} = \varepsilon f + \varepsilon^2 g$, $x(t_0) = x_0$*) and the initial value problem*

$$\dot{y} = \varepsilon f^0(y), \; y(t_0) = x_0, \tag{5.3}$$


F. Verhulst, *Averaging for Nonlinear Dynamics with Applications and Numerical Bifurcations*, Applied Mathematical Sciences 223,
https://doi.org/10.1007/978-3-032-12745-7_5

*with $f^0(x)$ existing and defined by Eq. (5.1). Assume for $x, y, x_0 \in D \subset \mathbb{R}^n, t_0 \leq t \leq \infty, 0 \leq \varepsilon \ll 1$:*

1. *$f, g, Df$ are defined, continuous, and bounded by a constant $M$ independent of $\varepsilon$ in $[t_0, \infty] \times D$.*
2. *$g$ is Lipschitz-continuous with respect to $x \in D$.*
3. *$y(t)$ belongs to a bounded interior subset of $D$ on the timescale $1/\varepsilon$.*

*Then we have for the solutions of the initial value problems the estimate:*

$$x(t) - y(t) = O(\delta(\varepsilon)) \text{ as } \varepsilon \to 0 \text{ on the timescale} 1/\varepsilon.$$

**Remark 5.1** The proof is rather subtle, see [123] ch. 4. An early proof was given by Bogoliubov and Mitropolsky in [20].

The more general version of averaging is highly relevant for the treatment of the so-called almost-periodic functions. We will not give an explicit definition, but see [23] and typical expression (5.4). We note that for almost-periodic functions the average $f^0$ in the sense of Theorem 5.3 always exists. It is also important to know that there exists an analogue of Fourier theory so that a vector field $f(t, x)$ that is almost-periodic in $t$ can be written as a uniformly convergent series:

$$f(t, x) = \sum_{n=0}^{\infty} (A_n(x) \cos \lambda_n t + B_n(x) \sin \lambda_n t). \tag{5.4}$$

If $f(t, x)$ is $2\pi$-periodic, we would have $\lambda_n = n$, but in this more general case the $\lambda_n$ can be any sequence of real numbers. The collection of these numbers is called the generalised Fourier spectrum. As stated above for an almost-periodic vector function, the following average exists:

$$f^0(x) = \lim_{T \to \infty} \frac{1}{T} \int_0^T f(t, x) dt.$$

A strong contrast with periodic Fourier theory is that if the average $f^0(x)$ vanishes, *the primitives of $f(t, x)$ need not be bounded for all time*. This weakens the error estimate, as we shall see in an exercise.

**Remark 5.2** A useful lemma was proved in [123]. Suppose that there exists a positive number $\alpha$ such that we have for the spectrum of the generalised Fourier expansion (5.4) of $f(t, x)$, $\lambda_n \geq \alpha, n = 0, 1, 2, \ldots$, then the primitive of $f(t, x) - f^0(x)$ is bounded, and the error by first order general averaging is $O(\varepsilon)$.

**Remark 5.3** If the expansion (5.4) is finite, we call $f(t, x)$ a quasi-periodic function. The error by first order averaging is $O(\varepsilon)$ in this case. See also Theorem 2.2.

## 5.2 Second Order General Averaging

Consider the slowly varying system (2.1) and assume that the general average (5.1) exists. In Theorem 5.1 we have assumed that $Df$ is continuous. It is easy to formulate a second order approximation result if the vector field $f(t, x)$ is even smoother, at least $C^2(D)$. We transform

$$x(t) = y(t) + \delta(\varepsilon)u^1(t, y(t)),$$

with $u^1$ defined as for periodic averaging. Then we can construct for a large number of systems, dependent on the choice of $f(t, x)$, an $O(\delta^2)$ approximation of $x(t)$ on time intervals of size $1/\varepsilon$.

That we cannot obtain a second order approximation in some cases depends on the rate of convergence of the limit (5.1). For details and proofs, see Lemma 4.5.1 in [123]. We will discuss some aspects of second order general averaging in subsequent exercises.

## 5.3 Exercises

In physics and engineering, the conditions for general averaging are quite natural. Still, the number of applications is restricted. If we are considering equations with time-dependent terms that are not periodic and even show limiting behaviour, we may come across the concept of *adiabatic invariant*. This invariant plays usually the part of approximate time-dependent integral of motion in such dissipative equations.

### 5.3.1 *Van der Pol Equation with Changing Friction*

Consider the equation:

$$\ddot{x} + x = \varepsilon(1 - ae^{-t} - x^2)\dot{x},\ a > 0. \tag{5.5}$$

Transforming to amplitude-phase with (1.6), we find the variational equations:

$$\begin{aligned}
\dot{r} &= \varepsilon \sin(t + \phi)(1 - ae^{-t} - r^2 \cos^2(t + \phi))r \sin(t + \phi),\\
\dot{\phi} &= \tfrac{\varepsilon}{r} \cos(t + \phi)(1 - ae^{-t} - r^2 \cos^2(t + \phi))r \sin(t + \phi).
\end{aligned}$$

We have

$$0 < \int_0^t ae^{-s} \sin^2(s + \phi)ds \le \int_0^t ae^{-s}ds = a(1 - e^{-t}).$$

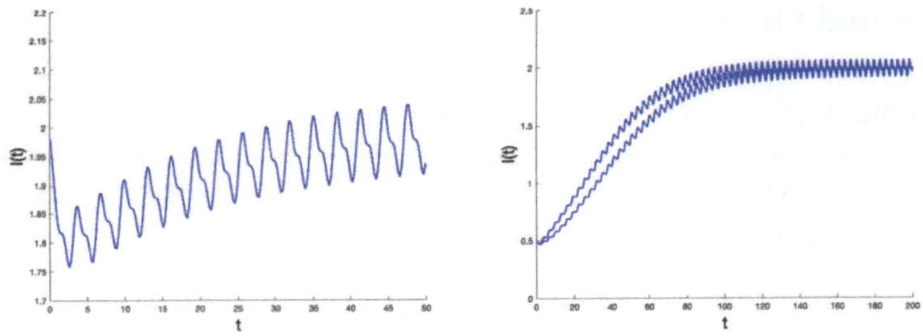

**Fig. 5.1** Behaviour of $I = 0.5(\dot{x}^2 + x^2)$ with time of the Van der Pol equation, with changing damping in Eq. (5.5), $\varepsilon = 0.05, a = 2, x(0) = 0, \dot{x}(0) = 2$. Right the start at $x(0) = 0, \dot{x}(0) = 1$ and both exponential and polynomial variation of the damping

We conclude that

$$\lim_{t \to \infty} \frac{1}{T} \int_0^T ae^{-s} \sin^2(s + \phi)ds = 0.$$

A similar estimate holds for the equation for $\phi$. We conclude that the estimate for the approximations of the standard Van der Pol equation carries over to this case with changing damping, see Fig. 5.1 where initially the action $I(t)$ moves away from the end stage near $I = 2$. For comparison we show in Fig. 5.1 (right) the times series starting at $x(0) = 0, \dot{x}(0) = 1$ with $-a^{-t}$ in the equation, and this term replaced by $-a/(1 + t)$ moving slower to $I = 2$.

### 5.3.2 *Oscillations with Increasing Friction, Adiabatic Invariant*

Consider a linear oscillator with friction that, through wear and tear, increases slowly with time. The equation is

$$\ddot{x} + \varepsilon f(t)\dot{x} + x = 0. \tag{5.6}$$

Assume that $f(0) = 1$ and $f(t)$ increases monotonically with time. We discuss two examples:

$$f_1(t) = 2 - e^{-t^2},\ f_2(t) = 2 - \frac{t}{1 + t^2}.$$

In amplitude-phase variables $r, \psi$, we have with transformation (1.6)

$$\begin{aligned}
\dot{r} &= -\varepsilon r \sin^2(t + \phi) f(t), \\
\dot{\psi} &= -\varepsilon \sin(t + \phi) \cos(t + \phi) f(t).
\end{aligned}$$

We can perform averaging in the more general sense, and for both choices $f = f_1(t)$ and $f = f_2(t)$, we find the approximating system:

$$\dot{r} = -\varepsilon r, \ \dot{\psi} = 0.$$

In both cases, we have on the timescale $1/\varepsilon$ the approximate solution:

$$x(t) = r(0)e^{-\varepsilon t} \cos(t + \psi(0)) + O(\delta(\varepsilon)). \tag{5.7}$$

However, the error is different. For the function $f_1(t)$, the integral $\int_0^t (f(s, x) - f^0(x))ds$ is uniformly bounded, so the error is $O(\varepsilon)$. Choosing $f_2(t)$, the integral is logarithmically growing as $\ln(1 + t^2)$; we have on the timescale $1/\varepsilon$: $\delta(\varepsilon) = O(\varepsilon |\ln \varepsilon|)$.

Such a difference is not unexpected. Note that the approximation corresponds with a constant friction coefficient 2, which is the limit for $t \to \infty$ of $f_1(t)$ and $f_2(t)$. The function $f_1(t)$ reaches this limit exponentially fast, but $f_2(t)$ increases polynomially to the value 2.

The non-autonomous equation (5.6) has the time-varying amplitude

$$r(t) = r(0)e^{-\varepsilon t}$$

as approximate quantity; this is called an *adiabatic invariant* of the non-autonomous equation (5.6).

If the variation in time of the friction is even slower, for instance by taking

$$f(t) = 2 - \frac{1}{(1+t)^\alpha}, 0 < \alpha < 1,$$

then a first order approximation exists with an $O(\varepsilon^\alpha)$ approximation, but a second order approximation with the construction mentioned above does not exist. See also the related example 4.5.1 in [123].

### 5.3.3 *Quasi-periodic Forcing of the Duffing Equation*

Although we can handle many quasi-periodic problems with the basic techniques of Chap. 2, the behaviour of quasi-periodic systems is of special interest as it represents relative simple almost-periodic problems. Consider the forced Duffing equation:

$$\ddot{x} + x + \varepsilon \gamma x^3 = \varepsilon f(t, x), \tag{5.8}$$

with $\gamma \neq 0$ a constant, and $f(t, x)$ is almost-periodic in $t$. Transforming with (1.6), we find

$$\dot{r} = \varepsilon \sin(t+\phi)(\gamma r^3 \cos^3(t+\phi) - f(t,x)),$$
$$\dot{\phi} \;\; = \varepsilon \tfrac{\cos(t+\phi)}{r}(\gamma r^3 \cos^3(t+\phi) - f(t,x))).$$

Assume that $f(t,x)$ is quasi-periodic of the form:

$$f(t,x) = \sum_{n=1}^{N}(A_n(x)\cos(\lambda_n t) + B_n(x)\sin\lambda_n t), \tag{5.9}$$

with independent (incommensurable) frequencies $\lambda_n$.

As a first example, we take for $f(t,x)$: $f_1(t,x) = x(\cos t + \cos\sqrt{2}t)$. Applying the basic quasi-periodic averaging Theorem 2.2 (Remark 5.9), we find for the averaged equations:

$$\dot{r} = 0,\; \dot{\psi} = \frac{3}{8}\varepsilon\gamma r^2. \tag{5.10}$$

We conclude that on the timescale $1/\varepsilon$ the amplitude is constant with error $O(\varepsilon)$, and the phase $\psi$ changes slowly. A numerical approximation confirms this analysis for a long time.

Consider an almost-periodic function of the form:

$$f_2(t,x) = \sum_{n=0}^{\infty}\frac{x}{(2n+1)^2}\sin\left(\frac{t}{2n+1}\right). \tag{5.11}$$

The spectrum accumulates near zero. The approximating system (5.10) will be the same, but the error will be larger than $O(\varepsilon)$. This is a consequence of the fact that the integral over time of the function $f_2(t,x)$ is not bounded.

But in the case of for instance the choice:

$$f_3(t,x) = \sum_{n=1}^{\infty}\frac{x}{(2n+1)^2}\sin\left(\sqrt{n}t\right),$$

we would have the same approximation with error $O(\varepsilon)$ (see Remark 5.2, $\alpha = 1$).

### 5.3.4 Quasi-periodic Forcing of a Van der Pol Limit Cycle

Consider the Van der Pol equation (2.12) but with $O(1)$ forcing added. A simple case is

$$\ddot{x} + x = \varepsilon\dot{x}(1-x^2) + f(t), \tag{5.12}$$

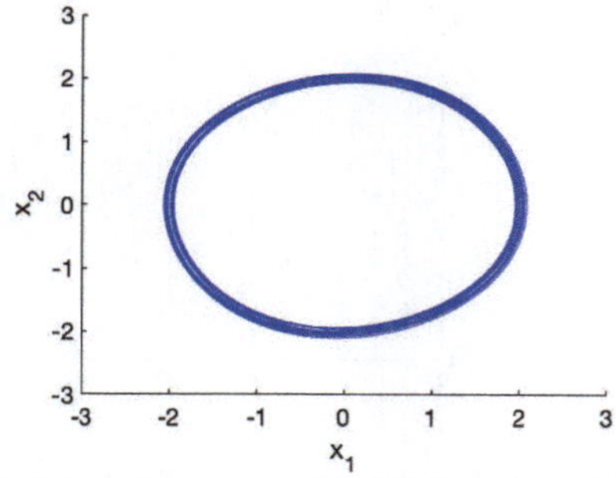


**Fig. 5.2** Projection on the $(x_1, x_2) = (x, \dot{x})$ phase-plane for the forced Van der Pol equation (5.12) in the case of small forcing $\varepsilon \cos \sqrt{2}t$ with $\varepsilon = 0.05$

with $f(t)$ a quasi-periodic function; put

$$f(t) = \sum_{n=1}^{N} a_n \cos \omega_n t, \tag{5.13}$$

with $N$ a natural number, the frequencies $\omega_n$, $n = 1, \ldots, N$ are irrational numbers, incommensurable and not $\varepsilon$-close to 1.

If we replace $f(t)$ in Eq. (5.12) by $\varepsilon f(t)$, first order averaging produces the result of the standard Van der Pol equation except that we have no existence of a periodic solution. We conjecture for $\varepsilon$ small the existence of a quasi-periodic solution moving on a torus in the phase space with time extended; see Fig. 5.2.

Consider $f(t) = O(1)$, $t \geq 0$. If $\varepsilon = 0$, a particular solution $\eta(t)$ of Eq. (5.12) is

$$\eta(t) = \sum_{n=1}^{N} \frac{a_n}{1 - \omega_n^2} \cos \omega_n t. \tag{5.14}$$

Transforming with (1.27) for forced equations, we put

$$x = r\cos(t + \phi) + \eta(t),\ \dot{x} = -r sin(t + \phi) + \dot{\eta}(t)$$

to find for $r, \phi$:

$$\dot{r} = -\varepsilon \sin(t + \phi)(-r sin(t + \phi) + \dot{\eta}(t))(1 - (r\cos(t + \phi) + \eta(t))^2),$$
$$\dot{\phi} = -\frac{\varepsilon}{r} \cos(t + \phi)(-r sin(t + \phi) + \dot{\eta}(t)) \left(1 - (r\cos(t + \phi) + \eta(t))^2\right).$$

The right-hand side is quasi-periodic in $t$; as an example we choose

$$f(t) = a \cos \sqrt{\omega} t. \tag{5.15}$$

Averaging produces the system:

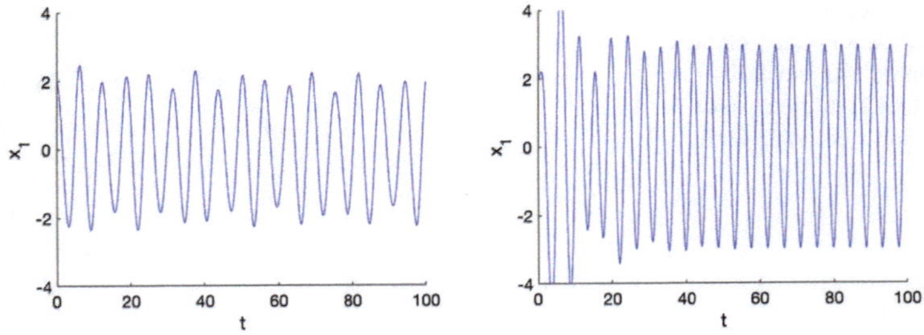


**Fig. 5.3** Two time series $x_1(t)$ for the forced Van der Pol equation (5.12) with $f(t) = a\cos\sqrt{2}t$. Left the case $a = 0.3$ where $r(t)$ does not tend to zero and is modulated by the quasi-periodic forcing. Right the case $a = 3$ where $r(t)$ vanishes and the motion $x(t)$ tends to the particular solution $-3\cos\sqrt{2}t$

$$\dot{r} = \frac{\varepsilon}{2}r(1 - \frac{1}{4}r^2 - \frac{1}{2}\frac{a^2}{(1-\omega^2)^2}),\ \dot{\phi} = 0.$$

The forcing causes an important change of the standard Van der Pol equation. If $\omega^2 = 2$, the approximate solution with constant $r$ vanishes if $1 - a^2/2$ becomes negative so if $a \geq \sqrt{2}$. Experiments confirm that in such cases the forcing dominates and produces quasi-periodic solutions (Fig. 5.3). In various parameter cases, the dynamics of the forced Van der Pol equation can be quite complicated, see also Chap. 12.

# Chapter 6
# Approximations on Timescales Longer Than $1/\varepsilon$

**Abstract** In some cases it is possible to establish a longer timescale of validity of approximation. If the average over time of the system $\dot{x} = \varepsilon f(t, x)$ vanishes, the approximation can in the periodic case be extended to $1/\varepsilon^2$. Another important case arises if the averaged equation contains a solution that is exponentially attracting. In such a case we can extend the timescale to the whole time axis.

For non-linear variational systems of the form $\dot{x} = \varepsilon f(t, x)$, we have a priori existence of solutions on the timescale $1/\varepsilon$. With special assumptions, it is possible to extend the timescale of existence and of approximations. We will formulate theorems in the case of two rather different sets of assumptions.

## 6.1 If First Order Averaging Gives Trivial Results

If in the case of periodic averaging (notation Chap. 2) we have $f^0(x) = 0$, it follows that with $x(0) = x_0$, $|x(t) - x_0| = O(\varepsilon)$ on the timescale $1/\varepsilon$. It is possible that significant qualitative phenomena take place on longer intervals of time. If $f^0(x) = 0$, we can improve the approximation and extend the timescale of validity. We formulate the theorem.

**Theorem 6.1** *Consider the initial value problems*

$$\dot{x} = \varepsilon f(t, x) + \varepsilon^2 g(t, x) + \varepsilon^3 R(t, x, \varepsilon), x(t_0) = x_0, \tag{6.1}$$

*with* $f^0(x) = 0$ *and*

$$\dot{y} = \varepsilon^2 f_1^0(y) + \varepsilon^2 g^0(y), y(t_0) = x_0, \tag{6.2}$$

*with*



F. Verhulst, *Averaging for Nonlinear Dynamics with Applications and Numerical Bifurcations*, Applied Mathematical Sciences 223,
https://doi.org/10.1007/978-3-032-12745-7_6

$$f_1(t,x) = Df(t,x).u^1(t,x) - Du^1(t,x).f^0(x),$$

*and*

$$u^1(t,x)) = \int_{t_0}^{t_0+t} f(s,x)ds.$$

*Assume for $x, y, x_0 \in D \subset \mathbb{R}^n$, $t_0 \leq t \leq \infty$, $0 \leq \varepsilon \ll 1$:*

1. *$f, g, R, Df$ are defined, continuous, and bounded by a constant $M$ independent of $\varepsilon$ in $[t_0, \infty] \times D$.*
2. *$R$ is Lipschitz-continuous with respect to $x \in D$.*
3. *$f$ and $g$ are smooth and $T$-periodic in $T$ with $T$ a constant independent of $\varepsilon$, and $f^0$, $f_1^0$, $g^0$ are the averages of $f$, $f_1$, $g$.*
4. *$y(t)$ belongs to a bounded interior subset of $D$ on the timescale $1/\varepsilon$.*

*Then we have for the solutions of the initial value problems the estimate:*

$$x(t) - (y(t) + \varepsilon u^1(t,y)) = O(\varepsilon) \text{ as } \varepsilon \to 0 \text{ on the timescale } 1/\varepsilon^2.$$

**Remark 6.1** Theorem 6.1 was formulated and proved in [139], see also [140] and [123] ch. 2.9.

## 6.2 The Case of Attraction

Another type of extension of timescales happens when attraction is involved. We present a theorem in a simple form. Essential is that we have attraction in linear approximation to an equilibrium or critical (fixed) point.

**Theorem 6.2** *Consider the initial value problem* (2.1) *(abbreviated: $\dot{x} = \varepsilon f + \varepsilon^2 g$):*

$$\dot{y} = \varepsilon f^0(y), \; y(t_0) = x_0. \tag{6.3}$$

*Assume for $x, y, x_0, a \in D \subset \mathbb{R}^n$, $t_0 \leq t \leq \infty$, $0 \leq \varepsilon \ll 1$:*

1. *$f, g, Df$ are defined, continuous and bounded by a constant $M$ independent of $\varepsilon$ in $[t_0, \infty] \times D$.*
2. *$g$ is Lipschitz-continuous with respect to $x \in D$.*
3. *$f$ is $T$-periodic in $T$ with $T$ a constant independent of $\varepsilon$.*
4. *$y(t)$ belongs to a bounded interior subset of $D$ on the timescale $1/\varepsilon$.*
5. *$f^0(a) = 0$ with $x = a$ an asymptotically stable point in linear approximation of Eq.* (6.3).
6. *$x_0$ is in an interior subset of the domain of attraction of $x = a$.*

*Then we have for the solutions of the initial value problems the estimate:*

$$x(t) - y(t) = O(\varepsilon) \text{ as } \varepsilon \to 0 \text{ for } t \geq t_0.$$

**Remark 6.2** Extensions of the theorem are useful in the case of limit cycles of autonomous systems and in dynamical systems described by maps with attracting manifolds. For a proof of theorem 6.2 and discussion, see [123] ch. 5.

## 6.3 Exercises

As mentioned before interesting new phenomena can take place at higher order approximation and on longer timescales. More complicated applications will be studied in Chaps. 9 and 11.

### 6.3.1 *The Amplitude of Periodic Solutions of Autonomous Equations*

We noted in remark (3.3) that for periodic solutions of autonomous ODEs we will always find at least one eigenvalue zero (Lyapunov exponent 1). For the Van der Pol equation (2.12) and its modified versions, this means that the approximations obtained imply asymptotic stability, but that Theorem 6.2 cannot be applied. However, the attraction takes place to the orbits in linear approximation. The consequence is that the amplitude(s) has approximations valid for all time.

### 6.3.2 *The Anharmonic Equation with Damping*

We return to the anharmonic equation (1.4) and add linear damping. Consider the oscillator described by the non-linear equation:

$$\ddot{x} + \varepsilon\mu\dot{x} + x - \varepsilon x^n = 0, \tag{6.4}$$

with initial values $x(0), \dot{x}(0)$ given, $n$ is a natural number ($n = 2, 3, 4, \ldots$) but as we noted at first order there is a difference between odd and even $n$. As for Eq. (1.4) we transform $x, \dot{x} \mapsto r, \phi$ by Eq. (1.6); It is not necessary to use comoving variables as the singularity at $r = 0$ is removed. For $r, \phi$ we find from transformation (1.8) the variational equations:

$$\begin{cases} \dot{r} &= -\varepsilon(\mu r \sin^2(t + \phi) + r^n \sin(t + \phi)\cos^n(t + \phi)), \\ \dot{\phi} &= -\varepsilon(\mu \sin(t + \phi)\cos(t + \phi) + r^{n-1}\cos^{n+1}(t + \phi)). \end{cases} \tag{6.5}$$

Averaging the right-hand sides over $t$ keeping $r$ and $\phi$ constant, we find

$$\frac{dr}{dt} = -\frac{1}{2}\mu r,\ \frac{d\phi}{dt} = 0 \text{ if n is even,} \tag{6.6}$$

whereas $d\phi/dt \neq 0$ if $n =$ odd. For instance if $n = 3$ we find:

$$\frac{dr}{dt} = -\frac{1}{2}\mu r,\ \frac{d\tilde{\phi}}{dt} = -\varepsilon\, \frac{3}{8} r(0)^2.$$

If $\mu > 0$ the result produces asymptotic stability of the origin in both cases and validity of the $O(\varepsilon)$ approximation for $t \geq 0$. The phase changes non-trivially if $n$ is odd depending on the initial amplitude $r(0)$ as in Chap. 1.

### 6.3.3 The Damped Duffing Equation with $O(1)$ Forcing

We consider another example but with dissipation and forcing. It is interesting and realistic to modify the Duffing equation to allow $O(1)$ forcing. Forced oscillations with a non-linear restoring force present basic problems in non-linear mechanics leading to much attention. See for instance the book by Stoker [126] ch. 4 with a detailed analysis and emphasis on the use of the harmonic balance method; this choice of analysis makes the results of Stoker mathematical formal, but in [126] the mathematically rigorous Poincaré-Lindstedt method is also partly used.

Consider the Duffing equation with small viscous damping and $O(1)$ forcing in the form:

$$\ddot{x} + \varepsilon\mu\dot{x} - \varepsilon x^3 + x = h \sin\omega t,\ \omega \neq 1, h, \mu > 0. \tag{6.7}$$

The restoring force is softening. Following Sect. 1.5, we start with a particular inhomogeneous solution $x_p(t)$ of the reduced equation for $\varepsilon = 0$:

$$x_p(t) = \frac{h}{1-\omega^2} \sin\omega t,\ \ \omega \neq 1. \tag{6.8}$$

With homogeneous solution $r_0 \cos(t + \phi_0)$, this induces us to transform $x, \dot{x} \to r, \phi$ by

$$\begin{cases} x(t) & = r\cos(t+\phi) + \frac{h}{1-\omega^2} \sin\omega t, \\ \dot{x}(t) & = -r\sin(t+\phi) + \frac{h\omega}{1-\omega^2} \cos\omega t. \end{cases} \tag{6.9}$$

Replacing $-\mu\dot{x} + x^3$ by $F$, we find the variational equations:

$$\dot{r} = -\varepsilon \sin(t + \phi)F, \ \dot{\phi} = -\frac{\varepsilon}{r}\cos(t + \phi)F, \tag{6.10}$$

with the assumption that a neighbourhood of the origin in polar coordinates $r = 0$ is excluded, and where the arguments of $F(x, \dot{x})$ have to be replaced using transformation (6.9).

In a neighbourhood of $r = 0$, we should use the transformation $x, \dot{x} \to A, B$ by

$$\begin{cases} x(t) &= A\cos t + B\sin t + \frac{h}{1-\omega^2}\sin\omega t, \\ \dot{x}(t) &= -A\sin t + B\cos t + \frac{h\omega}{1-\omega^2}\cos\omega t, \end{cases} \tag{6.11}$$

leading to the variational system:

$$\dot{A} = -\varepsilon \sin(t)F, \ \dot{B} = \varepsilon\cos(t)F. \tag{6.12}$$

For this variational system the arguments of $F(x, \dot{x})$ have to be replaced using transformation (6.11).

We use $F = -\mu\dot{x} + x^3$ and consider first a neighbourhood of $x_p(t)$, requiring us to use variational system (6.12). We will use for the averaged quantities again $A, B$ or $r, \phi$.

**The case** $\omega \neq 1, \frac{1}{3}, 3$. We find after averaging over $t$:

$$\begin{cases} \dot{A} &= -\varepsilon\frac{\mu}{2}A - \varepsilon\frac{3B}{8}(A^2 + B^2 + 2\frac{h^2}{(1-\omega^2)^2}), \\ \dot{B} &= -\varepsilon\frac{\mu}{2}B + \varepsilon\frac{3A}{8}(A^2 + B^2 + 2\frac{h^2}{(1-\omega^2)^2}). \end{cases} \tag{6.13}$$

Multiplying the first equation by $A$, the second one by $B$, and adding the equations, we find

$$\frac{d}{dt}(A^2 + B^2) = -\varepsilon\mu(A^2 + B^2),$$

so $A, B$ decrease and the phase flow shrinks exponentially as $\exp(-\varepsilon\mu t)$. If $\omega \neq 1, \frac{1}{3}, 3$, the origin $A = B = 0$ of the averaged system is exponentially stable. We conclude with Theorem 6.2:

*If* $\omega \neq 1, \frac{1}{3}, 3$, *the solution* $x_p(t)$of (6.8) *approximates the solutions of* Eq. (6.7) *to* $O(\varepsilon)$ *for all time on a domain containing the origin with size independent of* $\varepsilon$.

**The case** $\omega = 3$. After averaging over $t$: Using transformation (6.11), we find that $A = B = 0$ is stable in the averaged system. The results are more transparent using amplitude-phase variables with transformation (6.9) for $r > 0$; we checked that the results are equivalent. We find the averaged system:

$$\dot{r} = \varepsilon r(-\frac{1}{2}\mu + \frac{3}{64}hr\cos 3\phi), \ \dot{\phi} = -\frac{3}{8}\varepsilon(r^2 + \frac{h}{8}r\sin 3\phi + \frac{1}{32}h^2).$$

Critical points for $r > 0$ arise if:

$$\frac{3}{32}hr\cos 3\phi = \mu,\ r^2 + \frac{h}{8}r\sin 3\phi + \frac{1}{32}h^2 = 0.$$

As the discriminant of the second equation will be negative, we find at this order of approximation no additional subharmonic ($\omega = 3$) periodic solution of size $O(\varepsilon)$ of Eq. (6.7). We conclude with Theorem 6.2: *If* $\omega = 3$, *the solution* $x_p(t)$ *of* (6.8) *approximates the solutions of* Eq. (6.7) *starting in a neighbourhood of* $A = B = 0$ $(r = 0)$ *to* $O(\varepsilon)$ *for all time. No subharmonic periodic solutions of system* (6.7) *exist in this neighbourhood apart from* $x_p(t)$.

**The case** $\omega = \frac{1}{3}$. As for the case $\omega = 3$, we can average over $t$ using transformation (6.9); again we checked that the results are equivalent. Interestingly the origin is not a solution of the averaged system in this case.

Using $(r, \phi)$ coordinates, we find after averaging:

$$\dot{r} = \frac{\varepsilon}{2}(-\mu r + \frac{729}{2048}h^3\cos\phi),\ \dot{\phi} = -\frac{3}{8}\frac{\varepsilon}{r}(r^3 + \frac{81}{32}h^2 r + \frac{243}{512}h^3\sin\phi), \tag{6.14}$$

with critical points $r_0, \phi_0$. We find for these stationary solutions $r_0, \phi_0$:

$$\cos\phi_0 = \frac{2048}{729}\frac{\mu}{h^3}r_0,\ \sin\phi_0 = -\frac{512}{243}\frac{r_0^3}{h^3} - \frac{16}{3}\frac{r_0}{h}. \tag{6.15}$$

Using the formula $\cos^2\phi_0 + \sin^2\phi_0 = 1$, we can derive a cubic equation for $r_0^2 = z$. This equation is of the form:

$$\frac{a}{h^6}z^3 + \frac{b}{h^4}z^2 + \frac{c\mu^2}{h^6}z + \frac{d}{h^2}z = 1,\ a, b, c, d > 0.$$

It is easy to see that for fixed damping coefficient $\mu$ and $h$ large enough we will have a unique positive solution $z = r_0^2$ with $|\cos\psi_0| \leq 1, |\sin\psi_0| \leq 1$. Using expression (6.9) and $x_p(t) = \frac{9}{8}h\sin(t/3)$, we have found an $O(\varepsilon)$ approximation of a periodic solution.

The stability of this periodic solution is obtained by linearising system (6.14) at $r_0, \phi_0$. We find the matrix:

$$\begin{pmatrix} -\frac{\varepsilon}{2}\mu & -\frac{1}{2}\varepsilon\frac{729}{2048}h^3\sin\phi_0 \\ -\frac{3}{8}\frac{\varepsilon}{r_0}(3r_0^2 + \frac{81}{32}h^2) & -\frac{3}{8}\frac{\varepsilon}{r_0}\frac{243}{512}h^3\cos\phi_0 \end{pmatrix}. \tag{6.16}$$

The eigenvalues are $\lambda_1, \lambda_2$; as $\cos\psi_0 > 0$ the trace of the matrix $(\lambda_1 + \lambda_2)$ is negative. As $\sin\phi_0 < 0$, the product $\lambda_1\lambda_2$ is positive. We conclude that the critical point and so the periodic solution are asymptotically stable.

Figure 6.1 (left) presents the $x(t)$ time series if $\omega = 1/3$ with initial transient in the case $\mu = h = 1, \varepsilon = 0.1$. Note that we did not study the domains of attraction

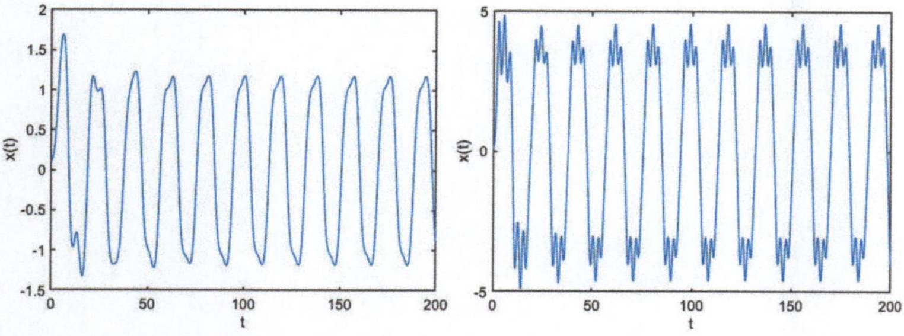


**Fig. 6.1** Time series $x(t)$ (200 timesteps) with initial transient in the case of the forced Duffing equation with damping (6.7) starting at $x(0) = 0.1, \dot{x}(0) = 0.1$ tending to a periodic solution $r_0 \cos(t + \psi_0) + \frac{9}{8} h \sin(t/3)$. We have chosen $\varepsilon = 0.1$, $\mu = h = 1$, $\omega = 1/3$. For comparison we present (right figure) the time series in the case of hardening force field $+\varepsilon x^3$ with the same data except $h = 10$

of the periodic solutions as dependent on $\mu$ and $h$, and there may be other types of solutions. A number of numerical experiments confirm the possibility of unbounded solutions, for instance changing in Fig. 6.1 $h$ from 1 to 2 produces these. On the other hand, with hardening force $+\varepsilon x^3$, we still find periodic solutions for larger forcing amplitudes; in Fig. 6.1 (right) we show the periodic solution in the hardening case with $h = 10$.

# Chapter 7
# Averaging over Angles

**Abstract** A natural perturbation formulation involves amplitudes and angles. Averaging over the angles produces valid approximations, but special problems and new phenomena in certain parts of phase space, the so-called resonance manifolds, will arise. A full description of the dynamics in resonance manifolds usually requires the construction of a second order approximation. Several examples illustrate the dynamics, and in Chap. 12 the theory is applied to the dynamics of a flywheel on an elastic foundation.

Consider again the perturbed harmonic Eq. (1.5) or systems of such equations and more general equations describing oscillations that will involve amplitudes and angles. As we will show in exercises and in the applications, we can often transform to other natural variational systems as will be presented in the following theorems. In this context it can be useful to average over one or more angles instead of over time, but they have to be timelike; for this concept see Remark 7.2 in the sequel.

## 7.1 Averaging over One Angle

**Theorem 7.1** *Consider the* $(n+1)$*-dimensional system:*

$$\dot{x} = \varepsilon X(\phi, x), \quad \dot{\phi} = \omega(x) + \varepsilon \ldots \tag{7.1}$$

*with* $x, x_0 \in D \subset \mathbb{R}^n, 0 \leq \phi \leq 2\pi$ *(*$\phi$ *1-dimensional) and* $X$ $2\pi$*-periodic in* $\phi$*. Introduce averaging over the angle* $\phi$ *by*

$$X^0(y) = \frac{1}{2\pi} \int_0^{2\pi} X(s, y) ds.$$

*Assume:*




F. Verhulst, *Averaging for Nonlinear Dynamics with Applications and Numerical Bifurcations*, Applied Mathematical Sciences 223,
https://doi.org/10.1007/978-3-032-12745-7_7

1. *The vector functions $X$, $\omega$ and the higher order terms indicated by dots are continuously differentiable on their respective domains of definition.*
2. *The frequency function $\omega(x)$ is bounded away from zero by a constant ($\phi$ is timelike).*
3. *The solutions of the system averaged over the angle*

$$\dot{y} = \varepsilon X^0(y), \ \dot{\phi} = \omega(y), \tag{7.2}$$

*remain in an interior subset of $D$.*

*Then we have $x(t) - y(t) = O(\varepsilon)$ on the timescale $1/\varepsilon$.*

**Remark 7.1** A proof can be found in [123] ch. 7. The proof becomes more complicated if $\omega(x)$ depends explicitly on time. As we shall see, a domain in $x$-space where the frequency $\omega(x)$ vanishes will be called a *resonance manifold*.

**Remark 7.2** An alternative approach to the proof uses that the angle $\phi$ is timelike. We mean by this that $\phi$ is a monotonically increasing or decreasing function of time, and the time interval can be mapped one to one on the circle that $\phi$ travels around repeatedly. We can eliminate time in the case of one angle by dividing the equation for $x$ by the equation for $\phi$ and average over "time" $\phi$.

## 7.2 Averaging over More Angles

The general idea of averaging over more angles is not difficult to understand, but the details of the applications contain many difficulties and surprises. The reader is referred also to an extensive account in ch. 4 of [4].

Consider the system

$$\dot{x} = \varepsilon X(\phi, x), \ \dot{\phi} = \Omega(x), \tag{7.3}$$

with $x \in D \subset \mathbb{R}^n$, $\phi \in T^m$. $T^m$ is the $m$-dimensional torus described by $m$ angles. We assume that the vector field $X$ is periodic in the $m$ angles with in general different periods. Using the individual angles $\phi_i, i = 2, \ldots, m$ as timelike variables, we can average over the individual angles unless a component $\Omega_i(x)$ vanishes in $D$. However, this picture is not complete, with more than one angle the interaction can be more complicated. To show this consider a generalised Fourier analysis of $X(\phi, x)$ of the form:

$$X(\phi, x) = \sum c_k(x) e^{i(k_1\phi_1 + \ldots + k_m\phi_m)},$$

with index $k = (k_1, \ldots, k_m)$. If for certain $k$ with $c_k(x) \neq 0$, we have that there is a subdomain where the *combination angle* $\chi_k$ vanishes

$$\chi_k(x) = k_1\Omega_1(x) + k_2\Omega_2(x) + \ldots + k_m\Omega_m(x) = 0,$$

then the corresponding combination angle $\chi_k$ is not timelike in this subdomain. We will call such a subdomain a *resonance manifold* for system (7.3). Outside the resonance manifolds in $D$ Theorem 7.1 for more angles applies, inside and near these manifolds, we have to consider system (7.3) separately.

**Remark 7.3 (Nonautonomous Equations)** In practice, it happens quite often that time $t$ enters explicitly into the equations. Consider the system

$$\dot{x} = \varepsilon X(\phi, t, x),$$
$$\dot{\phi} = \Omega(x),$$

with the vector function $X$ periodic in $t$. Suppose that $\phi$ is $m$-dimensional; we add the angle $\phi_{m+1}$ with:

$$\phi_{m+1} = t,\ \dot{\phi}_{m+1} = 1,$$

and consider averaging over $m + 1$ angles.

## 7.3 Exercises

We start with an artificial problem.

### 7.3.1 An Example with Two Angles

To illustrate the concepts, we consider a system with two angles.

$$\begin{cases} \dot{x} &= \varepsilon(c_0(x) + c_1(x)\cos\phi_1 + c_2(x)\sin 2\phi_2 + c_3(x)\cos(\phi_1 - \phi_2)), \\ \dot{\phi}_1 &= \omega_1(x), \\ \dot{\phi}_2 &= \omega_2(x). \end{cases} \tag{7.4}$$

The functions $c_0(x), \ldots, c_3(x), \omega_1(x), \omega_2(x)$ are smooth and nontrivial on $\mathbb{R}$. Suppose first that the frequency functions are constant: $\omega_1(x) = k, \omega_2(x) = l$. If $k \neq l$, there is no resonance manifold, and the angles $\phi_1, \phi_2$ and the combination angle $(\phi_1 - \phi_2)$ are timelike so we can average to find $\dot{x} = \varepsilon c_0(x)$.

Suppose now that the frequency functions are not constant but positive, there exists an $x_0$ such that $\omega_1(x_0) = \omega_2(x_0)$. So if $x(t) = x_0$, $\dot{\phi}_1 - \dot{\phi}_2 = 0$, we cannot average over $(\phi_1 - \phi_2)$ in a neighbourhood of $x_0$. A neighbourhood of $x_0$ is called a resonance manifold $M$, and outside this neighbourhood, we find again by averaging

$\dot{x} = \varepsilon c_0(x)$. In subsequent examples, we will characterise the dynamics inside a resonance manifold. We will also present the case of solutions that are passing through a resonance manifold. In the present example this happens for instance if we put: $c_0(x) = 1, \omega_1(x) = 2, \omega_2(x) = x, x(0) = 1$. System (7.4) becomes

$$\begin{cases} \dot{x} &= \varepsilon(1 + c_1(x)\cos\phi_1 + c_2(x)\sin 2\phi_2 + c_3(x)\cos(\phi_1 - \phi_2)), \\ \dot{\phi}_1 &= 2, \\ \dot{\phi}_2 &= x. \end{cases} \tag{7.5}$$

The angle $\phi_1$ is timelike, and we can average over $\phi_1$; near the initial condition $x(0) = 1$ the angle $\phi_2$ is also timelike, and we can average over $\phi_2$ unless at some point $x$ tends to zero. For the combination angle $\psi = \phi_1 - \phi_2$ we have

$$\dot{\psi} = \dot{\phi}_1 - \dot{\phi}_2 = 2 - x.$$

So near $x = 2$, we have a resonance manifold $M$. Initially we have after averaging over $\phi_1, \phi_2, \psi$ the approximation $x(t) = 1+\varepsilon t$, but we will move into the resonance manifold $M$ at $x = 2$. The questions are then, do we pass $M$ and how, or are the solutions caught in $M$? A similar question would arise if we start at $x(0) < 0$. Such questions will be raised and answered in the applications. To answer the question in the case of Eq. (7.5), we have to be more explicit about the coefficients $c_1(x), c_2(x), c_3(x)$.

In the case of two-dimensional autonomous systems like the Van der Pol equation and the Duffing equation, we can easily apply theorem 7.1; we leave such applications to the reader. It is dynamically interesting to study what happens near resonance manifolds. In the case of more angles, this turns out to be quite complicated, and we will develop the necessary theory while discussing examples.

### 7.3.2 A Pendulum with Slow Frequency Variation

The transformations we will use in this problem offer a different perspective on the perturbation behaviour. Consider the equation:

$$\ddot{x} + \omega^2(\varepsilon t)x = 0. \tag{7.6}$$

We put $\dot{x} = \omega(\varepsilon t)y, \tau = \varepsilon t$; and using a change of transformation (1.21) by computing $\ddot{x} = \dot{\omega}y + \omega\dot{y}$ so that

$$\dot{y} = \frac{\ddot{x}}{\omega} - \frac{\dot{\omega}}{\omega}y = -\omega x - \frac{\dot{\omega}}{\omega}y,$$

we find with $x = r\sin\phi,\ y = r\cos\phi$

$$\dot{r} = -\varepsilon \frac{1}{\omega(\tau)} \frac{d\omega}{d\tau} r \cos^2 \phi,$$

$$\dot{\phi} = \omega(\tau) + \varepsilon \frac{1}{\omega(\tau)} \frac{d\omega}{d\tau} \sin \phi \cos \phi,$$

$$\dot{\tau} = \varepsilon.$$

If we assume that $0 < a < \omega(\tau) < b$ (with $a, b$ constants independent of $\varepsilon$), $\phi$ is timelike. We have a three-dimensional system periodic in $\phi$ with two equations slowly varying. The function $\omega(\tau)$ is smooth, and its derivative is bounded. Interestingly, we have made no other assumptions on $\omega(\varepsilon t)$.

Averaging the two slowly varying equations over $\phi$, we obtain

$$\dot{r} = -\varepsilon \frac{1}{2\omega(\tau)} \frac{d\omega}{d\tau} r,$$

$$\dot{\tau} = \varepsilon.$$

The equation for $\tau$ does not change. From this system, we get

$$\frac{dr}{d\tau} = -\frac{1}{2\omega(\tau)} \frac{d\omega}{d\tau} r,$$

which can be integrated to find

$$r(\tau) = \frac{r(0)\sqrt{\omega(0)}}{\sqrt{\omega(\tau)}}.$$

Note that the quantity $r(\varepsilon t)\sqrt{\omega(\varepsilon t)}$ is conserved in time, and it is an *adiabatic invariant* of Eq. (7.6). It is clear that the same technique can be applied to equations of the form

$$\ddot{x} + \varepsilon \mu \dot{x} + \omega^2(\varepsilon t) x + \varepsilon f(x) = 0.$$

### 7.3.3 Frequency Variations with a Zero

Interesting problems may arise when $\omega(x)$ is not bounded away from zero, and this will become an important issue in higher-dimensional problems with more angles. To prepare for this, we consider an artificial example.

Consider the two scalar equations

$$\dot{x} = \varepsilon(1 - \cos \phi + \sin 3\phi),$$

$$\dot{\phi} = x - 1.$$

If $x(0) \neq 1$ $\phi$ is timelike and averaging over $\phi$ produces

$$\dot{x} = \varepsilon, \; x(0) = x(0),$$

so that $x(t) = x(0) + \varepsilon t$ and $x(t) - (x(0) + \varepsilon t) = O(\varepsilon)$ on the timescale $1/\varepsilon$ as long as $x(0) + \varepsilon t$ remains outside a neighbourhood of $x = 1$. However, if $x(0) < 1$, the approximate solution will pass into a neighbourhood of resonance manifold $x = 1$. Such a neighbourhood is called a *resonance zone*.

As we shall see, a resonance zone is really a boundary layer in the sense that its size tends to zero as $\varepsilon$ tends to zero. We can see this when introducing a local variable $\xi$ by rescaling near $x = 1$. We put

$$\xi = \frac{x - 1}{\delta(\varepsilon)},$$

with $\delta(\varepsilon)$ a function of $\varepsilon$ to be determined. Transforming $x, \phi \rightarrow \xi, \phi$, the system becomes

$$\begin{aligned} \dot{\xi} &= \frac{\varepsilon}{\delta(\varepsilon)}(1 - \cos\phi + \sin 3\phi), \\ \dot{\phi} &= \delta(\varepsilon)\xi. \end{aligned}$$

The equations are balanced (sometimes called a significant degeneration of the differential operator) if the terms on the right-hand side are of the same order in $\varepsilon$ or

$$\frac{\varepsilon}{\delta(\varepsilon)} = \delta(\varepsilon).$$

We conclude that $\delta(\varepsilon) = \sqrt{\varepsilon}$, which is the size of the resonance zone. Differentiating the equation for $\phi$, we can eliminate $\xi$ to find the equation:

$$\ddot{\phi} + \varepsilon(\cos\phi - \sin 3\phi) = \varepsilon.$$

This conservative equation governs the dynamics in the resonance zone with timescale $1/\sqrt{\varepsilon}$.

### 7.3.4 Behaviour in a Resonance Manifold

The preceding example of a conservative equation is more typical than it seems. Consider, for instance, the two-dimensional system consisting of two coupled scalar equations:

$$\dot{x} = \varepsilon X(\phi, x),$$
$$\dot{\phi} = \omega(x),$$

with $X$ and $\omega$ smooth scalar functions and $X(\phi, x)$ $2\pi$-periodic in $\phi$. Averaging over $\phi$ is possible outside domains determined by the zeros of $\omega(x)$. Suppose $\omega(c) = 0$ with $c$ an isolated zero of $\omega(x)$ and rescale:

$$\xi = \frac{x - c}{\delta(\varepsilon)}.$$

The equations become

$$\delta(\varepsilon)\dot{\xi} = \varepsilon X(\phi, c + \delta(\varepsilon)\xi),$$
$$\dot{\phi} = \omega(c + \delta(\varepsilon)\xi).$$

Expanding we find:

$$\delta(\varepsilon)\dot{\xi} = \varepsilon X(\phi, c) + O(\delta(\varepsilon)\varepsilon),$$
$$\dot{\phi} = \delta(\varepsilon)\frac{d\omega}{dx}(c)\xi + O(\delta^2(\varepsilon)).$$

Again, a suitable balancing (significant degeneration) arises if we choose $\delta(\varepsilon) = \sqrt{\varepsilon}$, which is the size of the resonance zone. The equations in the resonance zone are to first order

$$\dot{\xi} = \sqrt{\varepsilon} X(\phi, c),$$
$$\dot{\phi} = \sqrt{\varepsilon}\frac{d\omega}{dx}(c)\xi,$$

which is again to first order a conservative system (one can show this by differentiating the equation for $\phi$ and eliminating $\xi$). We demonstrate this explicitly in a pendulum system.

### 7.3.5 A Basic Pendulum Equation

Consider the system:

$$\begin{cases} \dot{x} = & \varepsilon(-x - \sin\phi), \\ \dot{\phi} = & x. \end{cases} \tag{7.7}$$

From the preceding analysis, we know that a neighbourhood of $x = 0$ is a resonance zone. Outside the resonance zone, we average over $\phi$ to obtain an approximation of $x(t)$ described by

$$\dot{x} = -\varepsilon x.$$

With given $x(0)$, we find the approximation $x(0)\exp(-\varepsilon t)$. In the resonance zone, we rescale $\xi = x/\sqrt{\varepsilon}$, which leads to the system:

$$\begin{aligned} \sqrt{\varepsilon}\dot{\xi} &= \varepsilon(-\sqrt{\varepsilon}\xi - \sin\phi), \\ \dot{\phi} &= \sqrt{\varepsilon}\xi. \end{aligned}$$

We find

$$\dot{\phi} = \sqrt{\varepsilon}\xi,\ \dot{\xi} = -\sqrt{\varepsilon}\sin\phi - \varepsilon\xi,$$

which is to order $O(\sqrt{\varepsilon})$ again conservative, but not to order $O(\varepsilon)$. More insight is obtained by differentiating the equation for $\phi$ in system (7.7), producing

$$\ddot{\phi} + \varepsilon\dot{\phi} + \varepsilon\sin\phi = 0.$$

The equation describes the motion of a mathematical pendulum with linear damping. The equilibria at $(2n\pi, 0), n = 1, 2, \ldots$, are positive attractors, but this is described asymptotically by a second order calculation in the resonance zone, not by first order.

### 7.3.6 A Three-Dimensional System with Two Angles

As indicated earlier, when more angles are present, subtle problems and interesting phenomena may arise. We start again with an example. Consider the system with given initial value $x(0)$:

$$\begin{aligned} \dot{x} &= \varepsilon x(\cos\phi_1 + \cos(2\phi_1 - \phi_2)), \\ \dot{\phi}_1 &= x^2 + 1, \\ \dot{\phi}_2 &= 1. \end{aligned}$$

In the equation for $x$, we have angle $\phi_1$ and combination angle $2\phi_1 - \phi_2$. Both are timelike as $\dot{\phi}_1 > 0$ and $2\dot{\phi}_1 - \dot{\phi}_2 = 2x^2 + 1 > 0$. Averaging the equation for $x$ over $\phi_1$ and the combination angle, we find the averaged equation $\dot{x} = 0$. We conclude that averaging over more angles yields that $x(0)$ is an $O(\varepsilon)$ approximation of $x(t)$ on the timescale $O(1/\varepsilon)$.

### 7.3.7 A Modification

We consider a modification of the last example. Consider the system

$$\begin{cases} \dot{x} &= \varepsilon + \varepsilon x(\cos\phi_1 - 2\cos(2\phi_1 - \phi_2)), \\ \dot{\phi_1} &= x^2 + 1, \\ \dot{\phi_2} &= 4. \end{cases} \tag{7.8}$$

We have still the same two angle combinations in the equation for $x$ as in the preceding example. Again $\dot{\phi}_1 > 0$ so $\phi_1$ is timelike, but we find $\dot{\psi} = 2\dot{\phi}_1 - \dot{\phi}_2 = 2x^2 - 2$, so the second combination angle is timelike only outside two resonance zones which are neighbourhoods of $x = \pm 1$. Outside the resonance zones, we have the approximation $x(0) + \varepsilon t$ for $x(t)$, so if $x(0) < 1$ we will enter at least one resonance zone. To describe the dynamics near $x = 1$, we rescale as before

$$\xi = \frac{x-1}{\sqrt{\varepsilon}},$$

to find in the resonance zone near $x = 1$:

$$\begin{aligned} \dot{\xi} &= \sqrt{\varepsilon}(1 + \cos\phi_1 - 2\cos\psi) + O(\varepsilon), \\ \dot{\phi_1} &= 2 + O(\sqrt{\varepsilon}), \\ \dot{\psi} &= 4\sqrt{\varepsilon}\xi + O(\varepsilon). \end{aligned}$$

Averaging over $\phi_1$ and differentiating the equation for the combination angle $\psi$, we find to first order

$$\ddot{\psi} = 4\varepsilon(1 - 2\cos\psi).$$

The first order equation in the resonance zone is again conservative, but it is interesting to note that we have a centre critical point (equilibrium) ($\psi = -\pi/2, \dot{\psi} = 0$) and a saddle ($\psi = \pi/2, \dot{\psi} = 0$) in the resonance zone. The saddle will persist as an unstable solution in higher order approximation. The nature of the perturbed centre will show at higher order, see Fig. 7.1.

In some of the preceding examples, we encountered the phenomenon of solutions that are attracted to an equilibrium in a resonance zone. We call this *resonance locking*.

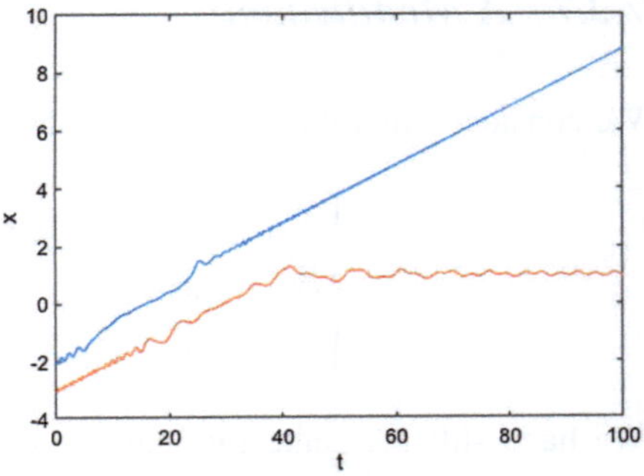


**Fig. 7.1** Passing into resonance for system (7.8) with $\varepsilon = 0.1$, initial conditions $x(0) = -2, \phi_1(0) = 1, \phi_2(0) = 0$ and $x(0) = -3, \phi_1(0) = 1, \phi_2(0) = 0$. In the second case, the attraction is into the resonance zone $x = 1$

### *7.3.8 Intersection of Resonance Manifolds*

We analyse a more complicated example with two slowly varying components and two angles where we have resonance domains that may intersect.

Consider the system:

$$\begin{cases} \dot{x}_1 &= \varepsilon(0.5 + 2x_1 \cos\phi_1 + x_2 \cos\phi_2), \\ \dot{x}_2 &= \varepsilon(0.5 - x_1 \cos\phi_1 + 1.5x_2 \cos\phi_2), \\ \dot{\phi}_1 &= x_2 - x_1^2, \\ \dot{\phi}_2 &= x_1 - 1. \end{cases} \tag{7.9}$$

The angle $\phi_1$ is timelike outside the set of points where $x_2 = x_1^2$; $\phi_2$ is timelike outside the set of points where $x_1 = 1$. Averaging outside the two resonance manifolds produces the approximations $x_1(t) = x_1(0) + \varepsilon t$ and $x_2(t) = x_2(0) + \varepsilon t$. This follows from averaging theory, and it is for initial values outside the resonance zones confirmed by numerical computation.

A double resonance arises at the intersection $x_1 = x_2 = 1$.

The analysis at the separate resonance manifolds runs as before. Leaving out the location of the double resonance, we put respectively

$$\frac{x_1 - 1}{\sqrt{\varepsilon}} = \xi, \text{ and } \frac{x_2 - x_1^2}{\sqrt{\varepsilon}} = \eta$$

to perform the rescalings of the equations at the two sets. We discuss the dynamics in the separate manifolds.

**The Resonance Manifold** $x_1 = 1$

Assume $x_2 \neq 1$. The equations for $x_1, x_2$ can be averaged over $\phi_1$ producing

$$\dot{x}_1 = \varepsilon(0.5 + x_2 \cos\phi_2), \; \dot{x}_2 = \varepsilon(0.5 + 1.5x_2 \cos\phi_2),$$

with integral

$$\frac{3}{2}x_1(t) - x_2(t) = \frac{\varepsilon}{4}t + \frac{3}{2}x_1(0) - x_2(0).$$

If $x_1(t)$ remains in the resonance set, we should have

$$x_2(t) = -\frac{\varepsilon}{4}t + x_2(0) + \dots,$$

where the dots stand for higher order terms. This is not compatible with the equation for $x_2$ so we conclude that $x_1(t)$ leaves the resonance manifold. This is confirmed by a number of numerical experiments.

**The Resonance Manifold $x_2 = x_1^2$**

As we leave out a neighbourhood of $(x_1, x_2) = (1, 1)$, averaging over $\phi_2$ yields that the terms with $\phi_2$ in system (7.9) vanish at first order. Choosing $x_1 = x_r \neq 1, x_2 = x_r^2$, we obtain after rescaling

$$\dot{\eta} = \sqrt{\varepsilon}(\frac{1}{2}(1 - 2x_r) - (x_r + 4x_r^2)\cos\phi_1) + O(\varepsilon), \dot{\phi}_1 = \sqrt{\varepsilon}\eta.$$

Differentiating the equation for $\phi_1$, we get

$$\ddot{\phi}_1 + \varepsilon(x_r + 4x_r^2)\cos\phi_1 = \varepsilon\frac{1}{2}(1 - 2x_r) + O(\varepsilon^{\frac{3}{2}}). \tag{7.10}$$

If $x_r < -\frac{1}{4} - \frac{1}{4}\sqrt{3}$ or $x_r > -\frac{1}{4} + \frac{1}{4}\sqrt{3}, x_r \neq 1$, Eq. (7.10) has a centre and a saddle equilibrium. See Fig. 7.2 where the behaviour of $\eta(t)$ and $x_2 - x_1^2$ is presented. The first order equation (7.10) becomes with the initial condition $x_1(0) = 0.5, x_2(0) = 0.25$

$$\ddot{\phi}_1 + \frac{3}{2}\varepsilon\cos\phi_1 = 0, \text{ critical points } \phi_1 = \pm\pi/2.$$

It is remarkable that with the initial conditions of Fig. 7.2 in the resonance zone, the solutions follow the resonance zone to end up in the double resonance zone. Keeping the initial values $x_1(0) = 0.5, x_2(0) = 0.25, \phi_1(0) = -1.57$, the same pattern of ending up in the double resonance zone is obtained when choosing $\phi_2(0)$ in the interval $I$ given by $-0.05 \leq \phi_2(0) \leq 0.88$.

Solutions that are starting close to the trapping interval $I$ are interacting for some time in the double resonance zone followed by escape from this zone; see Fig. 7.3.

**The Double Resonance**

Near the double resonance, we use the rescaling (blowup):

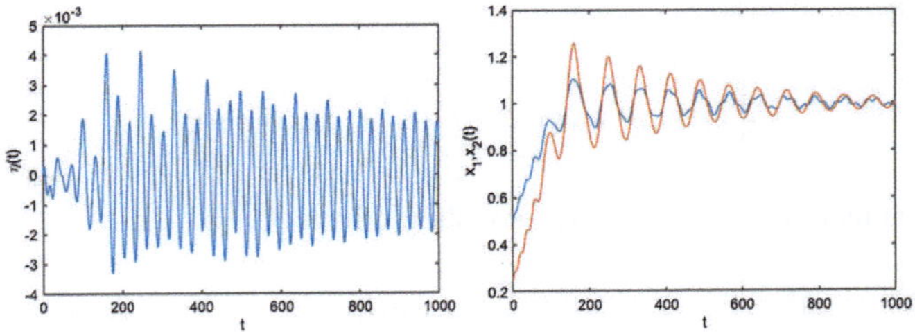


**Fig. 7.2** Dynamics starting in the resonance set $x_2 = x_1^2$. Time series $\eta(t)$ (1000 timesteps) for system (7.9) starting left at $x_1(0) = 0.5, x_2(0) = 0.25, \phi_1(0) = -1.57, \phi_2(0) = 0.5$ with $\varepsilon = 0.01$. The solutions are starting near an unstable solution and tend to a stable solution in the double resonance set near $x_1 = x_2 = 1$ (right picture)

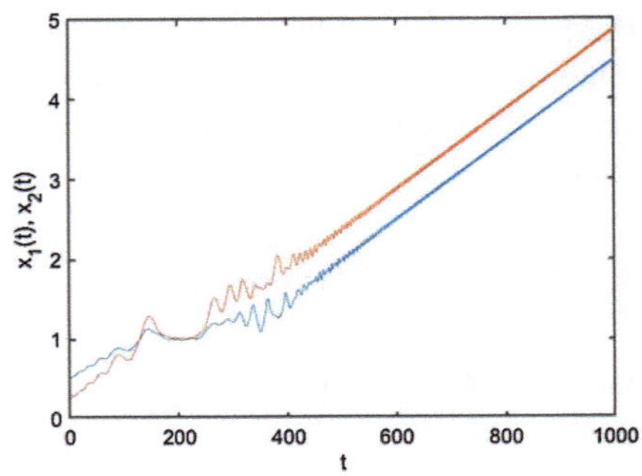


**Fig. 7.3** Dynamics starting in the resonance manifold $x_2 = x_1^2$. Time series (1000 timesteps) for $x_1(t), x_2(t)$ of system (7.9) starting at $x_1(0) = 0.5, x_2(0) = 0.25, \phi_1(0) = -1.57, \phi_2(0) = -0.06$ with $\varepsilon = 0.01$. The solutions are close to being trapped in the double resonance zone near $x_1 = x_2 = 1$ but are leaving this zone after about 200 timesteps

$$\frac{x_1 - 1}{\sqrt{\varepsilon}} = \xi, \ \frac{x_2 - 1}{\sqrt{\varepsilon}} = \eta,$$

producing with system (7.9):

$$\begin{cases} \dot{\xi} &= \sqrt{\varepsilon}(0.5 + 2\cos\phi_1 + \cos\phi_2) + \varepsilon \ldots, \\ \dot{\eta} &= \sqrt{\varepsilon}(0.5 - \cos\phi_1 + 1.5\cos\phi_2) + \varepsilon \ldots, \\ \dot{\phi}_1 &= \sqrt{\varepsilon}(\eta - 2\xi) + \varepsilon\xi^2, \\ \dot{\phi}_2 &= \sqrt{\varepsilon}\xi. \end{cases} \tag{7.11}$$

Differentiation of the equations for $\phi_1, \phi_2$ yields

$$\ddot{\phi}_1 = -\varepsilon(0.5+5\cos\phi_1+0.5\cos\phi_2)+O(\varepsilon^{3/2}), \ \ddot{\phi}_2 = \varepsilon(0.5+2\cos\phi_1+\cos\phi_2)+O(\varepsilon^{3/2}). \tag{7.12}$$

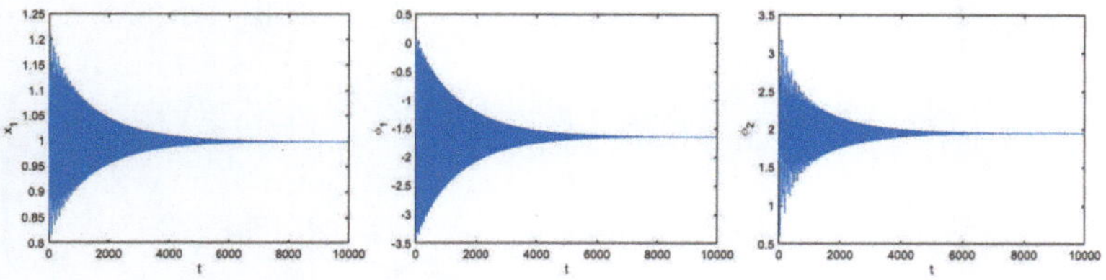

**Fig. 7.4** Time series (10,000 timesteps horizontally) for system (7.9) starting at $x_1(0) = 1$, $x_2(0) = 0.95$, $\phi_1(0) = 0.1$, $\phi_2(0) = 1.6$ with $\varepsilon = 0.01$. These solutions tend to a stable solution in the double resonance zone near $x_1 = x_2 = 1$ (left), $\phi_1 \approx -1.7$ (middle), $\phi_2 \approx 1.9$ (right)

The first order equations generate a volume-preserving flow in phase space; in Fig. 7.4, a numerical calculation of system (7.9) shows convergence to a stable solution in the double resonance zone.

An independent check is the analysis of critical points (equilibria) of autonomous system (7.9). The four critical points are in the double resonance zone and are determined by

$$(x_1, x_2) = (1, 1),\ \cos\phi_1 = -\frac{1}{16}, \cos\phi_2 = -\frac{3}{8}. \tag{7.13}$$

We abbreviate $S_1 = \sin\phi_1$, $S_2 = \sin\phi_2$. The matrix determining the stability at the four critical points of system (7.9) is

$$\begin{pmatrix} -\frac{1}{8}\varepsilon & -\frac{3}{8}\varepsilon & -2\varepsilon S_1 & -\varepsilon S_2 \\ \frac{1}{16}\varepsilon & -\frac{9}{16}\varepsilon & \varepsilon S_1 & -\frac{3}{2}\varepsilon S_2 \\ -2 & 1 & 0 & 0 \\ 1 & 0 & 0 & 0 \end{pmatrix}.$$

We find one equilibrium with negative real eigenvalues and so stability if $\sin\phi_1 < 0$, $\sin\phi_2 > 0$; the eigenvalues depend on $\varepsilon$, and we find for instance

$$\varepsilon\ = 0.1;\ \text{eigenvalues}\ -0.0065 \pm 0.7217i\ \ -0.0279 \pm 0.2651i,$$

$$\varepsilon = 0.01;\ \text{eigenvalues}\ -0.0007 \pm 0.2282i\ \ -0.0028 \pm 0.0843i\ ,$$

$$\varepsilon = 0.001;\ \text{eigenvalues}\ -0.0001 \pm 0.0722i\ \ -0.0003 \pm 0.0267i.$$

This settles the question of positive attraction in the double resonance zone.

# Chapter 8
# Averaging for Partial Differential Equations

**Abstract** For formal quantitative analysis of PDEs, many approaches are known. A number of averaging techniques are considered tin this chapter that can be validated mathematically. Under certain conditions, one can average an equation where time dependence and a linear operator are present. A major problem is that the "unperturbed" problem can sometimes be understood qualitatively but is difficult to be used to formulate slowly varying variational equations; however, this is possible in a number of cases. Another approach, suitable for boundary value problems, is using projection methods. This will be illustrated in Chap. 14.

The qualitative and quantitative theory of weakly non-linear partial differential equations is still fragmented, and it is too early to present a coherent picture of the theory, but we will discuss a number of promising methods and interesting problems. For more details and references, see the appendix of [123] and [145] and for applications Chap. 14.

Apart from averaging methods, formal approximation methods, as for example multiple timing, have sometimes been successful and stimulating, for both PDEs on bounded and unbounded domains. Another formal method that attracted a lot of interest is Whithams approach to combine averaging and variational principles; see for these formal methods [148]. The validation for certain parameter values and side conditions has to be numerical for these methods. One of the reasons of the relatively many open problems for perturbed PDEs is that the unperturbed problem is often already quite difficult and no easy start for a perturbation approach. To study linear perturbation, problems can already be very instructive.

The mathematical analysis of asymptotic approximations of PDEs with proofs of validity moves at a slow pace, and it rests firmly on the qualitative theory of weakly non-linear partial differential equations. Existence and uniqueness results are available, which involve typically contraction, or other fixed point methods, and maximum principles; one can also use projection methods in Hilbert spaces where eigenfuction expansions are used. If a functional analytic variational principle is available, one can use Galerkin methods, possibly in combination with averaging.



F. Verhulst, *Averaging for Nonlinear Dynamics with Applications and Numerical Bifurcations*, Applied Mathematical Sciences 223,
https://doi.org/10.1007/978-3-032-12745-7_8

Various forms of averaging techniques are being used in the literature. Some of them are indicated by terms as "homogenisation" or "regularisation" methods. Their main purpose is to stabilise numerical integration schemes for partial differential equations. They open the possibility of hybrid methods where asymptotic analysis is used to obtain numerical improvements of a PDE problem.

## 8.1 Metric Spaces and Norms, a Reminder

For metric spaces and norms, also leading to Banach and Hilbert spaces, we do not aim at an introduction, but we remind the reader only of concepts that are useful in the analysis of PDEs. There are many good introductions to functional analysis directed at applications, see, for instance, ch. 2 in [83]; the reader is referred to such introductions.

A set of elements $a, b, c, \ldots$ is called a vector space (or linear vectorspace) if in the set multiplication and addition have been defined according to certain rules, for instance with respect to addition the set has to be an Abelian group. The elements can be vectors in the vector space $\mathbb{R}^n$ or for instance the continuous functions or continuously differentiable ($C^1$) functions on the interval [0, 1].

We can provide the vector space with a norm that means that we assign to each element $a$ of the vector space a unique real number $\geq 0$; the norm of $a$ is indicated as $\|a\|$. In $\mathbb{R}^n$ the Euclidean distance for $a = (a_1, a_2, \ldots, a_n)$ as defined for recurrence in Chap. 9 is a norm:

$$\|a\| = \sqrt{a_1^2 + a_2^2 + \ldots + a_n^2}.$$

For ODEs we usually have estimates in the so-called max- or sup norm. This means that the estimate of an error is given as follows. Consider two vector functions $\phi(t), \psi(t)$ on the time interval $L : 0 \leq t \leq T_0$. With Euclidean norm $\|.\|$, we have sup norm $\|.\|_{sup}$

$$\|\phi - \psi\|_{sup} = \sup_{t \in L} \|\phi(t) - \psi(t)\|.$$

In the case of PDEs, one has to estimate functions depending on space and time variables. In Fourier theory, one uses the so-called $L_2$ norm. We put for functions $u : \mathbb{R} \mapsto \mathbb{C}$ the $L_2$ norm:

$$\|u\|_2 = \left( \int_{\mathbb{R}} u(x)^2 dx \right)^{1/2}.$$

The $L_2$ norm is natural in Fourier theory but weaker than the sup norm. More in general we have for $p = 1, 2, \ldots$ the $L_p$ norm:

$$\|u\|_p = \left( \int_{\mathbb{R}} |u(x)|^p dx \right)^{1/p}.$$

In many problems we consider $x \in D \subset \mathbb{R}^n$, and we have to change the domain of integration accordingly.

For normed vector spaces that are Hilbert spaces, we can add for each two elements $f, g$ the inner product:

$$(f, g) = \int_{\mathbb{R}} f(x)g(x)dx$$

(if the elements are complex functions, we replace $g(x)$ by its complex conjugate).

When considering continuous differentiable functions $u(x)$ on a domain $D$, we use the Sobolev inner product:

$$\|u\|_{H^1}^2 = (u, u) + (u_x, u_x).$$

For $k$ times continuously differentiable functions on $D$, we have the Sobolev inner product:

$$\|u\|_{H^k}^2 = (u, u) + (u_x, u_x) + \ldots (u_k, u_k).$$

If $k = 0$, we have the $L_2$ inner product and norm. Note that estimates in Sobolev norms are for $k \geq 1$ stronger than the $L_2$ norm as apart from the function values in $D$ it also takes into account its derivatives.

## 8.2 Using Maximum Principles

Consider for function $u(t, x)$ a parabolic equation with one space dimension:

$$\frac{\partial^2 u}{\partial x^2} + \frac{\partial u}{\partial x} - \frac{\partial u}{\partial t} = f(t, x) + \varepsilon f_1(t, x), \tag{8.1}$$

with initial-boundary values $u(0, x) = g(x), 0 \leq x \leq 1, u(0, t) = h_1(t), u(t, 1) = h_2(t), 0 \leq t \leq T$. The time bound $T$ will be chosen as needed, for instance $t = 1, 1/\varepsilon, \infty$.

The initial-boundary value problem of Eq. (8.1) can be seen as a model for the heat flow in a metal rod with prescribed initial temperature distribution $g(x)$ and temperatures at the ends determined by $h_1(t), h_2(t)$. External heating is described by $f(t, x) + \varepsilon f_1(t, x)$; assume that $f(t, x) > \mu > 0$ in the domain $[0, T] \times [0, 1]$ with $\mu$ a constant. We assume that there exists a positive constant $\alpha$ such that

$$-\alpha < f_1(t, x) < \alpha, \; (t, x) \in [0, T] \times [0, 1].$$

We will use maximum principles based on [112]; Eq. (8.1) is the simplest formulation for our result, and we can add coefficients dependent on $t$, $x$, coupled parabolic equations, non-linearities and also consider more dimensional spatial problems. It follows from ch. 3 in [112] that the maximum and minimum values of the solutions of Eq. (8.1) are assumed on the boundary.

Consider $u_\varepsilon(t, x)$ the solution with the initial-boundary values of Eq. (8.1), $u_0(t, x)$ the solution in the case $\varepsilon = 0$. Consider the function:

$$w(t, x) = u_\varepsilon(t, x) - \varepsilon\alpha x - u_0(t, x). \tag{8.2}$$

. For $w(t, x)$, we have the initial-boundary value problem:

$$\frac{\partial^2 w}{\partial x^2} + \frac{\partial w}{\partial x} - \frac{\partial u}{\partial t} = \varepsilon(\alpha + f_1(t, x)). \tag{8.3}$$

The right-hand side of the equation for $w$ is positive; the initial-boundary conditions are $u(0, x) = -\varepsilon\alpha x, 0 \leq x \leq 1, u(0, t) = u(t, 1) = 0, 0 \leq t \leq T$. The maximum and minimum values of $w(t, x)$ are assumed on the boundary, so we conclude

$$u_\varepsilon(t, x) = u_0(t, x) + O(\varepsilon),\ 0 \leq t \leq T, 0 \leq x \leq 1.$$

This result can be useful as we have no Poincaré expansion theorem for PDEs.

## 8.3 Averaging a Linear Operator

A typical problem formulation would be to consider the Gauchy problem or an initial-boundary value problem for equations like

$$u_t + Lu = \varepsilon f(u),\quad t > 0, u(0) = u_0. \tag{8.4}$$

L is a linear operator like the Laplace $\Delta$ or the wave operator $\Delta - \partial^2/\partial t^2$, $f(u)$ represents the perturbation terms, possibly non-linear.

To obtain a standard form, $u_t = \varepsilon F(t, u)$, suitable for averaging in the case of a PDE, can already pose a formidable technical problem, even in the case of simple geometries. However it is reasonable to suppose that one can solve the "unperturbed" ($\varepsilon = 0$) problem in some explicit form before proceeding to the perturbation problem.

A number of authors, in particular in the former Soviet Union, have addressed problem (8.4). For a survey of results, see [106] and [124].

We shall follow the theory developed by Krol in [91] which has some interesting applications on periodic flow in tidal basins. Consider the problem (8.4) with two spatial variables $x$, $y$ and time $t$, and the coefficients are $T$-periodic; we choose $f(u)$ linear; from [91] we have:

**Theorem 8.1** *Assume that, after solving the unperturbed problem* (8.4), *by a variation of constants procedure, we can write the problem in the form*

$$\frac{\partial F}{\partial t} = \varepsilon L(t)F, \quad F(x, y, 0) = \gamma(x, y). \tag{8.5}$$

*We have on the domain* $D \in \mathbb{R}^2$

$$L(t) = L_2(t) + L_1(t), \tag{8.6}$$

*where*

$$L_2(t) = b_1(x, y, t)\frac{\partial^2}{\partial x^2} + b_2(x, y, t)\frac{\partial^2}{\partial x \partial y} + b_3(x, y, t)\frac{\partial^2}{\partial y^2}, \tag{8.7}$$

$$L_1(t) = a_1(x, y, t)\frac{\partial}{\partial x} + a_2(x, y, t)\frac{\partial}{\partial y} \tag{8.8}$$

*in which* $L_2(t)$ *is a uniformly elliptic operator on the domain* $D$, $L_1, L_2$ *and so* $L$ *are* $T$*-periodic in* $t$*; the coefficients* $a_i$, $b_i$ *and* $\gamma$ *are* $C^\infty$ *and bounded with bounded derivatives.*

*We average the operator* $L$ *by averaging the coefficients* $a_i, b_i$ *over* $t$*:*

$$\bar{a}_i(x, y) = \frac{1}{T}\int_0^T a_i(x, y, t)dt, \quad \bar{b}_i(x, y) = \frac{1}{T}\int_0^T b_i(x, y, t)dt, \tag{8.9}$$

*producing the averaged operator* $\bar{L}$*. As an approximating problem for (8.5), we now take*

$$\frac{\partial \bar{F}}{\partial t} = \varepsilon \bar{L}\bar{F}, \quad \bar{F}(x, y, 0) = \gamma(x, y). \tag{8.10}$$

*A rather straightforward analysis shows existence and uniqueness of the solutions of problems (8.5) and (8.10) on the timescale* $1/\varepsilon$*. We have:*

*Let* $F$ *be the solution of initial value problem (8.5) and* $\bar{F}$ *the solution of initial value problem (8.10), and then we have the estimate* $\|F - \bar{F}\| = O(\varepsilon)$ *on the timescale* $1/\varepsilon$*.*

This is a strong result as the norm $\|.\|$ is the sup norm on the spatial domain and on the timescale $1/\varepsilon$. By Ben Lemlih and Ellison [18] and, independently in [91], a near-identity transformation is applied to $\bar{F}$, which is autonomous and on which we have explicit information; see for technical details [18].

## 8.4 Wave Equations, Projection Methods

Consider on a bounded spatial domain hyperbolic, weakly non-linear equations of the form:

$$u_{tt} + Au = \varepsilon g(u, u_t, t, \varepsilon), \tag{8.11}$$

where $A$ is a positive, self-adjoint linear differential operator on a separable real Hilbert space, think of the Laplace operator $\Delta$. Equation (8.11) can be studied by expanding the solutions in eigenfunctions of the operator $A$, but for equivalence of the solutions with Eq. (8.11) this involves solving a system corresponding of an infinite number of ODEs. In many cases, resonance will make this virtually impossible, the averaged (normalised) system is too large, and we have to use truncation techniques. We discuss results from [90] on the asymptotic validity of truncation methods; this yields at the same time interesting information on the timescale of physical interaction of modes.

An example where we can apply expansion in eigenfunctions is the perturbed wave equation

$$u_{tt} - u_{xx} = \varepsilon f(u, u_x, u_t, t, x, \varepsilon), \ \ t \geq 0, 0 < x < \pi, \tag{8.12}$$

where

$$u(0, t) = u(\pi, t) = 0, u(x, 0) = \phi(x), u_t(x, 0) = \psi(x), 0 \leq x \leq \pi.$$

The eigenfunctions are $\sqrt{2}\sin nx, n = 1, 2, \ldots$ However, a difficulty is that apart from being infinite the system has a resonant spectrum, without the possibility of reduction unless we choose special initial conditions consisting of a number of finite modes. A typical example in the subsequent applications is $f = u^3$; see [148] and Sect. 14.4.

An example which is easier to handle is the Klein-Gordon equation with dispersion:

$$u_{tt} - u_{xx} + a^2 u = \varepsilon u^3, \ \ t \geq 0, 0 < x < \pi, a > 0. \tag{8.13}$$

We can apply almost-periodic averaging, and the averaged system splits into finite-dimensional parts, see the applications in Chap. 14. Similar phenomena arise in applications to rod and beam equations.

We will summarise the results for projection methods from [90] and [137].

Consider again the initial-boundary value problem for the non-linear wave equation (8.12). The normalised eigenfunctions of the unperturbed ($\varepsilon = 0$) problem are $v_n(x) = \sqrt{2}\sin(nx)$, $n = 1, 2, \cdots$, and we expand the solution of the initial-boundary value problem for equation (8.12) in a Fourier series with respect to these eigenfunctions of the form:

$$u(t,x) = \sum_{n=1}^{\infty} u_n(t)v_n(x). \tag{8.14}$$

As announced above, by taking inner products, this yields an infinite system of ordinary differential equations, which is equivalent to the original problem. The next step is then to truncate this infinite-dimensional system (a projection) and apply averaging to the truncated system. The truncation is related to Galerkin's method; this method uses as approximation a finite expansion of the solutions in suitable functions. For our problem one has to *estimate the combined error of truncation and averaging*.

The modes $u_n(t)v_n(x)$ will also be called eigenfunctions. As we will see in Chap. 14, the functions $u_n(t)$ will often satisfy a system of equations of the form:

$$\ddot{u}_n + \omega_n^2 u_n = \varepsilon F(u, t, x),\ n = 1, 2, \ldots$$

with in the example of Eq. (8.13) $u = (u_1, u_2, \ldots, u_N)$ if we truncate to $N$ modes. The truncation is called a wave packet. We will see that a first order averaging approximation $u_N$ of the system Eq. (8.13) contains $u_n$ functions of the form

$$u_n = r_n(\varepsilon t)\cos(\omega_n t + p_n \varepsilon t)$$

with the $p_n$ real numbers. This produces the approximation:

$$u_N(t,x) = \frac{1}{2}\sqrt{2}\sum_{n=1}^{N}\left(\sin(\omega_n t + p_n\varepsilon t + nx) + \sin(\omega_n t + p_n \varepsilon t - nx)\right). \tag{8.15}$$

The phase velocity $W_n$ of each eigenfunction in the wave packet is $W_n = (\omega_n + \varepsilon p_n)/n$.

Equation (8.12) with its initial-boundary values has exactly one solution in a suitably chosen Hilbert space $\mathcal{H}_k = H_0^k \times H_0^{k-1}$, where $H_0^k$ are the well-known Sobolev spaces consisting of functions $u$ with derivatives $U^{(k)} \in L_2[0,1]$ and $u^{(2l)}$ zero on the boundary whenever $2l < k$. Sobolev spaces are natural in physics and engineering as they *measure both positions and derivatives*. It also makes sense to relate the estimate to the energy content of the modes in the system.

It is rather standard to establish existence and uniqueness of solutions *on the timescale* $1/\varepsilon$ under certain mild conditions on $f$; examples are right-hand sides $f$ like $u^3$, $uu_t^2$, $\sin u$, $\sinh u_t$ or $u^k$, $k = 1, 2, \ldots$, etc. Moreover we note that:

1. If $k \geq 3$, $u$ is a classical solution of equation (8.12).
2. If $f = f(u)$ is an odd function of $u$, one can find an even energy integral. If such an integral represents a positive-definite energy integral, we are able to prove existence and uniqueness for all time.

In Galerkin's truncation method, one considers only $N$ modes of the expansion (8.14), which we shall call the projection $u_N$ of the solution $u$ on a $N$-dimensional space. To find $u_N$, we have to solve a $2N$-dimensional system of ordinary differential equations for the expansion coefficients $u_n(t)$ with appropriate (projected) initial values. The estimates for the error $\|u - u_N\|$ depend strongly on the smoothness of the right-hand side $f$ of equation (8.12) and the initial values $\phi(x)$, $\psi(x)$ but, remarkably enough, not on $\varepsilon$.

Periodic averaging of the truncated system produces a time-dependent approximation $\bar{u}_N$ of $u_N$ from the Galerkin-averaging theorem, see Krol [90]. We have:

**Theorem 8.2** *Consider the initial-boundary value problem*

$$u_{tt} - u_{xx} = \varepsilon f(u, u_x, u_t, t, x, \varepsilon), \quad t \geq 0, 0 < x < \pi,$$

*where*

$$u(0, t) = u(\pi, t) = 0, u(x, 0) = \phi(x), u_t(x, 0) = \psi(x), 0 \leq x \leq \pi.$$

*Suppose that $f$ is $k$-times continuously differentiable and satisfies the existence and uniqueness conditions on the timescale $1/\varepsilon$, $(\phi, \psi) \in \mathcal{H}_k$; if the solution of the initial-boundary problem is $(u, u_t)$ and the approximation by the Galerkin-averaging procedure $(\bar{u}_N, \bar{u}_{Nt})$, we have on the timescale $1/\varepsilon$*

$$\|u - \bar{u}_N\|_\infty = O(N^{\frac{1}{2}-k}) + O(\varepsilon), \quad N \to \infty, \varepsilon \to 0 \tag{8.16}$$

$$\|u_t - \bar{u}_{Nt}\|_\infty = O(N^{\frac{3}{2}-k}) + O(\varepsilon), \quad N \to \infty, \varepsilon \to 0. \tag{8.17}$$

**Remark 8.1**

- Taking $N = O(\varepsilon^{-\frac{2}{2k-1}})$ we obtain an $O(\varepsilon)$-approximation on the timescale $1/\varepsilon$. So, the required number of modes decreases when the regularity of the data and the order up to which they satisfy the boundary conditions increase.
- However, this decrease of the number of required modes is not uniform in $k$. So it is not obvious for which choice of $k$ the estimates are optimal at a given value of $\varepsilon$.
- An interesting case arises if the non-linearity $f$ satisfies the regularity conditions for all $k$. This happens for instance if $f$ *is an odd polynomial in* $u$ and with analytic initial values. In such cases the results can be improved by introducing Hilbert spaces of analytic functions (the so-called Gevrey classes). The estimates in [90] for the approximations on the timescale $1/\varepsilon$ obtained by the Galerkin-averaging procedure become in this case

$$\|u - \bar{u}_N\|_\infty = O(N^{-1}a^{-N}) + O(\varepsilon), \quad N \to \infty, \varepsilon \to 0 \tag{8.18}$$

$$\|u_t - \bar{u}_{Nt}\|_\infty = O(a^{-N}) + O(\varepsilon), \quad N \to \infty, \varepsilon \to 0, \tag{8.19}$$

where the constant $a$ arises from the bound one has to impose on the size of the strip around the real axis on which analytic continuation is permitted in the initial-boundary value problem.

The important implication is that, because of the $a^{-N}$-term we need only $N = O(|log\varepsilon|)$ terms to obtain an $O(\varepsilon)$ approximation on the timescale $1/\varepsilon$.

- It is not difficult to improve the result in the case of finite-modes initial values, i.e., the initial values can be expressed in a finite number of eigenfunctions $v_n(x)$. In this case the error becomes $O(\varepsilon)$ on the timescale $1/\varepsilon$ if $N$ is taken large enough.
- Here and in the sequel we have chosen Dirichlet boundary conditions. It is stressed that this is by way of example and not a restriction. We can also use the method for Neumann conditions, periodic boundary conditions, etc.
- It is possible to generalise these results to higher-dimensional (spatial) problems; see [90] for remarks and [108] for an analysis of a two-dimensional non-linear Klein-Gordon equation with Dirichlet boundary conditions on a rectangle. In the case of more than one spatial dimension, many more resonances may be present. See Chap. 14.
- Related proofs for Galerkin-averaging were given in [56] and [57]. These papers also contain extensions to difference and delay equations.

# Part II
# Averaging and Numerical Bifurcation Analysis

# Chapter 9
# Hamiltonian Systems

**Abstract** To understand the dynamics of Hamiltonian systems, we need basic concepts like first integrals, integrability, special aspects of stability of periodic solutions and resonance. The use of the Poincaré recurrence theorem will play an important part to characterise the dynamics between regular and complex or chaotic. This is illustrated for the famous Hénon-Heiles system in the regular and in the chaotic domain. A systematic analysis of two- and three-dof systems shows interesting differences. For two-dof systems, the averaged-normal form is always integrable, and for three dof this is not the case. For two-dof systems we consider 1 : 2 and 1 : 1 resonances and higher order resonance; as an application, the spring pendulum is analysed at various resonances values. For three dof the first order resonances in normal form are given together with their integrability. The chapter closes with a discussion of adiabatic invariants and interaction of first and higher order resonance.

The development of mechanics and dynamical systems cannot be understood without knowledge of Hamiltonian systems. This sounds like a paradox as this field is the main part of conservative dynamics where dissipation is neglected. However, the iconic features of resonance, invariant manifolds, and integrability are the basic elements of Hamiltonian systems that carry over to many other parts of dynamical systems.

This chapter demonstrates the use of averaging for Hamiltonian systems in some important cases. However, we omit in this chapter the interesting case of systems with gyroscopic terms. A simple formulation of a time-independent Hamiltonian system starts with introducing a smooth function $H(p, q)$ of $2n$ variables defined on a part of $\mathbb{R}^{2n}$ or a manifold in Euclidean $2n$-space. We have for the positions $q \in \mathbb{R}^n$, for the momenta $p \in \mathbb{R}^n$. The equations of motion induced by the Hamiltonian $H$ written out in components are

$$\dot{q}_i = \frac{\partial H}{\partial p_i},\ \dot{p}_i = -\frac{\partial H}{\partial q_i}, i = 1, \ldots, n. \tag{9.1}$$




F. Verhulst, *Averaging for Nonlinear Dynamics with Applications and Numerical Bifurcations*, Applied Mathematical Sciences 223,
https://doi.org/10.1007/978-3-032-12745-7_9

A critical point $x_0$ of the Hamiltonian vector field describing the equations of motion is an equilibrium of the phase flow; it is a stationary point of the Hamiltonian function: $\partial H/\partial p = \partial H/\partial q = 0$. The function $H$ will often be chosen analytic with possibly excluding a few singular points, for instance, in gravitational and electromagnetic theory. The quadratic terms obtained by expansion of $H(p, q)$ near equilibrium will be indicated by $H_2$.

It is a restriction of the general theory to use perturbation methods near equilibria, but the results are impressive. More general introductions to Hamiltonian dynamics can be found in [4], [5], [27], and [163]. In this chapter we emphasise quantitative aspects, but qualitative aspects cannot be ignored to obtain general insight. In dynamics, in particular Hamiltonian dynamics, symmetries play an important part; a nice introduction is ch. 10 in [62]. We will pay attention to symmetry aspects.

## 9.1 Concepts

The analysis and approximation of Hamiltonian dynamics are often carried out by computing *normal forms*. Canonical normalisation of Hamiltonian vector fields is closely related to averaging (see [4] and [123]), and normalisation is often preferred as the process can be formulated to conserve the Hamiltonian character at each approximation step. Canonical variables are actions $\tau$ and angles $\phi$ given by

$$p_i = \sqrt{2\tau_i}\,\cos\phi_i,\; q_i = \sqrt{2\tau_i}\,\sin\phi_i,\; i = 1, 2, \ldots, n. \tag{9.2}$$

However, using averaging the conservative character of the dynamics can be saved so there is no qualitative difference between normal forms and averaging results; see [123] for a detailed comparison. In any case, averaging has the added advantage of producing validity of approximations on certain timescales.

This is of great importance because of the issue of timescales in Hamiltonian systems where many different timescales may play a part. In Sect. 9.5 we consider a Hamiltonian with two degrees of freedom (dof); for the 1 : 2 resonance, the dominant timescale is $1/\varepsilon$, and the timelike variables are $t, \varepsilon t$. Also in 9.5 we have the 1 : 1 Hamiltonian resonance and timescale $1/\varepsilon^2$, and timelike variables are $t, \varepsilon^2 t$. Chapter 7 showed us resonances with timelike variables $\varepsilon^q t$, where $q$ is a rational number. We will meet these again for higher order Hamiltonian resonance.

We will discuss various aspects of Hamiltonian examples (see also [123] ch. 10) with attention to different variational equations. It is important to keep in mind that in Hamiltonian systems we have no attraction but *recurrent* motion if the energy manifold is bounded. Roughly stated this means that for a Hamiltonian system on a bounded domain nearly all orbits return an infinite number of times arbitrarily close to the initial conditions; see Sect. 9.3. Passage of resonance zones will be characterised by complicated dynamics involving motion around tori and passage near saddle points but not by capture of orbits. In such complicated dynamics, one expects delay of recurrence, see [146], and [155].

The recurrence theorem, Sect. 9.3, is important but has not been used much in practical problems. The reason is that many interacting particles and complex dynamics may extend the recurrence time enormously, but for a few dof and to characterise complexity it is a useful concept. In Sect. 9.5, we will discuss two-dof systems including higher order resonance; we show the use of the recurrence theorem and other aspects of three-dof problem in Sect. 9.6.

### 9.1.1 First Integrals and Integral Manifolds

For time-independent Hamiltonian systems, the Hamiltonian function $H(p, q)$ is a first integral of motion; the interpretation of this integral is usually "energy" of the system. Near stable equilibrium $H(p, q) = E_0$ (constant) describes a family of closed manifolds (surfaces) around the equilibrium as $E_0$ takes different positive values. A classical question in the literature for $n$ dof is whether more first integrals independent of $H$ exist for certain important systems with two or more dof. If apart from $H$ $n - 1$ mutually independent integrals exist, the Hamiltonian system is called integrable. In the integrable case we can describe the dynamics completely by motion on invariant manifolds embedded in the energy manifolds; there will be no chaos. An example is the gravitational two-body problem.

A remarkable result in the perturbation theory of normalisation and averaging is described and explained in ch. 10 of [123]. After canonical normalisation or averaging near equilibrium, $H_2(p, q)$ will be a second independent integral of the averaged system. The implication is that for two dof the perturbation results will not show chaos. This is not in contradiction with the fact that generally Hamiltonian systems with $n \geq 2$ dof are non-integrable and often chaotic. The perturbation result gives an $O(\varepsilon^m)$, $m \geq 1$ approximation on a certain timescale; this means that chaotic behaviour in two-dof systems near equilibrium is limited to sets of $O(\varepsilon^m)$ measure.

An interesting insight was developed by Chechin et al. in [36] with applications to Fermi-Pasta-Ulam chains that will be discussed in the next chapter; the idea is to identify in such Hamiltonian problems lower-dimensional invariant manifolds, called "bushes" in [36], that determine to a large extent the dynamics.

### 9.1.2 Why Are Integrable Systems of Interest?

Poincaré proved in [110] that Hamiltonian systems with two or more dof are generally non-integrable. Why would one bother then to study integrable systems? One reason is that for Hamiltonian systems with the Hamiltonian function dependent on parameters there are no general results on the sets of parameter values where one has integrability or not. Usually only for special parameter values integrability is known.

Secondly, for iconic examples like the harmonic equation, the mathematical pendulum or the advanced theory of water waves and specific problems in non-linear mechanics integrability emerges often and plays an inspiring role.

A third reason, connected with the second, is that in models of mathematical physics often symmetry assumptions play a part. The spherical symmetry of gravitational and electromagnetic forces produces examples, but also there is an advantage in assuming symmetries as this simplifies the models. We have some examples in Sect. 9.55, see also ch. 10 of [123].

### 9.1.3 *Stability of Periodic Solutions*

As discussed in Chap. 3 perturbing around a periodic solution of an autonomous system, one of the characteristic Floquet exponents will always be zero. Poincaré proved in [110] vol. 1 that the characteristic exponents have pairwise the same size and have opposite sign. This means a periodic solution produced by a time-independent Hamiltonian will have two characteristic exponents zero. Geometrically, this is tied in with the fact that periodic solutions in Hamiltonian systems are present in continuous families parameterised by the energy.

It follows that for a periodic solution in a two-dof Hamiltonian system we have at most two characteristic exponents non-zero; with $a \neq 0$ real they will be either $\pm\, a$ (indicated by H, hyperbolic) or imaginary $\pm\, ai$ (indicated by E, elliptic). An important consequence is that for two dof stability of the periodic solutions by linear analysis is definitive, and higher order approximation and non-linear terms cannot destroy in two dof the linear result.

For three dof the situation is different. We have at most four characteristic exponents non-zero with possibly four exponents real (indicated by HH), four exponents purely imaginary (indicated by EE), mixed real and purely imaginary (EH), and four exponents symmetrically complex (indicated by C). If the purely imaginary exponents are double, higher order perturbations may move the exponents into the general complex domain; this is called the Hamiltonian-Hopf bifurcation. It can be associated with complex, chaotic dynamics (see Fig. 9.1).

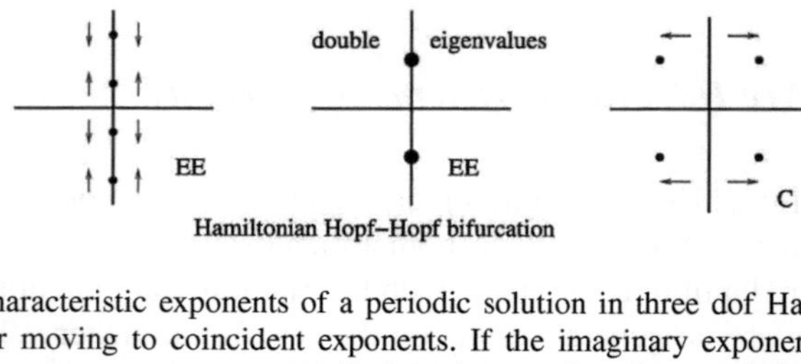


**Fig. 9.1** Four characteristic exponents of a periodic solution in three dof Hamiltonian systems with a parameter moving to coincident exponents. If the imaginary exponents are coincident, the exponents may move next symmetrically into the complex plane, and the periodic solution is unstable. This is the Hamiltonian-Hopf bifurcation

## 9.2 Frequencies and Resonances

Consider time-independent Hamiltonian systems $H$ near stable equilibrium. With $n$ dof (degrees of freedom) and basic frequencies $\omega_i, i = 1, \ldots, n$, we write

$$H = \sum_{i=1}^{n} \frac{1}{2}(p_i^2 + \omega_i^2 x_i^2) + H_3(p_1, \ldots p_n, x_1, \ldots, x_n)$$
$$+ H_4(p_1, \ldots p_n, x_1, \ldots, x_n) + \ldots, \qquad (9.3)$$

with $H_3$ a cubic homogeneous polynomial in its arguments, $H_4$ a quartic homogeneous polynomial, etc. The sum of quadratic terms is called $H_2$. If $H_3, H_4, \ldots$ depend on $x$ only, we call the Hamiltonian "derived from a potential". Note that instead of variables $p_i, x_i, i = 1, \ldots, n$, standard use is $p$ for the momenta and $q$ for the positions. In the literature, all these notations are mixed, also canonical variables $\tau, \phi$ are used, and we have to follow this slightly confusing custom on and off to keep contact with the referenced material.

An equivalent form for $H_2$ is

$$H_2 = \sum_{i=1}^{n} \frac{1}{2}\omega_i(p_i^2 + x_i^2).$$

If one wants to study the dynamics near equilibria with mixed positive and negative real parts of the eigenvalues (unstable equilibria) or cases with negative and positive $\omega_i$, the last form is preferred. In $H_2$ the masses $m_i$ are not dependent on $\varepsilon$ and were scaled to 1. One can introduce the masses explicitly by rescaling $p_i$.

The equations of motion induced by Hamiltonian (9.3) are given by Eq. (9.1). If we have a potential problem for Hamiltonian (9.3), the equations of motion near stable equilibrium correspond with coupled harmonic equations of the form

$$\ddot{x}_i + \omega_i^2 x_i = \cdots$$

with the dots standing for quadratic, cubic, etc., polynomial terms in $x_i, i = 1, \ldots, n$.

In the sequel, we will consider an $\varepsilon$-neighbourhood of equilibrium by scaling $p \to \varepsilon p, x \to \varepsilon x$ and dividing Hamiltonian (9.3) by $\varepsilon^2$ to produce $H = H_2 + \varepsilon H_3 + \varepsilon^2 \ldots$

**The Order of Resonance**

As we will show explicitly in the examples, to be "near equilibrium at the origin" is expressed by rescaling $p_i, x_i \to \varepsilon p_i, \varepsilon x_i, i = 1, \ldots, n$ followed by dividing the equations by $\varepsilon$. We will formulate variational equations. In the case of coupled harmonic equations, the polynomial right-hand sides produce Fourier expansions, resonance zones with periodic solutions and possibly invariant manifolds will emerge for rational relations of the frequencies $\omega_i$. We summarise a few results from ch. 10 of [123] for prominent resonances. For the frequencies $\omega_1, \ldots, \omega_n$, we have a resonance if for some $k_1, k_2, \ldots, k_n$ with $k_i \in \mathbb{Z}$, we have

$$k_i\omega_1 + k_2\omega_2 + \ldots + k_n\omega_n = 0. \tag{9.4}$$

It makes sense to choose the numbers $k_i$ relative prime.

2 dof: low order resonance at $H_3$ if $\omega_1 : \omega_2 = 1 : 2$; resonance at $H_4$ if $\omega_1 : \omega_2 = 1 : 1$ or $1 : 3$.

2 dof: higher order resonance if $\omega_1 : \omega_2 = k_1 : k_2, k_1 + k_2 > 4$ (see Sect. 9.5).

3 dof: 4 resonances at $H_3$ if $(\omega_1 : \omega_2 : \omega_3) = (1 : 2 : 1), (1 : 2 : 2), (1 : 2 : 3), (1 : 2 : 4)$.

3 dof: 12 resonances at $H_3 + H_4$ if $(\omega_1 : \omega_2 : \omega_3) = (1 : 1 : 1), (1 : 1 : 3), (1 : 2 : 5), (1 : 2 : 6), (1 : 3 : 3), (1 : 3 : 4), (1 : 3 : 5), (1 : 3 : 6), (1 : 3 : 7), (1 : 3 : 9), (2 : 3 : 4), (2 : 3 : 6)$.

Applications in Sect. 9.5 contain examples of low order 2-dof resonances. They can be handled by first and second order averaging. Note that symmetries in the equations of motion may change the part played by resonances drastically. If the frequencies are constant (as assumed in this chapter), it will become clear that the use of angles instead of phases makes little difference. The combination angles that arise in the resonance zones can in this case be replaced by combination phases.

We will also discuss a few embedded double resonances in three dof, the $(2 : 2 : 3)$ and $(1 : 1 : 4)$ resonances; they are a mix of low and higher order resonances.

## 9.3 The Poincaré Recurrence Theorem

This section is based on Poincaré's original text [110] and on [153]. It is included as it presents a tool to characterise conservative dynamical systems. The phase flow induced by a time-independent Hamiltonian is volume-preserving, see [146]. As stated before, the recurrence theorem implies that for Hamiltonian systems on a bounded energy manifold, nearly all solutions return after a finite time $T_r$ arbitrarily close to their original position in phase space. The recurrence theorem applies quite generally to systems with measure-preserving flow, but we restrict ourselves to time-independent Hamiltonian systems. More precise:

**Theorem 9.1** *The Poincaré Recurrence Theorem.*

*Consider the phase flow $F$ of a dynamical system induced by the Hamiltonian function $H(p, q)$ with $n$ dof ($2n$ equations of motion). A bounded energy manifold $M$ determined by $H(p, q) = E_0$ (positive constant) has a subset $D_0 \subset M$ with positive measure. Consider solutions of Hamiltonian* (9.3) *with solution $(p(t), x(t))$. Nearly all initial points $P = (p(0), x(0))$ in $D_0$ (in the sense of measure theory) will return under the map $F$ after a finite time arbitrarily close to initial $P$ in $D_0$.*

**Remark 9.1** A set consisting of points that do not return near the initial points under the map $F$ has measure zero. Examples of initial points that do not return are starting points of homoclinic and heteroclinic solutions. Other exceptional solutions with this property will be called *wandering*.

For one-dof systems on a bounded domain, recurrence is trivial in the phase-plane. For two-dof systems that are integrable, recurrence behaviour is relatively simple near a stable periodic solution, but in general this is already not so easy for chaotic two-dof systems. Consider the Euclidean norm for $n$ dof; for the phase distance $d(t)$ of solution $(p(t), x(t))$ to the initial point $(p(0), x(0))$ in phase space, we put

$$d(t) = \sqrt{\sum_{i=1}^{n} \left((p_i(t) - p_i(0))^2 + (x_i(t) - x_i(0))^2\right)}. \tag{9.5}$$

A recurrence time $T_r$ arises if we choose a fixed distance $d_0 > 0$ with $d(T_r) \leq d_0$. The Euclidean distance $d(t)$ does not produce a fixed time of recurrences for a fixed value of $d_0$, and we will find a number of times for which the domain around $(p(0), x(0))$ with size $d_0$ is traversed. So the recurrence time $T_r$ for a given $n$ dof Hamiltonian system is not a fixed number, and it depends on its specific phase flow and the chosen initial position in phase space.

In [153] the following (rough) upper bound is derived for the recurrence time $T_r$:

**Theorem 9.2** *Suppose that we consider an energy manifold of system* (9.3) *that is contained in a sphere with radius $R = \sqrt{E_0}$, and $d_0$ is the chosen distance from $(p(0), x(0))$ in phase space. Then we have for the recurrence time*

$$T_r \leq \left(\frac{E_0^{n-1/2}}{d_0^{2n-1}}\right) \text{ as } d_0 \to 0.$$

Suppose that for a problem with three dof we have $E_0 = 1, d_0 = 0.1$ (fairly large), and then we have already $T_r \leq 10^5$ timesteps. Considering the time series (9.5) numerically will give an indication of the complexity of the phase flow.

## 9.4 Recurrence in the Hénon-Heiles System

A two-dof Hamiltonian system that changed our view on Hamiltonian dynamics completely is the paper [75] with the cubic Hamiltonian:

$$H = \frac{1}{2}(\dot{x}^2 + x^2 + \dot{y}^2 + y^2) + x^2 y - \frac{1}{3}y^3. \tag{9.6}$$

At this point we will globally describe the system without detailed orbit calculations, see also 10.1, to illustrate the meaning of the recurrence theorem. The equations of motion are

$$\begin{cases} \ddot{x} + x &= -2xy, \\ \ddot{y} + y &= -x^2 + y^2. \end{cases} \tag{9.7}$$

Analysing the Hamiltonian geometrically, we find that the energy manifold obtained from putting $H(\dot{x}, x, \dot{y}, y) = E$ (constant) is bounded for $0 < E < 1/6$ ( $=$ 0.1667). At the origin of phase space, the energy manifold is a slightly distorted sphere, and we have a Lyapunov stable equilibrium. The manifold contains unstable equilibria if $E = 1/6$. On the bounded energy manifold, we find families of stable and unstable periodic solutions. We have three exact families of periodic solutions that we can study globally for $0 < E < 1/6$. First the $y$ normal mode:

$$\ddot{y} + y = y^2,\ \dot{x} = x = 0. \tag{9.8}$$

Two families in general position are given by

$$\ddot{y} + y = -2y^2,\ x = \pm\sqrt{3}\, y. \tag{9.9}$$

The three families of exact periodic solutions are linking the origin of phase space with the energy manifold that loses boundedness at $E = 1/6$.

The remarkable conclusion of [75] is that increasing the energy $E$ slowly from $E = 0$ to $E = 1/12$ the dynamics looks regular with families of stable periodic solutions surrounded by invariant tori together with unstable periodic solutions bounding the families of tori. The dynamics can for $0 < E < 1/12$ be described by averaging-normalisation as formulated in the next section.

For larger values of the energy, for instance $E = 1/10$, some of the stable periodic solutions vanish or become unstable and chaotic regions appear. Increasing the energy to a neighbourhood of $E = 1/6$, the phase space seems to become dominated completely by chaotic motion. It is known now that the presence of the invariant tori is understood from KAM theory, see for details Chap. 12. We also know that the apparent regularity of the dynamics for $0 < E < 1/12$ is caused by the exponential smallness of the chaotic sets.

In Fig. 9.2 we present the Euclidean distance $d(t)$ to the initial conditions of a few orbits. Left we show the distance $d$ for the fairly regular case of $E = 1/12$, and right we have $E = 1/6.5$ ($= 0.1538\cdots$) making the Hénon-Heiles system very chaotic. In both cases the recurrence theorem applies.

## 9.5 Two Degrees of Freedom

The literature contains many applications of Hamiltonian systems with two dof. In the case of more than two dof, one can sometimes obtain a reduction of the

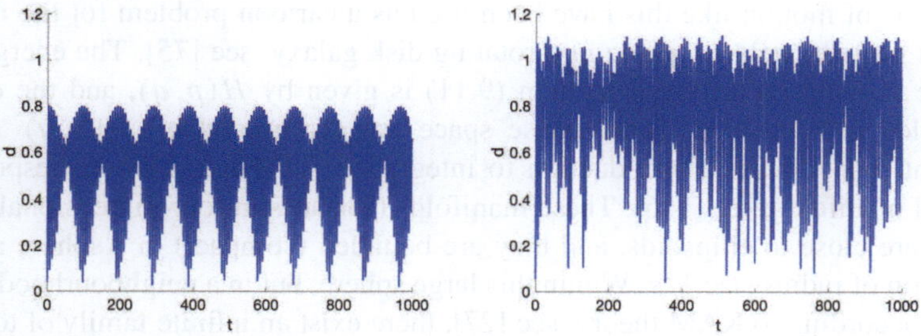


**Fig. 9.2** Euclidean distance $d(t)$ for two orbits starting at $(0, \sqrt{E}, 0, \sqrt{E})$ with left $E = 1/12$ where the dynamics is regular and right $E = 0.1538$ in the chaotic region. As expected, the recurrence is very different in the two cases

problem by using an integral of the system independent of the energy. A classical case of reduction is the gravitational problem of two point masses that has in relative coordinates three dof (six dimensions). The system has apart from the energy integral a vector angular momentum integral caused by the assumption of spherical symmetry of the potential. The four integrals enable a reduction that makes the gravitational two-body problem super-integrable.

### 9.5.1 A Hamiltonian System with Cubic Term

The results in this subsection are based on [144]. Consider a two-degrees-of-freedom (dof) Hamiltonian system generated by the Hamiltonian:

$$H(p,q) = \frac{1}{2}(p_1^2 + \omega^2 q_1^2) + \frac{1}{2}(p_2^2 + q_2^2) - \varepsilon q_1 q_2^2. \tag{9.10}$$

The variables $p_1, p_2$ are called the momenta (here identified with velocities), and the variables $q_1, q_2$ are called positions. The quadratic part of the Hamiltonian is abbreviated as $H_2(p,q)$, and $\omega$ and 1 are frequencies in the linear part of the equations of motion; in this example we choose $\omega = 2$. The equations of motion are derived from

$$\dot{q}_i = \frac{\partial H}{\partial p_i},\ \dot{p}_i = -\frac{\partial H}{\partial q_i}, i = 1, 2.$$

The equations of motion are for this Hamiltonian in quasi-harmonic form:

$$\begin{cases} \ddot{q}_1 + 4q_1 &= \varepsilon q_2^2, \\ \ddot{q}_2 + q_2 &= 2\varepsilon q_1 q_2. \end{cases} \tag{9.11}$$

Equations of motion like this have been used as a cartoon problem for the motion of stars near the galactic plane in a rotating disk galaxy, see [75]. The energy of a particle moving according to system (9.11) is given by $H(p, q)$, and the energy manifolds in four-dimensional phase space are determined by $H(p, q) = E_0$ constant. For a general introduction to integrals of ODEs and the corresponding integral manifolds, see [146]. These manifolds (surfaces in four-dimensional phase space) are close to ellipsoïds, and they are bounded (compact) in a sphere around the origin of radius size $1/\varepsilon$. Within this large sphere, but in a neighbourhood of the origin, according to KAM theory, see [27], there exist an infinite family of tori that surround an infinite number of periodic solutions. For larger values of the energy but still on bounded energy manifolds, chaos will emerge.

The problem for system (9.11) looks simple, but the dynamics is complicated. We will see what first order averaging-normalisation has to say about this problem. Slowly varying equations arise from transformation (1.6) ($q_i, \dot{q}_i \mapsto r_i, \psi_i, i = 1, 2$):

$$\begin{cases} \dot{r}_1 &= -\frac{1}{2}\varepsilon \sin(2t + \psi_1) r_2^2 \cos^2(t + \psi_2), \\ \dot{\psi}_1 &= -\varepsilon \frac{\cos(2t+\psi_1)}{2r_1} r_2^2 \cos^2(t + \psi_2), \\ \dot{r}_2 &= -2\varepsilon \sin(t + \psi_2) r_1 r_2 \cos(t + \psi_2) \cos(2t + \psi_1), \\ \dot{\psi}_2 &= -2\varepsilon r_1 \cos(2t + \psi_1) \cos^2(t + \psi_2). \end{cases} \tag{9.12}$$

Note that we have to exclude regions where the amplitudes are small or zero. The $2\pi$-periodicity of the right-hand sides suggests averaging over t. To avoid too many indices, we use as before the original variables for the approximate quantities. Averaging yields

$$\begin{cases} \dot{r}_1 &= -\frac{\varepsilon}{8} r_2^2 \sin\chi, \ \dot{\psi}_1 = -\frac{\varepsilon}{8} \frac{r_2^2}{r_1} \cos\chi, \\ \dot{r}_2 &= -\frac{\varepsilon}{2} r_1 r_2 \sin\chi, \ \dot{\psi}_2 = -\frac{\varepsilon}{2} r_1 \cos\chi. \end{cases} \tag{9.13}$$

We introduced the combination angle $\chi = \psi_1 - 2\psi_2$. Our aim is to find periodic solutions and invariant manifolds embedded in the energy manifolds characterising the phase flow.

**Periodic solutions**

Note that putting $q_2 = \dot{q}_2 = 0$ in system (9.11) (a $q_1$ normal mode) produces a family of periodic solutions, parameterised by the energy. This $q_1$-normal mode family is not described by the averaged equations in amplitude-phase form. An exact $q_2$-normal mode family does not exist but is there such an $\varepsilon$-close family? From system (9.13) we find that $r_1$ is forced by $r_2 \neq 0$ so this seems not the case. The amplitude-phase transformation is of course not suitable for normal mode analysis, but repeating the calculations with transformation (1.16), we find again that there is not a $q_2$-normal mode family. This is surprising as one expects usually continuation from the normal modes of the linearised system to the non-linear case (also called Lyapunov continuation).

Periodic solutions in *general position* (not $\varepsilon$-close to coordinate planes) arise from system (9.13) if $\sin\chi = 0$ or $\chi = 0, \pi$. These solutions with constant amplitude exist if $\dot{\chi} = 0$ so that

$$\dot{\chi} = \dot{\psi}_1 - 2\dot{\psi}_2 = -\varepsilon\left(\frac{r_2^2}{8r_1} - r_1\right)\cos\chi = 0,$$

leading to the requirement

$$r_2^2 = 8r_1^2.$$

This yields two families of periodic solutions; we can make them more explicit by considering them on invariant manifolds.

**Invariant manifolds**

Multiplying in system (9.13) the equation for $r_1$ with $r_1$, the equation for $r_2$ with $r_2$, adding the equations, and integrating, we find the approximate energy integral:

$$4r_1^2 + r_2^2 = 2E_0, \tag{9.14}$$

with $E_0$ a positive constant, the initial energy. This is in fact an $O(\varepsilon)$ approximation of the energy integral $H$ = constant. In the case of bounded solutions of the Hamiltonian system, we do not need averaging to obtain result (9.14) from Hamiltonian (9.10). Combining the condition for general position periodic solutions with integral (9.14), we find

$$r_1 = \sqrt{\frac{E_0}{6}},\ r_2 = \sqrt{\frac{4E_0}{3}}.$$

The two families are parameterised by the energy.

Remarkably enough we find a second integral from system (9.13):

$$\frac{1}{2}r_1r_2^2\cos\chi = I, \tag{9.15}$$

with $I$ a constant. The two integrals and the three families of periodic solutions are useful to obtain a picture of the geometry of phase space to $O(\varepsilon)$. The analysis for the following summary is based on the study of the averaged system (9.13); we summarise:

The general position orbits are Lyapunov stable on each energy manifold, and the $q_1$-normal mode family is unstable. For definitions of various types of stability, see [146]. One can visualise the flow on a fixed energy manifold, but this flow is still three-dimensional. It helps to consider a transversal of this three-dimensional flow for a fixed value of one of the variables; this corresponds with a plane that is punctured by the oscillating orbits on the energy manifold and is called a Poincaré map. The $q_1$ normal mode produces one point, a saddle, see Fig. 9.3. Because of the

**Fig. 9.3** Poincaré map of the 2 : 1 resonance Hamiltonian system described by (9.13); the energy is fixed, and the orbits move on a fixed energy manifold puncturing the plane $\dot{q}_2 = 0$. The fixed point in the centre corresponds with the unstable $q_1$ normal mode. The closed orbits consist of points where the orbits are passing the transversal plane, and the orbits are filling up tori

frequency 2 of the $q_1$ mode, the periodic orbits produce two fixed points each in the transversal map. The closed curves around the fixed points consist actually of points from orbits on the tori surrounding the stable periodic solutions that puncture the transversal plane. See for more details [144].

### 9.5.2 The Cubic System with Detuning

It is not realistic to assume in experiments exact values 2 and 1 for the frequencies. In physical experiments, one expects a small detuning of the frequencies. We shall keep the frequency 1 and put

$$\omega^2 = 4 + \delta\varepsilon. \tag{9.16}$$

The parameter $\delta$ can be positive or negative. It is remarkable that we can show that this small change has both qualitative and quantitative consequences. The equations of motion (9.11) change to

$$\begin{cases} \ddot{q}_1 + 4q_1 & = -\varepsilon\delta q_1 + \varepsilon q_2^2, \\ \ddot{q}_2 + q_2 & = 2\varepsilon q_1 q_2. \end{cases} \tag{9.17}$$

As in the preceding subsection the $q_1$ normal mode ($q_2 = \dot{q}_2 = 0$) is an exact solution. We repeat the averaging process to obtain

$$\begin{cases} \dot{r}_1 & = -\frac{\varepsilon}{8} r_2^2 \sin\chi,\ \dot{\psi}_1 = \varepsilon\frac{\delta}{4} - \frac{\varepsilon}{8}\frac{r_2^2}{r_1}\cos\chi, \\ \dot{r}_2 & = -\frac{\varepsilon}{2} r_1 r_2 \sin\chi,\ \dot{\psi}_2 = -\frac{\varepsilon}{2} r_1 \cos\chi. \end{cases} \tag{9.18}$$

Again, two periodic solutions may exist if $\sin\chi = 0$ ($\chi = 0, \pi$) for $t \geq 0$. These solutions with constant amplitude exist if $\dot{\chi} = 0$ so that

$$\dot{\chi} = \dot{\psi}_1 - 2\dot{\psi}_2 = \varepsilon\frac{\delta}{4} - \varepsilon\left(\frac{r_2^2}{8r_1} - r_1\right)\cos\chi = 0,$$

leading to the requirement:

$$\delta - 4\left(\frac{r_2^2}{8r_1} - r_1\right)\cos\chi = 0.$$

System (9.18) admits the same approximate integral (9.14) as in the case with $\delta = 0$ so we have the condition $r_1^2 \leq E_0/2$, and by eliminating $r_2$:

$$6r_1^2\cos\chi + \delta r_1 - E_0\cos\chi = 0.$$

If $\chi = 0$, the limit value $r_1^2 = E_0/2$ produces the condition $\delta > -2\sqrt{2E_0}$ for existence of this periodic solution.

If $\chi = \pi$, the limit value $r_1^2 = E_0/2$ produces the condition $\delta < 2\sqrt{2E_0}$ for existence of this periodic solution.

We conclude that for detuning $-2\sqrt{2E_0} < \delta < 2\sqrt{2E_0}$ we have three families of periodic solutions parameterised by $E_0$. The values $\delta = \pm 2\sqrt{2E_0}$ are bifurcation values of the Hamiltonian system. This result also elucidates the origin of these general position periodic solutions. The 2 : 1 resonance produces these periodic solutions for values of $\delta$ between the bifurcation values, for $\delta < -2\sqrt{2E_0}$ and $\delta > 2\sqrt{2E_0}$ the $q_2$ normal mode exists, and at the bifurcation values, the general position periodic solutions emerge from the normal mode solutions.

### 9.5.3 A General Cubic Potential

The cubic potential we have considered thus far consists of one term only. We will look at a more general potential; this will also illustrate the importance of symmetries. Consider the two-dof potential problem with Hamiltonian:

$$H = \frac{1}{2}(\dot{x}^2 + 4x^2 + \dot{y}^2 + y^2) - \varepsilon(\frac{1}{3}a_1x^3 + a_2xy^2 + \frac{1}{3}a_3y^3 + a_4x^2y), \tag{9.19}$$

with $a_1, a_2, a_3, a_4$ constants, and the system is in 2 : 1 resonance. The equations of motion are

$$\begin{cases} \ddot{x} + 4x &= \varepsilon(a_1x^2 + a_2y^2 + 2a_4xy), \\ \ddot{y} + y &= \varepsilon(2a_2xy + a_3y^2 + a_4x^2). \end{cases} \tag{9.20}$$

The preceding subsections used the cubic $a_2$-term only, and such a potential is discrete symmetric (mirror symmetry) in the second position ($y$). Are there new phenomena in this more general system? The system was discussed by several authors, for an amplitude-phase treatment if $a_2 \neq 0$, see [144]; we summarise these results. After this summary, we will repeat the analysis partly with amplitude-angle transformation. Using transformation (1.6) in the form

$$x = r_1 \cos(2t+\psi_1), \dot{x} = -2r_1 \sin(2t+\psi_1), y = r_2 \cos(t+\psi_2), \dot{y}=-r_2 \sin(t+\psi_2),$$

we obtain from [144] with $\chi = \psi_1 - 2\psi_2$ the first order averaged system:

$$\begin{cases} \dot{r}_1 &= -\varepsilon \frac{a_2}{8} r_2^2 \sin\chi\,, \dot{\psi}_1 = -\varepsilon \frac{a_2}{8} \frac{r_2^2}{r_1} \cos\chi\,, \\ \dot{r}_2 &= \varepsilon \frac{a_2}{2} r_1 r_2 \sin\chi\,, \dot{\psi}_1 = -\varepsilon \frac{a_2}{2} r_1 \cos\chi\,. \end{cases} \tag{9.21}$$

It is remarkable that the coefficients $a_1, a_3, a_4$ play no part in an $O(\varepsilon)$ approximation on the timescale $1/\varepsilon$. System (9.20) has to first order a hidden symmetry. The system (9.21) has two independent integrals of motion and three periodic solutions for each value of the energy: the $x$-normal mode corresponding with $y = \dot{y} = 0$(unstable), the in-phase solution for $\chi = 0$ (stable), and the out-phase solution for $\chi = \pi$ (stable). As the energy is a free parameter, we have three families of periodic solutions. Again, the $y$-normal mode cannot be continued.

We take now a different approach of [144] to show both the similarity but also slight difference with averaging over angles. Using the amplitude-angle transformation (1.21)

$$x = r_1 \sin\phi_1, \dot{x} = 2r_1 \cos\phi_1, y = r_2 \sin\phi_2, \dot{y} = r_2 \cos\phi_2,$$

we obtain the variational system:

$$\begin{cases} \dot{r}_1 &= \frac{\varepsilon}{2} \cos\phi_1 (a_1 r_1^2 \sin^2\phi_1 + a_2 r_2^2 \sin^2\phi_2 + 2a_4 r_1 r_2 \sin\phi_1 \sin\phi_2), \dot{\phi}_1 = 2+\varepsilon \ldots, \\ \dot{r}_2 &= \varepsilon \cos\phi_2 (2a_2 r_1 r_2 \sin\phi_1 \sin\phi_2 + a_3 r_2^2 \sin^2\phi_2 + a_4 r_1^2 \sin^2\phi_1), \dot{\phi}_2 = 1+\varepsilon \ldots. \end{cases} \tag{9.22}$$

It is easy to see that the angles in the system for $(r_1, r_2)$ are $\phi_1, \phi_2, \phi_1+2\phi_2, \phi_1 - 2\phi_2$. The only angle that is not timelike is $\chi = \phi_1 - 2\phi_2$ (see for "timelike" Chap. 7). Averaging produces the approximating system:

$$\dot{r}_1 = -\frac{\varepsilon}{8} a_2 r_2^2 \cos(\phi_1 - 2\phi_2), \dot{r}_2 = \frac{\varepsilon}{2} a_2 r_1 r_2 \cos(\phi_1 - 2\phi_2).$$

If we use the $O(\varepsilon)$ terms of the equations for $\phi_1, \phi_2$ in system (9.22), we obtain after averaging the same results as before: periodic solutions and integrals. The concept of timelike and not timelike angles helps us to obtain a slowly varying equation for $\chi$. Domains in space where $d\chi/dt = 0$ correspond with resonance manifolds; this is where we located periodic solutions.

This calculation would be more complicated if the $O(1)$ variation of the angles $\phi_1, \phi_2$ would depend on $r_1, r_2$. In near stable equilibrium in phase space, one can often safely assume constant frequencies but not always for larger values of the energy.

We conclude that the advantage of using amplitude-angle variables is that we know that the size of the resonance zone containing the periodic solutions in this case is $O(\sqrt{\varepsilon})$. Apart from that, the different variational approaches produce similar results here.

We observe that the Hamiltonian (9.19) is a potential problem with four free parameters. If $a_1 = a_2 = 0$, the potential is discrete (mirror) symmetric in position variable $x$. In addition, only one of them, $a_2$, plays a part in the first order approximation. If $a_2 = 0$, the first order approximation is trivial. Such degenerations arise often when introducing symmetries; such assumptions can lead to interesting results, but they are not always generic.

**The Case** $a_2 = 0$

In this case Hamiltonian (9.19) leads to the equations of motion:

$$\begin{cases} \ddot{x} + 4x &= \varepsilon(a_1 x^2 + 2a_4 xy), \\ \ddot{y} + y &= \varepsilon(a_3 y^2 + a_4 x^2). \end{cases} \tag{9.23}$$

As noted above, first order averaging leads to trivial results. Second order averaging (see Chap. 4) leads to

$$\dot{r}_1 = O(\varepsilon^3), \dot{r}_2 = O(\varepsilon^3), \tag{9.24}$$

but to a nontrivial equation for the combination angle $\chi = 2\phi_1 - 4\phi_2$:

$$\frac{d\chi}{dt} = \varepsilon^2 f(r_1, r_2) + O(\varepsilon^3).$$

In this case we have to consider the potential problem as a $4 : 2$ higher order resonance. The angle $\chi$ is timelike except in the resonance manifold determined by $f(r_1, r_2) = 0$.

More explicit details to handle such higher order resonances are given in later subsections for instance 9.5.8.

### 9.5.4 *A Cubic Hamiltonian System in* 1 : 1 *Resonance*

In time-independent Hamiltonian systems and in general conservative systems, the problem of extension of the timescale of validity is more difficult as these systems do not contain asymptotically stable solutions. For two-dof Hamiltonian systems, we can obtain long time approximations by using geometric arguments based on the theory of area-preserving maps. See [5] and [27].

The perturbation results in this subsection are based on [144]. Consider the two-degrees-of-freedom Hamiltonian system generated by the cubic Hamiltonian: (9.10) but with $\omega = 1$:

$$H(p,q) = \frac{1}{2}(p_1^2 + q_1^2) + \frac{1}{2}(p_2^2 + q_2^2) - \varepsilon q_1 q_2^2.$$

The equations of motion are

$$\begin{cases} \ddot{q}_1 + q_1 &= \varepsilon q_2^2, \\ \ddot{q}_2 + q_2 &= 2\varepsilon q_1 q_2. \end{cases} \tag{9.25}$$

In Subsection 10.1 we will consider a related cubic Hamiltonian producing the so-called Hénon-Heiles system. Note that for system (9.25), as in subsection 9.5.1, putting $q_2 = \dot{q}_2 = 0$ produces a family of periodic solutions (harmonic functions), parameterised by the energy. This is the $q_1$-normal mode family. Slowly varying equations arise from transformation (1.6) ($q_i, \dot{q}_i \mapsto r_i, \phi_i,\ i = 1, 2$). We can average over time $t$, but this produces (not very interesting) zero right-hand sides. Again, to avoid too many indices and tildes, we use with some abuse of notation the original variables for the approximate quantities. We will apply Theorem 6.1, so we have to compute $f_1 = Df.u^1$ (Theorem 6.1 notation). We omit the long computations involving vector fields of dimension 4 and $4 \times 4$ matrices. Averaging of $f_1(t, r_1, \phi_1, r_2, \phi_2)$ over time yields with combination angle $\chi = 2(\phi_1 - \phi_2)$:

$$\begin{cases} \dot{r}_1 &= -\frac{\varepsilon^2}{2} r_1 r_2^2 \sin\chi,\ \dot{\phi}_1 = -\varepsilon^2 r_2^2 (\frac{1}{3} + \frac{1}{2}\cos\chi), \\ \dot{r}_2 &= \frac{\varepsilon^2}{2} r_1^2 r_2 \sin\chi,\ \dot{\phi}_2 = -\varepsilon^2 (\frac{1}{3} r_1^2 + \frac{5}{12} r_2^2 + \frac{1}{2} r_1^2 \cos\chi). \end{cases} \tag{9.26}$$

As in application 9.5.1, we have only one combination angle $\chi$ in the averaged equations. From system (9.26) the equation for angle $\chi$ is

$$\dot{\chi} = \varepsilon^2 \left( \frac{2}{3} r_1^2 + \frac{1}{6} r_2^2 + (r_1^2 - r_2^2)\cos\chi \right). \tag{9.27}$$

**Invariant manifolds**

System (9.26) has the quadratic first integral

$$r_1^2 + r_2^2 = 2E_0, \tag{9.28}$$

with $E_0$ a positive constant. The family of integrals (9.28) approximates the family of energy manifolds around the origin of phase space. As the energy manifold is bounded in the neighbourhood of the origin, the approximate integral (9.28) is valid for all time with error $O(\varepsilon^2)$.

There exists a second integral of the averaged system (9.26) :

$$r_1^2\left(\frac{1}{4}r_1^2+\frac{1}{3}E_0-r_2^2\cos\chi\right)=I, \tag{9.29}$$

with integration constant determined by the initial conditions.

The integrals (9.28) and (9.29) describe together approximate families of tori embedded in the energy manifolds. The tori surround the stable periodic solutions listed below.

**Periodic solutions**

It turns out we find six families of periodic solutions. There are two normal mode families, the exact $q_1$ normal mode family and the $\varepsilon$-close $q_2$ normal mode family. For the second family we have to change the averaging scheme by using transformation (1.16); we omit these calculations.

As in the case of application (9.5.1), we obtain short-periodic solutions by considering the cases where $\sin\chi = 0$. The term "short" indicates usually that the frequencies of these periodic solutions are close to the basic frequencies of the linearised system near the origin. We have the possibilities:

- In-phase periodic solutions $\chi = 0, 2\pi$. Using the integrals and Eq. (9.27), we find $r_1^2=\frac{2}{3}E_0$, $r_2^2=\frac{4}{3}E_0$. The periodic solutions are stable.
- Out-phase periodic solutions $\chi=\pi, 3\pi$. Using the integrals and Eq. (9.27), we find again $r_1^2=\frac{2}{3}E_0$, $r_2^2=\frac{4}{3}E_0$. The periodic solutions are stable.

It is interesting that in this problem we can obtain some general position periodic solutions as exact solutions. We use as a guess $q_1 = aq_2$ with factor $a$ to be determined. We find

$$q_1=\pm\frac{1}{\sqrt{2}}q_2.$$

Substitution in system (9.25) produces two identical equations of the form:

$$\ddot{q}_2+q_2=\pm\varepsilon\sqrt{2}\,q_2^2. \tag{9.30}$$

The exact solutions determined by Eq. (9.30) are periodic elliptic functions, and they are parameterised by the initial conditions. Note that they also exist for larger values of the energy as a continuation of the asymptotic solutions; however, their stability may change when increasing $E_0$. The in- and out-phase periodic solutions were obtained by localising near the origin using parameter $\varepsilon$. The exact periodic solutions exist for arbitrary values of the parameter $\varepsilon$. It can be verified that they represent continuous families of periodic solutions connecting the origin of phase space with other critical points (corresponding with unstable equilibria) of the energy manifold.

### 9.5.5 Higher Order Resonance in Two Degrees of Freedom

Consider systems with two degrees of freedom (dof), frequencies $\omega_1, \omega_2$ and

$$k\omega_1 + l\omega_2 = 0,$$

with $k$ and $l$ relative prime. The case $|k| + |l| \geq 5$ is called a higher order resonance, the corresponding higher order theory was developed mainly for Hamiltonian systems, but the ideas apply also to dissipative systems. The theory of *higher order resonance* for two-dof Hamiltonian systems was started in [120], [131]; see also [123] ch. 10.6.4, for an introduction [121].

We consider two-dof Hamiltonian systems with basic $k : l$ higher order resonance. It happens in some cases because of symmetries that a lower order resonance is also described by the theory of higher order resonance as symmetries may produce certain degenerations in the averaged system, see the case of the cubic potential in 2 : 1 resonance and [131].

We will see that in the case of higher order resonance the essential perturbation terms are appearing at much higher order in the expansion of the Hamiltonian. We consider again the Hamiltonian $H(p, q)$ in homogeneous polynomials near stable equilibrium:

$$H(p,q) = \frac{1}{2}\omega_1(p_1^2 + q_1^2) + \frac{1}{2}\omega_2(p_2^2 + q_2^2) + \varepsilon H_3 + \varepsilon^2 H_4 + \dots \tag{9.31}$$

The averaging process removes the cubic terms to produce the normal form Hamiltonian $\bar{H}$. As in [123] ch. 10, we will often use the canonical action-angle coordinates

$$\tau = \frac{1}{2}(p^2 + q^2)$$

and $\phi$. It turns out that at higher order resonance the quartic terms, quadratic in $\tau$, do not depend on the angles. They arise at much higher order $(k + l - 2)$ after extended averaging:

$$\begin{cases} \bar{H} & = \omega_1\tau_1 + \omega_2\tau_2 + \varepsilon^2(A\tau_1^2 + 2B\tau_1\tau_2 + C\tau_2^2) + \dots + \\ & \varepsilon^{k+l-2}|D|\sqrt{2\tau_1}(2\tau_2)^{\frac{k}{2}}\cos(l\phi_1 - k\phi_2 + \alpha) + O(\varepsilon)^{k+l-1}, \end{cases} \tag{9.32}$$

with constants $|D|, \alpha, A, B, C$; the dots represent terms of size smaller than $O(\varepsilon^2)$ but not containing the combination angle $\chi = l\phi_1 - k\phi_2 + \alpha$. The first resonant term arrives from $H_{k+l}$ at $O(\varepsilon^{k+l-2})$ and depends on the actions $\tau_1, \tau_2$ and angle $\chi$.

In the analysis, both averaging over time $t$ and combination angle $\chi$ play a part. For the combination angle $\chi$, we have

$$\dot{\chi} = 2\varepsilon^2 \left(l(A\tau_1 + B\tau_2) - k(B\tau_1 + C\tau_2)\right) + \ldots + \varepsilon^{k+l-2} g(\tau_1, \tau_2) \sin \chi , \tag{9.33}$$

with $g$ a function of the actions.

Analysing the averaged equations, it turns out there are two domains in phase space where the dynamics is very different, and they are characterised by different timelike variables.

**Theorem 9.3** *Assume* $\omega_1, \omega_2 > 0$ *and smoothness of the Hamiltonian to an order as high as necessary as in Eq.* (9.32). *From the theory of averaging over angles, we identify a resonance manifold* $M$ *that is generically determined by the condition*

$$l(A\tau_1 + B\tau_2) - k(B\tau_1 + C\tau_2) = 0.$$

*If this equation has a solution on a bounded energy manifold, it determines a resonance manifold* $M$. *We have:*

- *The* resonance domain *(or zone)* $D_I$ *is a neighbourhood of the resonance manifold* $M$. *In terms of singular perturbations, the neighbouring resonance zone is the* inner boundary layer. *Introducing the distance* $d(P, M)$ *for a point* $P$ *in the resonance zone to the manifold* $M$, *we have*

$$D_I = P\{d(P, M) = O(\varepsilon^{\frac{k+l-4}{2}})\}, \quad k + l \geq 5.$$

- *The remaining part of phase space, outside the resonance zone, is* $D_0$, *the* outer *domain. In the domain* $D_0$, *there is, to a certain approximation, hardly any exchange of energy between the two degrees of freedom.*
- *From Hamiltonian* (9.33), *we find that variations of* $\chi$ *take place at* $O(\varepsilon^{k+l-2})$ *with factor* $\sin \chi$. *Periodic solutions, if present, can be found in the resonance manifold.*

**Remark 9.2** To compute the averaged system to such a high order is quite an effort. However, one can much easier recover some quantitative aspects by a result formulated in [131]: *If the Hamiltonian* (9.31) *is derived from a potential function, then we have* $\alpha = 0$ *for the constant in the combination angle* $\chi$ of averaged (normalised) Hamiltonian (9.32).

**Remark 9.3** Higher order resonance in three or more dof is largely unexplored. We will discuss a few applications in Sect. 9.6 where first order and higher order resonance arise in three-dof Hamiltonian systems. We will meet in this context the presence of embedded resonance manifolds.

### 9.5.6 A Cubic Potential in Higher Order Resonance

Consider again the cubic potential problem studied before. It represents a strongly simplified galactic potential problem outside the main resonances 2 : 1 and 1 : 1. So consider the Hamiltonian with higher order resonance:

$$H(p,q) = \frac{1}{2}(p_1^2 + k^2 q_1^2) + \frac{1}{2}(p_2^2 + l^2 q_2^2) - \varepsilon q_1 q_2^2. \tag{9.34}$$

The positive frequencies $k, l$ are relative prime and $k + l \geq 5$. The equations of motion are

$$\begin{cases} \ddot{q}_1 + k^2 q_1 &= \varepsilon q_2^2, \\ \ddot{q}_2 + l^2 q_2 &= 2\varepsilon q_1 q_2. \end{cases} \tag{9.35}$$

Following remark 9.2, we define $\chi = l\phi_1 - k\phi_2$. For transformation to slowly varying variables, one has several choices. Using the non-canonical but quantitatively equivalent amplitude-phase variables (1.6), we find by averaging to second order:

$$\begin{cases} \dot{r}_1 &= O(\varepsilon^3), \dot{r}_2 = O(\varepsilon^3), \\ \dot{\chi} &= \varepsilon^2 \frac{k^5}{l(k^2-4l^2)} \left( \frac{3k^2-4l^2}{4k^2} r_2^2 - r_1^2 \right). \end{cases} \tag{9.36}$$

Rescaling $\varepsilon^2 t = \tau$, we recognise the set-up of Theorem 7.1 for averaging over one angle. The resonance manifold, if it exists, is given by the equation:

$$\frac{3k^2 - 4l^2}{4k^2} r_2^2 - r_1^2 = 0. \tag{9.37}$$

As before we have the approximate energy integral:

$$k^2 r_1^2 + l^2 r_2^2 = 2E_0.$$

Using condition (9.37) and the approximate energy integral, we find for the location of the resonance manifold depending on the energy level $E_0$:

$$r_1^2 = \frac{3k^2 - 4l^2}{3k^4} 2E_0, \; r_2^2 = \frac{8}{3k^2} E_0. \tag{9.38}$$

The conditions that $k^2 r_1^2$ and $l^2 r_2^2$ have to be smaller than $2E_0$ lead with Eq. (9.38) to the existence condition of this resonance manifold $k/l > 2/\sqrt{3}$. In Fig. 9.4 we show projections of the stable (full line) and unstable (hashed) periodic solutions in two cases of $k/l$.

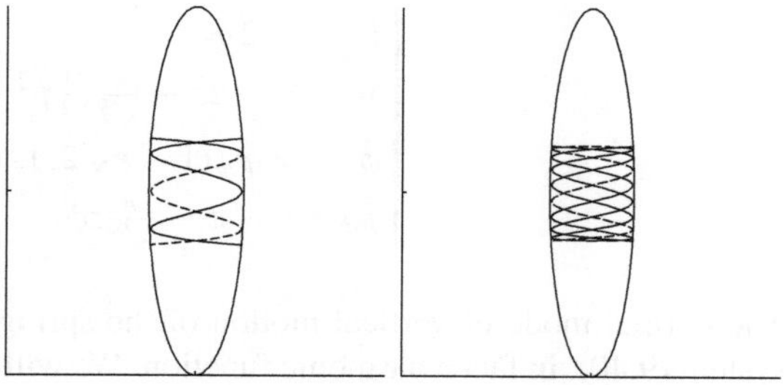

**Fig. 9.4** Projections of periodic solutions (stable full line, unstable hashed) on the $(q_1, q_2)$ plane of the $k/l = 4/1$ and $9/2$ resonances described for system (9.35); the figures are based on [121]

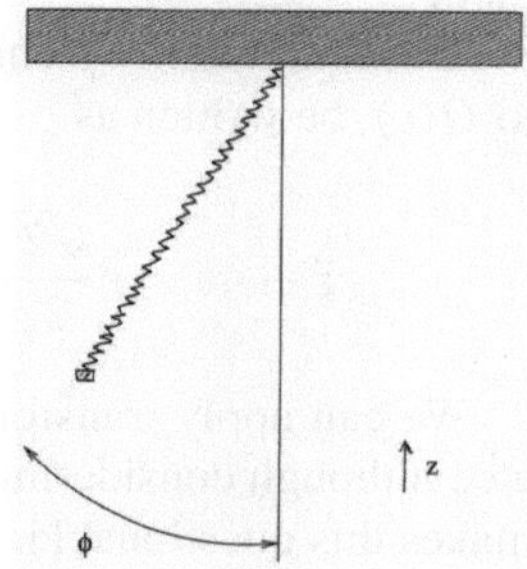


**Fig. 9.5** The spring pendulum described by system (9.40)

### 9.5.7 *The Spring Pendulum*

Consider a hanging spring attached at the top and with a weight attached to the lower end, see Fig. 9.5. The spring can oscillate without bending, and it can swing in a vertical plane as a pendulum making an angle $\phi$ with the vertical; without friction it can be described by a Hamiltonian system with two dof. A model extension would be the *spherical* spring pendulum requiring three dof. We will study the stability of the (normal mode) vertical oscillations in the two-dof case. In 1933 Gorelik and Vitt [63] showed experimentally that the vertical normal mode can be unstable resulting in strong exchanges of energy between the two dof. In [138] this phenomenon was explained by first order averaging. In section 10.5 of [62] the role of symmetries in the dynamics of the spring pendulum is analysed.

A local non-linear analysis involves expanding the Hamiltonian to a certain order and rescaling the variables by $\varepsilon$ to study the dynamics near equilibrium when the spring is hanging down at rest. For the expanded Hamiltonian $H = H_2 + \varepsilon H_3$ we have from [138] for small deflections to degree 3:

$$\begin{cases} H_2 &= \frac{1}{2}\omega_z(p_z^2 + z^2) + \frac{1}{2}\omega_\phi(p_\phi^2 + \phi^2), \\ H_3 &= \frac{\omega_\phi}{\sqrt{ml_0^2\omega_z}}\left(\frac{1}{2}z\phi^2 - zp_\phi^2\right). \end{cases} \tag{9.39}$$

Note that we have mirror (discrete) symmetry in $\phi$ as expected. To avoid too many parameters, we put $ml_0^2 = 1$. Consider the resonance ratio by choosing $\omega_z = 2, \omega_\phi = 1$. The equations of motion become to $O(\varepsilon)$:

$$\begin{cases} \dot{z} &= 2p_z, \\ \dot{p}_z &= -2z - \frac{\varepsilon}{\sqrt{2}}(\frac{1}{2}\phi^2 - p_\phi^2), \\ \dot{\phi} &= p_\phi(1 - \varepsilon\sqrt{2}z), \\ \dot{p}_\phi &= -\phi - \frac{\varepsilon}{\sqrt{2}}z\phi. \end{cases} \tag{9.40}$$

The normal mode of vertical motion of the spring ($\dot{\phi} = \dot{p}_\phi = 0$) is a solution of system (9.40), in fact a harmonic function. We will study the stability of this normal mode.

The equation for $\dot{\phi}$ enables us to eliminate $p_\phi$, and the equations of motion can, to $O(\varepsilon)$, be written as

$$\ddot{z} + 4z = \varepsilon\frac{\sqrt{2}}{2}(\dot{\phi}^2 - \frac{1}{2}\phi^2),\ \ddot{\phi} + \phi = \varepsilon\sqrt{2}(\frac{1}{2}z\phi - \dot{z}\dot{\phi}). \tag{9.41}$$

We can apply transformation (1.6): $z = r_1\cos(2t + \psi_1), \phi = r_2\cos(t + \psi_2)$, etc., although considering $\phi$ small to study the stability of the vertical normal mode makes this questionable. However, because of the non-linear terms in the equations for the $\phi$-mode, the polar coordinate singularity is in this case removed. Introducing the combination angle $\chi = \psi_1 - 2\psi_2$, we have by first order averaging for the variational equations:

$$\begin{cases} \dot{r}_1 &= \varepsilon\frac{3\sqrt{2}}{32}r_2^2\sin\chi,\ \dot{\psi}_1 = \varepsilon\frac{3\sqrt{2}}{32}\frac{r_2^2}{r_1}\cos\chi, \\ \dot{r}_2 &= -\varepsilon\frac{3\sqrt{2}}{8}r_1r_2\sin\chi,\ \dot{\psi}_2 = \varepsilon\frac{3\sqrt{2}}{8}r_1\cos\chi. \end{cases} \tag{9.42}$$

For the combination angle $\chi$, we have the equation:

$$\dot{\chi} = \varepsilon\frac{3}{4}\sqrt{2}(\frac{r_2^2}{8r_1} - r_1)\cos\chi. \tag{9.43}$$

System (9.42) has the first integral:

$$4r_1^2 + r_2^2 = 2E_0, \tag{9.44}$$

with $E_0$ a positive constant. Choosing $r_2^2 - 8r_1^2 = 0$, we have $\dot{\chi} = 0$; this relation holds for $t \geq 0$ if $r_1(t), r_2(t)$ are constant, so choose $\chi = 0, \pi$. This results in two families of periodic solutions. Using integral (9.44), we find with $\chi = 0, \pi$

$$r_1^2 = \frac{1}{6}E_0,\ r_2^2 = \frac{4}{3}E_0. \tag{9.45}$$

This makes with the vertical normal mode three families of periodic solutions. In Fig. 9.3 we used a Poincaré map for illustration of the cubic Hamiltonian flow on

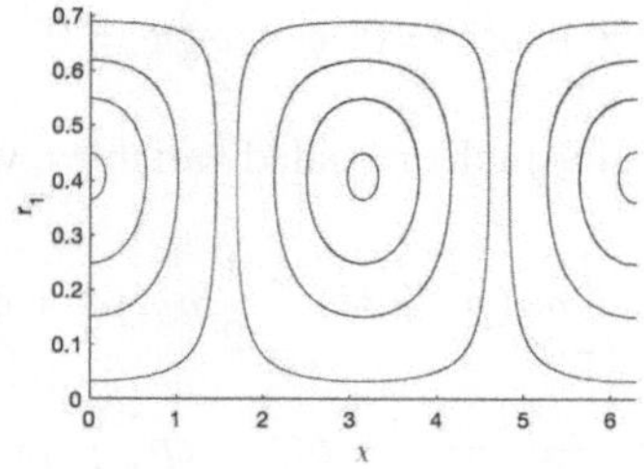


**Fig. 9.6** The $r_1, \chi$ phase-plane of the spring pendulum in 2 : 1 resonance described by system (9.42); we have chosen $E_0 = 1$ so we have from (9.44) that $0 \leq r_1 \leq 0.707$. The solutions corresponding with $\chi = 0, \pi$ are stable, and the vertical normal mode corresponding with $r_1 = 0.707, r_2(t) = 0$ is unstable

the energy manifold. As we can reduce the analysis in two dof (the case of the spring pendulum), we can also use a reduction to the $r_1, \chi$-plane; see Fig. 9.6.

For $\varepsilon = 0$ we have the pendulum normal mode in system (9.40), but this mode cannot be continued. The physical reasons for this are also quite clear as swinging the pendulum makes the string expand and contract.

In studies of mechanical oscillations with two dof, one often finds the 1 : 2 resonance, both in conservative and in dissipative systems. The main reason is that a first order approximation in $\varepsilon$ is usual sufficient in these problems. Other resonance ratios will be considered in the next subsection.

### 9.5.8 *The Spring Pendulum at Higher Order Resonance*

Consider again the spring pendulum, a pendulum where the suspending, inflexible string is replaced by a linear spring. We will look at the *higher order* resonances. As predicted by Theorem 9.3, if higher order resonance is active, there will be two domains in phase space where the dynamics is very different, and they are characterised by different timelike variables.

Let $r(t)$ be the length of the spring at time $t$ and $\phi$ the angular deflection of the spring with respect to the vertical. Neglecting friction, the Hamiltonian is

$$H = \frac{1}{2m}\left(p_r^2 + \frac{p_\phi^2}{r^2}\right) + \frac{s}{2}(r - l_0)^2 - mgr\cos\phi, \tag{9.46}$$

with $s$ a positive constant, $p_r = m\dot{r}$, $p_\phi = mr^2\dot{\phi}$. Using the notation of [131], a few rescalings are introduced, for the length of the spring $z = (r - l_0)/l_0$, also that the spring pendulum is at rest if $(p_z, z, p_\phi, \phi) = (0, 0, 0, 0)$. The basic frequencies are $\omega_z = \sqrt{s/m}$, $\omega_\phi = \sqrt{g/l_0}$ producing for the linearised motion in the vertical and horizontal directions, respectively:

$$\ddot{z} + \omega_z^2 z = 0, \; \ddot{\phi} + \omega_\phi^2 \phi = 0.$$

Expanding Hamiltonian (9.46) in the rescaled variables, we have

$$H_2 = \frac{1}{2}\omega_z(p_z^2 + z^2) + \frac{1}{2}\omega_\phi(p_\phi^2 + \phi^2), \tag{9.47}$$

$$H_3 = \frac{\omega_\phi}{\sqrt{ml_0^2\omega_z}}\left(\frac{1}{2}z\phi^2 - zp_\phi^2\right), \tag{9.48}$$

$$H_4 = \frac{1}{ml_0^2}\left(\frac{3}{2}\frac{\omega_\phi}{\omega_z}z^2 p_\phi^2 - \frac{1}{24}\phi^4\right). \tag{9.49}$$

For simplicity we replace $\omega_z, \omega_\phi$ by the natural numbers $k, l$. For averaging, we rescale the variables by $\varepsilon$, $z \mapsto \varepsilon\bar{z}$, etc., followed by dividing by $\varepsilon$ and omitting the bars.

We made a calculation in subsection 9.5.7 by putting $k = 2, l = 1$ and find that first order averaging using transformation (1.6) produces interesting results arising from the $H_3$ terms.

Assume now that $k/l \neq 2$ and also not $\varepsilon$-close to this ratio. Applying second order averaging, we find (even if $k/l = 1$)

$$\dot{r_1} = O(\varepsilon^3), \; \dot{r_2} = O(\varepsilon^3),$$

and for the combination angle $\chi$, see (9.33), following [131] an expression of the form

$$\dot{\chi} = l\dot{\psi}_1 - k\dot{\psi}_2 = f(r_1, r_2),$$

with $f$ depending on the frequencies $k, l$. We present the expression in the case of the 6 : 1 resonance:

$$\dot{\chi} = -\varepsilon^2\left(\frac{4.57}{\sigma}r_1^2 - 0.46r_2^2\right), \tag{9.50}$$

with $\sigma > 0$, so the right-hand side has a zero for fixed energy. The 6 : 1 resonance manifold exists. In Fig. 9.7 we present a Poincaré map for the 6 : 1 resonance with the resonance domain depicted as a small region $O(\varepsilon^{3/2})$.

Following [131], we summarise some results in Table 9.1. It is clear that the larger $k + l$, the smaller the resonance domain indicated by size $d_\varepsilon$. The interaction intervals of time between the two dof in the resonance zones increase with $k + l$.

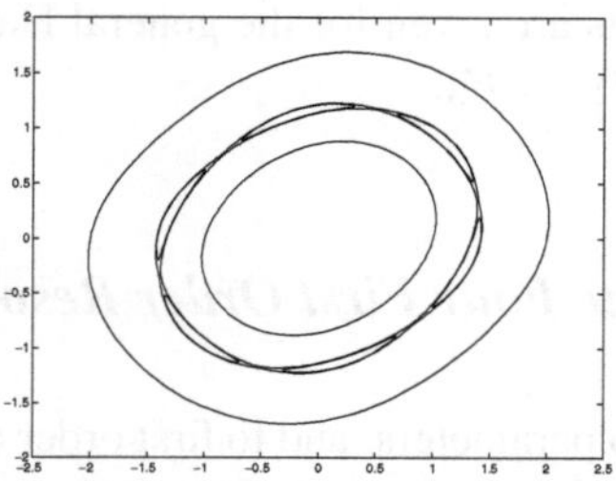

**Fig. 9.7** Poincaré map for the 6 : 1 resonance of the spring pendulum ($\varepsilon = 0.75$, large for illustration purposes). In the resonance zone, the saddles are connected by heteroclinic cycles and inside the cycles are centres, see [131], courtesy SIAP

**Table 9.1** The table from [131] presents the most prominent higher order resonances of the spring pendulum with lowest order resonant terms $O(\varepsilon^{k+l-2})$. The third column gives the size of the resonance zone in which the resonance manifold $M$ is embedded, while in the fourth column we find the timescale of interaction in the resonance zone

| *Resonance* | $k+l-2$ | $d_\varepsilon$ | Interaction timescale |
|---|---|---|---|
| 1 : 4 | 3 | $\varepsilon^{1/2}$ | $\varepsilon^{-5/2}$ |
| 3 : 4 | 5 | $\varepsilon^{3/2}$ | $\varepsilon^{-7/2}$ |
| 1 : 6 | 5 | $\varepsilon^{3/2}$ | $\varepsilon^{-7/2}$ |
| 2 : 6 | 6 | $\varepsilon^{2}$ | $\varepsilon^{-4}$ |
| 1 : 8 | 7 | $\varepsilon^{5/2}$ | $\varepsilon^{-9/2}$ |
| 4 : 6 | 8 | $\varepsilon^{3}$ | $\varepsilon^{-5}$ |

## 9.6 Three Degrees of Freedom

The case of three-dof Hamiltonian systems involves many new technical problems and new qualitative aspects. Technical, as the number of terms increases, qualitative as the complexity of periodic solutions and invariant manifolds increases enormously if the number of dof is larger than 2. This is partly illustrated by the integrability of the averaged (or normalised) Hamiltonian systems in two dof. Following the introduction of Chap. 9, a 3-dof time-independent Hamiltonian is called *integrable* if three independent first integrals exist foliating the bounded five-dimensional energy manifold in smooth manifolds on which the solutions are moving. For 2-dof Hamiltonian systems, the averaged Hamiltonian system is always integrable near stable equilibrium even if the original system is not integrable. As we shall see, the averaged Hamiltonian system with three dof is in general not integrable although special symmetries or coefficients in the Hamiltonian may change the picture. Another difference with two dof is the possibilities for stability of periodic solutions.

In [135] the four first low order resonance cases are presented, and in [123] a summary is given. We start with a summary of the first order resonance based on

[151] and [159]. The results are given for the general Hamiltonian (9.3) with three dof to cubic terms, $H = H_2 + \varepsilon H_3$.

### 9.6.1 Summary of the Four First Order Resonances

The cubic term contains 56 parameters, and to first order averaging, we have four or six parameters left. This is already remarkable as it shows that of the 56 parameters in the general $H_3$ case at least 50 play no prominent role at first order. Two or three combination angles arise in the first order averaged system: Normal forms of the Hamiltonian will be indicated by a bar, $\bar{H}_3$, $\bar{H}_4$, etc.

Normal forms (or averaging-normal forms) produce approximations. If the normal form is not integrable, the implication is that the dynamics shows more complexity than for instance in the case of two dof.

**The** 1 : 2 : 2 **Resonance**
The normal form is in action-angle coordinates $\tau$, $\psi$ to third order:

$$\bar{H}^3 = \tau_1 + 2\tau_2 + 2\tau_3 + 2\left[a_1\sqrt{2\tau_2}\cos(2\psi_1 - \psi_2 - a_2) + a_3\sqrt{2\tau_3}\cos(2\psi_1 - \psi_3 - a_4)\right], \tag{9.51}$$

with constants $a_1, \ldots, a_4$. For potential problems we have $a_2 = a_4 = 0$. The actions and the two combination angles are slowly varying.

The normal form $H_2 + \bar{H}^3$ in (9.51) is integrable, see for potential problems [101] and for the general Hamiltonian case [136]. The proofs are by inspection of the equations of motion.

For fixed energy there are two EE periodic solutions in general position. For $H_2 + \varepsilon\bar{H}_3$, $\tau_1 = 0$ is an invariant manifold. Extending to $H_2 + \bar{H}_3 + \bar{H}_4$, one finds that the $\tau_1 = 0$ manifold bifurcates into the unstable (HH) normal modes $\tau_2$, $\tau_3$ and four unstable (EH) periodic solutions at $\tau_1 = 0$; see [123].

It was shown in [43] by algebraic methods that for an open set of parameters the normal form $H_2 + \bar{H}_3 + \bar{H}_4$ is not integrable. The implication is that generically the 1 : 2 : 2 resonance behaves to $O(\varepsilon)$ integrable (leaving out the $\varepsilon$, integrability of $H_2 + \bar{H}_3$) and on smaller subsets to $O(\varepsilon^2)$ (so including $\bar{H}_4$) dynamically non-integrable. High precision numerics in [43] shows chaotic behaviour in Poincaré sections.

**The** 1 : 2 : 1 **Resonance**
The normal form is in action-angle coordinates $\tau$, $\psi$ to third order:

$$\begin{aligned}\bar{H}^3 =& \tau_1 + 2\tau_2 + \tau_3 + 2\sqrt{2\tau_2}\big[a_1\cos(2\psi_1 - \psi_2 - a_2) + \\ & a_3\sqrt{\tau_1\tau_3}\cos(\psi_1 - \psi_2 + \psi_3 - a_4) + a_5\tau_3\cos(2\psi_3 - \psi_2 - a_6)\big].\end{aligned} \tag{9.52}$$

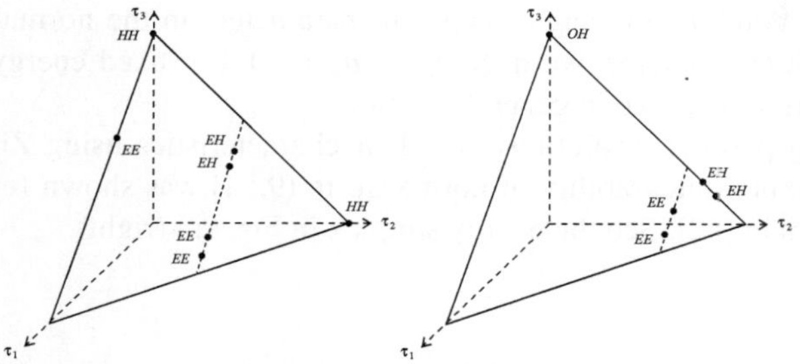


**Fig. 9.8** Energy simplices for the Hamiltonian resonances 1 : 2 : 3 (left) and 1 : 2 : 4 (right). Dots indicate periodic solutions. The instability of the $\tau_2$ normal mode of the 1 : 2 : 3 resonance can be HH or C dependent on the parameters (HH is indicated). Complex instability plays a part in the discussion of chaos in Sect. 9.12

We have three combination angles and 56 parameters in $H_2 + H_3$, reduced to six parameters in the normal form. In the case of potential problems, we have in (9.52) $a_2 = a_4 = a_6 = 0$. For fixed energy there exist seven periodic solutions with four in general position.

It was shown in [50] dat voor een open set of parameters in the Hamiltonian the normal form $H_2 + \bar{H}^3$ is non-integrable. The technique used in [50] is that in certain subsets one can identify solutions and continue time analytically into the complex domain. This leads to the presence of an essential singularity in the sense of complex analysis. The author of [50] remarks on the non-integrable dynamics of this resonance. More details will be given in Sect. 9.7.

**The** 1 : 2 : 3 **Resonance**

The normal form is in action-angle coordinates $\tau, \psi$ to third order $H_2 + \bar{H}^3$:

$$\begin{aligned}\bar{H}^3 =& \tau_1 + 2\tau_2 + 3\tau_3 + 2\sqrt{2\tau_1\tau_2}\big[a_1\sqrt{\tau_1}\cos(2\psi_1 - \psi_2 - a_2) + \\ & a_3\sqrt{\tau_3}\cos(\psi_1 + \psi_2 - \psi_3 - a_4)\big].\end{aligned} \tag{9.53}$$

We have two combination angles and four parameters in the normal form. In the case of potential problems, we have $a_2 = a_4 = 0$. We find for fixed energy seven periodic solutions with four in general position. See the energy simplex in Fig. 9.8 left. The 1 : 2 : 3 resonance in normal form is non-integrable, see Sect. 9.12.

**The** 1 : 2 : 4 **Resonance**

The normal form is in action-angle coordinates $\tau, \psi$ to third order:

$$\begin{aligned}\bar{H}^3 =& \tau_1 + 2\tau_2 + 4\tau_3 + 2\big[a_1\tau_1\sqrt{2\tau_2}\cos(2\psi_1 - \psi_2 - a_2) + \\ & a_3\, \tau_2\sqrt{2\tau_3}\cos(2\psi_1 - \psi_3 - a_4)\big].\end{aligned} \tag{9.54}$$

We have two combination angles and four parameters in the normal form. In the case of potential problems, we have $a_2 = a_4 = 0$. For fixed energy we find five periodic solution with two in general position.

Identifying particular solutions and their characteristics using Ziglin-Morales-Ramis theory, non-integrability of normal form (9.54) was shown for open sets of the parameters in [42]. See the energy simplex in Fig. 9.8 (right).

## 9.7 Details for the 1 : 2 : 1 Resonance

We know the normal form $H_2 + \bar{H}_3$ of Hamiltonian (9.52) in action-angle variables. We restrict the number of parameters to 10 by assuming that the Hamiltonian is a potential function. We start again to use averaging and to experiment with the parameters and symmetries. The assumption of potential dynamics leads to the general potential problem:

$$H(p,q) = \frac{1}{2}(p_1^2 + q_1^2) + \frac{1}{2}(p_2^2 + 4q_2^2) + \frac{1}{2}(p_3^2 + q_3^2) - \varepsilon H_3, \tag{9.55}$$

with 10 parameters, $H_3 =$

$$\frac{b_1}{3}q_1^3 + b_2q_1^2q_2 + b_3q_1^2q_3 + \frac{b_4}{3}q_2^3 + b_5q_2^2q_1 + b_6q_2^2q_3 + \frac{b_7}{3}q_3^3 + b_8q_3^2q_1 + b_9q_3^2q_2$$
$$+ b_{10}q_1q_2q_3.$$

The equations of motion are derived from $H_2$ and $H_3$:

$$\begin{cases} \ddot{q}_1 + q_1 &= b_1q_1^2 + 2b_2q_1q_2 + 2b_3q_1q_3 + b_5q_2^2 + b_8q_3^2 + b_{10}q_2q_3, \\ \ddot{q}_2 + 4q_2 &= b_2q_1^2 + b_4q_2^2 + 2b_5q_2q_1 + 2b_6q_2q_3 + b_9q_3^2 + b_{10}q_1q_3, \\ \ddot{q}_3 + q_3 &= b_3q_1^2 + b_6q_2^2 + b_7q_3^2 + 2b_8q_3q_1 + 2b_9q_3q_2 + b_{10}q_1q_2. \end{cases} \tag{9.56}$$

Excluding the normal mode planes, we use amplitude-phase variables (1.8) and find with the usual scaling $q_1 \mapsto \varepsilon q_i$, etc., and after averaging over time only the parameters $b_2, b_9, b_{10}$ are left:

$$\begin{cases} \dot{r}_1 &= -\varepsilon(\frac{b_2}{2}r_1r_2\sin(2\phi_1 - \phi_2) + \frac{b_{10}}{4}r_2r_3\sin(\phi_1 + \phi_3 - \phi_2)), \\ \dot{\phi}_1 &= -\varepsilon(\frac{b_2}{2}r_2\cos(2\phi_1 - \phi_2) + \frac{b_{10}}{4}\frac{r_2r_3}{r_1}\cos(\phi_1 + \phi_3 - \phi_2)), \\ \dot{r}_2 &= \frac{\varepsilon}{8}(b_2r_1^2\sin(2\phi_1 - \phi_2) + b_9r_3^2\sin(2\phi_3 - \phi_2) + b_{10}r_1r_3\sin(\phi_1 + \phi_3 - \phi_2)), \\ \dot{\phi}_2 &= -\frac{\varepsilon}{8r_2}(b_2r_1^2\cos(2\phi_1 - \phi_2) + b_9r_3^2\cos(2\phi_3 - \phi_2) + b_{10}r_1r_3\cos(\phi_1 + \phi_3 - \phi_2)) \\ \dot{r}_3 &= -\varepsilon(\frac{b_9}{2}r_3r_2\sin(2\phi_3 - \phi_2) + \frac{b_{10}}{4}r_1r_2\sin(\phi_1 + \phi_3 - \phi_2)), \\ \dot{\phi}_3 &= -\varepsilon(\frac{b_9}{2}r_2\cos(2\phi_1 - \phi_2) + \frac{b_{10}}{4}\frac{r_1r_2}{r_3}\cos(\phi_1 + \phi_3 - \phi_2)). \end{cases}$$

(9.57)

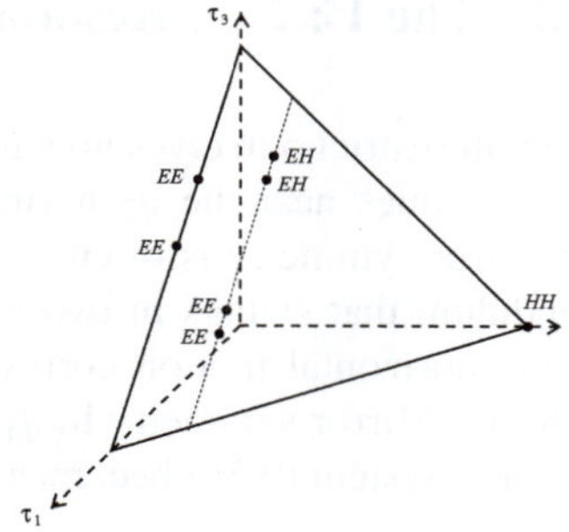


**Fig. 9.9** Energy simplex of system (9.56) showing the periodic solutions of the 1 : 2 : 1 resonance with stability characteristics

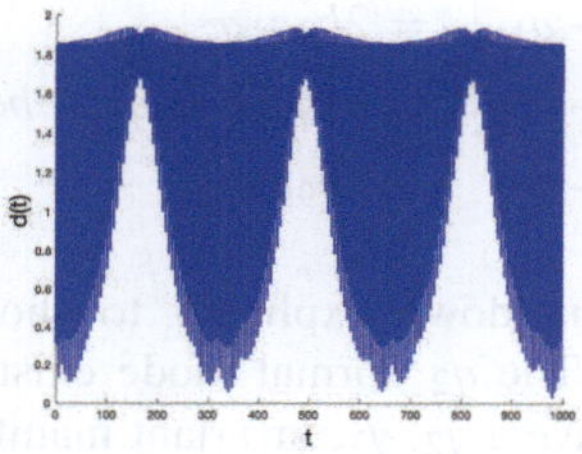


**Fig. 9.10** Phase distance $d(t)$ to show Poincaré recurrence of system (9.56) with initial values $q_1(0) = 0.1, q_2(0) = 0.8, q_3(0) = -0.1$ and initial velocities zero, so near the $q_2$ normal mode. Parameters: $\varepsilon = 0.05$; for $b_1, \ldots, b_{10}$, we have 0.2, 0.5, 0.2, 0.5, −0.03, 0.2, −0.4, 0.2, 0.6, −1 , $0 \leq t \leq 1000$. We have no symmetry assumptions. Courtesy Int. J. Bif. Chaos

The combination angles in the potential case are (of course) the same as in normal form (9.52). We put $\chi_1 = 2\phi_1 - \phi_2$, $\chi_2 = \phi_1 + \phi_3 - \phi_2$, $\chi_3 = 2\phi_3 - \phi_2$. Multiplying the equation for $\dot{r}_1$ by $r_1$, the equation for $\dot{r}_2$ by $4r_2$, the equation for $\dot{r}_3$ by $r_3$ and adding the equations, we have after integration that system (9.57) has the integral

$$\frac{1}{2}r_1^2 + 2r_2^2 + \frac{1}{2}r_3^2 = E_0, \tag{9.58}$$

with $E_0$ a positive constant. This is as expected. We can find periodic solutions in general position by putting the right-hand sides of the equations for $r_1, r_2, r_3$ zero. At the same time we have the conditions $\dot{\chi}_1 = \dot{\chi}_2 = \dot{\chi}_3 = 0$. We can solve the algebraic equations to obtain values for the amplitudes and combination angles dependent on parameter $E_0$. The analysis is elementary but requires too many details here, and the results are the same as summarised in Subsection 9.6.1. In Fig. 9.9 we show a so-called energy simplex for the actions and the periodic solutions with their stability.

In Fig. 9.10 we show the Poincaré recurrence by distance $d(t)$ for 1000 timesteps, and this gives an indication of the dynamics near the $q_2$ normal mode. The parameters are given in the caption. The instability of the $q_2$ mode is clear, and the recurrence is irregular and weak as on this interval of time the orbit does not approach the initial condition very close.

## 9.8 The 1 : 2 : 1 Resonance with Mirror Symmetry

We will restrict ourselves now to the physically interesting case of symmetries with initial values near the $q_2$ normal mode; the results are based on [159]. Discrete or mirror symmetry is often a typical characteristic in physical models, think of a pendulum that swings in two horizontal directions with respect to a vertical axis. If the horizontal motion corresponds with $q_1, q_3$, we have the case described as follows: Mirror symmetry in $q_1, q_3$ implies that only $b_2, b_4$ and $b_9$ can be unequal to zero. System (9.56) becomes

$$\begin{cases} \ddot{q}_1 + q_1 &= 2b_2 q_1 q_2, \\ \ddot{q}_2 + 4q_2 &= b_2 q_1^2 + b_4 q_2^2 + b_9 q_3^2, \\ \ddot{q}_3 + q_3 &= 2b_9 q_3 q_2. \end{cases} \tag{9.59}$$

We have written the system down explicitly to show the difference with the asymmetric system (9.56): The $q_2$ normal mode exists in system (9.59) exactly, in fact if $q_1$ vanishes, we have a $q_2, q_3$, invariant manifold, and if $q_3$ vanishes, we have a $q_1, q_2$ invariant manifold. The averaged system (9.57) reduces to

$$\begin{cases} \dot{r}_1 &= -\varepsilon \frac{b_2}{2} r_1 r_2 \sin(2\phi_1 - \phi_2), \dot{\phi}_1 = -\varepsilon \frac{b_2}{2} r_2 \cos(2\phi_1 - \phi_2), \\ \dot{r}_2 &= \frac{\varepsilon}{8}(b_2 r_1^2 \sin(2\phi_1 - \phi_2) + b_9 r_3^2 \sin(2\phi_3 - \phi_2)), \\ \dot{\phi}_2 &= -\frac{\varepsilon}{8r_2}(b_2 r_1^2 \cos(2\phi_1 - \phi_2) + b_9 r_3^2 \cos(2\phi_3 - \phi_2)), \\ \dot{r}_3 &= -\varepsilon \frac{b_9}{2} r_3 r_2 \sin(2\phi_3 - \phi_2), \dot{\phi}_3 = -\varepsilon \frac{b_9}{2} r_2 \cos(2\phi_1 - \phi_2). \end{cases} \tag{9.60}$$

System (9.60) has the integral (9.58). For the two combination angles $\chi_1 = 2\phi_1 - \phi_2$, $\chi_3 = 2\phi_3 - \phi_2$, we have

$$\begin{cases} \frac{d\chi_1}{dt} &= -\frac{\varepsilon}{8r_2}\left(b_2(8r_2^2 - r_1^2)\cos\chi_1 - b_9 r_3^2 \cos\chi_3\right), \\ \frac{d\chi_3}{dt} &= -\frac{\varepsilon}{8r_2}\left(b_9(8r_2^2 - r_3^2)\cos\chi_3 - b_2 r_1^2 \cos\chi_1\right). \end{cases} \tag{9.61}$$

The amplitudes are constant if $\chi_1 = 0, \pi$ and $\chi_3 = 0, \pi$ producing $\cos\chi_1 = \pm 1, \cos\chi_3 = \pm 1$. This produces four possibilities of families of periodic solutions. For instance, if we have $\cos\chi_1 = \cos\chi_3 = 1$, we have from the equations for $\chi_1, \chi_3$ the conditions for positive amplitudes:

$$b_2(8r_2^2 - r_1^2) - b_9 r_3^2 = 0,\ b_9(8r_2^2 - r_3^2) - b_2 r_3^2 = 0. \tag{9.62}$$

Together with integral (9.58), we will look for positive solutions of the amplitudes. With this procedure, we find at most four families of periodic soutions. We will leave the details to the reader, but we illustrate the complexity in Fig. 9.11 by using the recurrence theorem. We choose $b_2 = 0.5, b_4 = 0.5, b_9 = 0.6$; the recurrence is not very good on this interval of time, see Fig. 9.11. It was shown in [159] that

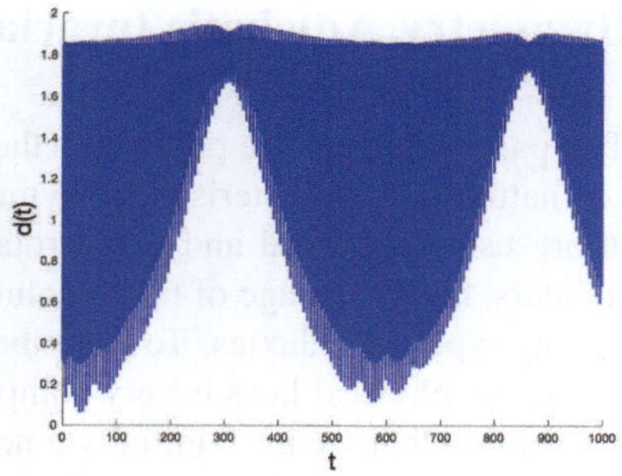


**Fig. 9.11** Phase distance $d(t)$ to show Poincaré recurrence of system (9.59) with initial values $q_1(0) = 0.1, q_2(0) = 0.8, q_3(0) = -0.1$ and initial velocities zero; $\varepsilon = 0.05, b_2 = 0.5, b_4 = 0.5, b_9 = 0.6$ and other $b$-parameters zero (mirror symmetry in the $q_1, q_3$ modes). Instability of the $q_2$ normal mode remains, and the recurrence has not improved. Courtesy Int. J. Bif. Chaos

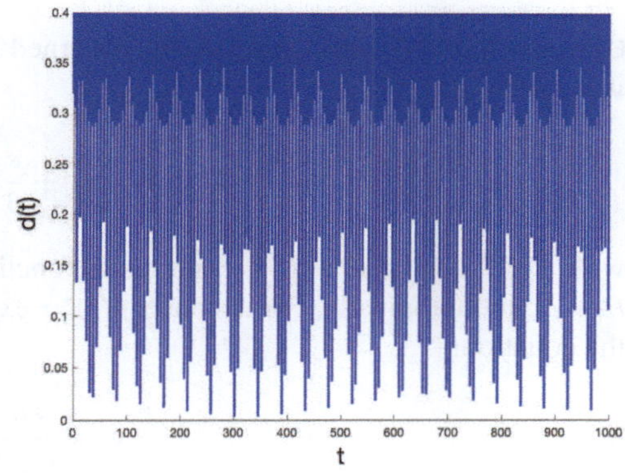


**Fig. 9.12** Phase distance $d(t)$ to show Poincaré recurrence of system (9.59) with initial values $q_1(0) = 0.1, q_2(0) = 0.8, q_3(0) = -0.1$ and initial velocities zero; $\varepsilon = 0.05, b_1 = 0.2, b_3 = 0.2, b_5 = -0.03, b_6 = 0.2, b_7 = -0.4$ and other $b$-parameters zero (mirror symmetry in the $q_2$ mode). The recurrence is much faster. Courtesy Int. J. Bif. Chaos

the system with mirror symmetry in $q_2, q_3$ producing the averaged system (9.59) is not integrable and contains the so-called Shilnikov-type chaotic behaviour. See the discussion in Sect. 9.12.

**Mirror Symmetry in $q_2$**

In this case we have $b_2 = b_4 = b_9 = b_{10} = 0$ in system (9.56). The system becomes

$$\begin{cases} \ddot{q}_1 + q_1 &= b_1 q_1^2 + 2b_3 q_1 q_3 + b_5 q_2^2 + b_8 q_3^2, \\ \ddot{q}_2 + 4q_2 &= 2b_5 q_2 q_1 + 2b_6 q_2 q_3, \\ \ddot{q}_3 + q_3 &= b_3 q_1^2 + b_6 q_2^2 + b_7 q_3^2 + 2b_8 q_3 q_1. \end{cases} \tag{9.63}$$

The first order averaging produces zero right-hand sides, and we have to average to second order in this case. This is left to the reader. The phase flow is to a larger extent determined by the linearised system, and the recurrence has strongly improved, see Fig. 9.12. Non-integrability behaviour will be at most of $O(\varepsilon^2)$ size.

## 9.9 Evolution to Symmetry, Adiabatic Invariants

This subsection is based on parts of [156]. The paper uses the observation that many macroscopic structures in nature are characterised by symmetries. One can think of planets and stars that are usual spherical and when rotating is flattened at the poles, globular clusters of stars, the final stage of tidal evolution in the gravitational two-body problem, and many types of galaxies. To describe the evolution of these systems with time in detail using physical laws is very complicated, so we choose a shortcut to evolution by assuming that we have initially a non-symmetric potential, whereas the asymmetries vanish with time. We start with a cartoon problem.

### 9.9.1 A Cartoon Model

Consider a two-dimensional system governed by a potential with asymmetric term vanishing:

$$\Phi(x) = \frac{1}{2}x^2 + \frac{\varepsilon}{3}f(t)x^3, \tag{9.64}$$

with $f(0) = 1$ and $f(t)$ a continuous function with the property $f(t) \to 0$ as $t \to \infty$ monotonically. For instance $f(t) = \exp(-t)$ or $f(t) = 1/(1+t)$. Consider the equation:

$$\ddot{x} + x = \frac{\varepsilon a}{1+t}x^2. \tag{9.65}$$

If $\varepsilon = 0$ or parameter $a = 0$, the action $I$

$$I = \frac{1}{2}(\dot{x}^2 + x^2) \tag{9.66}$$

will be constant, $I = 0.5(\dot{x}^2(0) + x^2(0))$. The non-linear term will vanish with time, and the question is whether the original dynamical state at $t = 0$ has still some influence after a long time. We use transformation (1.6) to obtain the variational equations:

$$\dot{r} = -\varepsilon \sin(t+\phi)\frac{a}{1+t}r^2\cos^2(t+\phi),\ \dot{\phi} = -\varepsilon\cos(t+\phi)\frac{a}{1+t}r\cos^2(t+\phi).$$

Using $\sin(t+\phi)\cos^2(t+\phi) = \frac{1}{4}(\sin(t+\phi) + \sin(3t+3\phi))$ and Theorem 5.1 we have to evaluate the integral for $\dot{r}$ (leaving out constants):

$$f_1^0(x) = \lim_{t\to\infty} \frac{1}{T}\int_0^T \frac{1}{1+t}(\sin(t+\phi)+\sin(3t+3\phi))dt.$$

Using partial integration the integral becomes before taking the limit:

$$\cos\phi(0) - \frac{\cos(t+\phi)}{1+t} + \cos 3\phi(0) - \frac{\cos(3t+3\phi)}{3(1+t)}$$
$$-\int_0^t \frac{1}{(1+t)^2}(\cos(t+\phi)+\frac{1}{3}\cos(3t+3\phi))dt.$$

The integral converges for $t \to \infty$ so we find the average $\dot{r} = 0$. In the same way we find the averaged equation $\dot{\phi} = 0$. We conclude that we have with $r(0) = r_0, \phi(0) = \phi_0$ the approximation:

$$x(t) = r_0 \cos(t+\phi_0) + \delta(\varepsilon) \text{ on the timescale } 1/\varepsilon. \tag{9.67}$$

The amplitude (or action $I$) is an adiabatic invariant of equation (9.65). We can estimate $\delta(\varepsilon) = \varepsilon|\ln\varepsilon|$. An improved first order estimate can be obtained by using convergent iteration of the corresponding integral equation:

$$x(t) = r_0\cos(t+\phi_0) + \varepsilon a\int_0^t \sin(t-s)\frac{x^2(s)}{1+s}ds.$$

Putting $\phi_0 = 0, x(s) = r_0\cos s$, we find after partial integration as a first step:

$$\tilde{x}(t) = r_0\cos t + \varepsilon\frac{a}{4}r_0^2\left(\sin t(3\frac{\sin t}{1+t} + \frac{\sin 3t}{3(1+t)} + \cos t(\frac{\cos t}{1+t} + \frac{\cos 3t}{3(1+t)} - \frac{4}{3})\right).$$

The convergence of the expansion is in $t$, not in $\varepsilon$. This is why we call $\tilde{x}(t)$ an improved first order approximation. It is clear from the approximations that the initial values play a part in the later dynamics; see also Fig. 9.13.

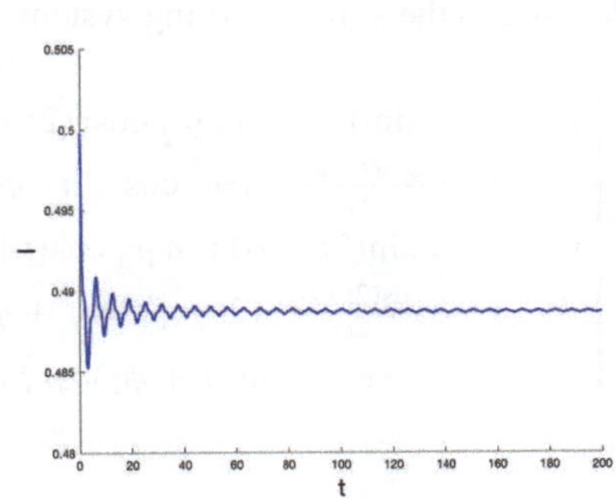


**Fig. 9.13** Time series for the action $I(t)$ of Eq. (9.65) with $\varepsilon = 0.05, a = 1, x(0) = 1, \dot{x}(0) = 0$. We have $I(0) = 0.5$ and $I(t)$ tending to 0.488

### 9.9.2 A Two-Dof Application to Astrophysics

In [156] the motion of stars in a galactic plane is described with $x(t)$ the epicyclic motion in the plane with respect to circular orbits, and $z(t)$ the vertical oscillation with respect to the galactic plane. The Hamiltonian and the equations of motion are derived with the assumption of axi-symmetry of the galaxy. The evolution results are more spectacular here as the evolutionary dynamics changes the stability of normal mode solutions. This is for one dof not so easy to achieve, but it may happen for two dof as we will show for the Hamiltonian:

$$H = \frac{1}{2}(\dot{x}^2 + x^2 + \dot{z}^2 + \omega^2 z^2) - \varepsilon(\frac{1}{3}a_1 x^3 + a_2 x z^2 + e^{-\delta t}\frac{1}{3}a_3 z^3 + e^{-\delta t} a_4 x^2 z). \tag{9.68}$$

The parameter $\delta > 0$ will govern the dynamics of evolution, and the terms asymmetric in $z$ will vanish with time producing a disk galaxy symmetric with respect to the galactic plane. We consider a case studied in [156] and keep the notation of that paper to make the connection easier. Consider the case $a_1 = a_3 = 0$, $\omega = 2$ producing the two-dof system:

$$\begin{cases} \ddot{x} + x &= \varepsilon a_2 z^2 + \varepsilon e^{-\delta t} 2a_4 x z, \\ \ddot{z} + 4z &= \varepsilon 2a_2 x z + \varepsilon e^{-\delta t} a_4 x^2. \end{cases} \tag{9.69}$$

We observe that the $x$-normal mode ($z = \dot{z} = 0$) exists and is harmonic, and the $z$-normal mode for motion perpendicular to the galactic plane does not exist.

Choosing $\delta = 1$, we have relatively fast evolution in the two-dof system (9.69). General averaging produces as in the preceding example no contribution of the asymmetric terms on the timescale $1/\varepsilon$, so the dynamical effects are small. Putting $\delta = \varepsilon$, the evolution takes longer and will produce less trivial results.

We transform to slowly varying polar coordinates $r, \phi$ by (1.6):

$$x = r_1\cos(t+\phi_1),\ \dot{x} = -r_1\sin(t+\phi_1),\ z = r_2\cos(2t+\phi_2),\ \dot{z} = -2r_2\sin(2t+\phi_2),$$

leading to the slowly varying system:

$$\begin{cases} \dot{r}_1 &= -\varepsilon\sin(t+\phi_1)\left(a_2 r_2^2\cos^2(2t+\phi_2) - e^{-\varepsilon t}2a_4 r_1\cos(t+\phi_1) r_2\cos(\omega t+\phi_2)\right), \\ \dot{\phi}_1 &= -\varepsilon\frac{\cos(t+\phi_1)}{r_1}\left(a_2 r_2^2\cos^2(2t+\phi_2) - e^{-\varepsilon t}2a_4 r_1\cos(t+\phi_1) r_2\cos(2t+\phi_2)\right), \\ \dot{r}_2 &= -\frac{\varepsilon}{2}\sin(2t+\phi_2)\left(2a_2 r_1\cos(t+\phi_1) r_2\cos(2t+\phi_2) - e^{-\varepsilon t}a_4 r_1^2\cos^2(t+\phi_1)\right), \\ \dot{\psi}_2 &= -\varepsilon\frac{\cos(2t+\phi_2)}{2r_2}\left(2a_2 r_1\cos(t+\phi_1) r_2\cos(2t+\phi_2)\right. \\ &\qquad \left. - e^{-\varepsilon t}\sin(2t+\phi_2) a_4 r_1^2\cos^2(t+\phi_1)\right). \end{cases} \tag{9.70}$$

Near the normal modes $r_1 = 0$ and $r_2 = 0$, we have to use a different coordinate transformation. We can put $\tau = \varepsilon t$ and treat $\tau$ as a new, independent variable. We add the equation:

$$\dot{\tau} = \varepsilon.$$

It will be useful to introduce the actions $E_1, E_2$ by

$$E_x = \frac{1}{2}(\dot{x}^2 + x^2) = \frac{1}{2}r_1^2,\ E_z = \frac{1}{2}(\dot{z}^2 + \omega^2 z^2) = \frac{\omega^2}{2}r_2^2. \tag{9.71}$$

First order averaging produces that in system (9.70) the terms with coefficients $a_2$ vanish. The implication is *that to first order in* $\varepsilon$ and on time intervals of size $1/\varepsilon$, system (9.69) is described by the intermediate normal form equations:

$$\begin{cases} \ddot{x} + x &= \varepsilon e^{-\tau} 2a_4 xz, \\ \ddot{z} + \omega^2 z &= \varepsilon e^{-\tau} a_4 x^2, \\ \dot{\tau} &= \varepsilon. \end{cases} \tag{9.72}$$

**The relation with dissipation**

As the non-linear terms are homogeneous in the coordinates, we can remove the time-dependent term by a transformation involving $e^{-\tau}$. We transform $x = e^{-\tau} y_1, z = e^{-\tau} y_2$. Such a time-dependent transformation exists for any positive sufficiently differentiable function of time that decreases monotonically to zero. Replacing $e^{-\varepsilon t}$ in system (9.69), system (9.72) transforms to

$$\begin{cases} \ddot{y}_1 + y_1 &= -2\varepsilon \dot{y}_1 - \varepsilon^2 y_1 + \varepsilon 2a_4 y_1 y_2, \\ \ddot{y}_2 + 4y_2 &= -2\varepsilon \dot{y}_2 - \varepsilon^2 y_2 + \varepsilon a_4 y_1^2. \end{cases} \tag{9.73}$$

So the time-dependent transformation transforms the asymmetric system to an equivalent dissipative system with friction coefficient $2\varepsilon$. We can study the resulting dissipative system, but we return to the original formulation.

We will average system (9.72) using transformation (1.6) keeping $r_1, r_2, \phi_1, \phi_2, \tau$ fixed. We find to first order with combination angle $\chi = 2\phi_1 - \phi_2$:

$$\begin{cases} \dot{r}_1 &= -\varepsilon e^{-\tau} \frac{a_4}{2} r_1 r_2 \sin\chi,\ \dot{\phi}_1 = -\varepsilon e^{-\tau} \frac{a_4}{2} r_2 \cos\chi, \\ \dot{r}_2 &= \varepsilon e^{-\tau} \frac{a_4}{8} r_1^2 \sin\chi,\ \dot{\phi}_2 = -\varepsilon e^{-\tau} \frac{a_4}{8} \frac{r_1^2}{r_2} \cos\chi. \end{cases} \tag{9.74}$$

For $\chi$ we have the equation:

$$\frac{d\chi}{dt} = \varepsilon a_4 e^{-\tau}\left(-r_2 + \frac{r_1^2}{8r_2}\right)\cos\chi. \tag{9.75}$$

It is surprising that to this first order system (9.74) admits a time-independent integral of motion (adiabatic invariant):

$$\frac{1}{2}r_1^2 + 2r_2^2 = E_0, \tag{9.76}$$

with constant $E_0 \geq 0$. In the original coordinates, we have

$$\frac{1}{2}(\dot{x}^2 + x^2) + \frac{1}{2}(\dot{z}^2 + 4z^2) = E_0.$$

We will see that the separate actions $E_x$, $E_z$ are not conserved. However, system (9.74) admits special families of solutions with constant amplitude on the timescale $O(1/\varepsilon)$ if

$$\chi = 0, \pi, r_1^2 = 8r_2^2,\ r_1^2 = \frac{4}{3}E_0, r_2^2 = \frac{1}{6}E_0. \tag{9.77}$$

The solutions with $\chi = 0$ are called in-phase, and the solutions with $\chi = \pi$ out-phase. The corresponding phases are slowly decreasing with rate $\exp(-\varepsilon t)$. These special solutions are found in a so-called resonance manifold, a geometric structure that was studied in detail in Chap. 7.

The solutions of system (9.74) have the error estimate $O(\varepsilon)$ on the timescale $1/\varepsilon$. On this long interval of time, we expect the terms $O(\varepsilon^2)$ to play a part as the solutions of system (9.74) with coefficients $a_4$ will vanish and maybe terms with coefficient $a_2$ arise.

**Second Order Averaging**

To improve the approximations by second order averaging [123] of system (9.70), we borrow the result from [156]:

$$\begin{cases} \dot{r}_1 &= -\varepsilon e^{-\tau}\frac{a_4}{2}r_1r_2\sin\chi,\ \dot{r}_2 = \varepsilon e^{-\tau}\frac{a_4}{8}r_1^2\sin\chi, \\ \dot{\phi}_1 &= -\varepsilon e^{-\tau}\frac{a_4}{2}r_2\cos\chi - \varepsilon^2 e^{-2\tau}\frac{1}{64}a_4^2(9r_1^2+4r_2^2), \\ \dot{\phi}_2 &= -\varepsilon e^{-\tau}\frac{a_4}{8}\frac{r_1^2}{r_2}\cos\chi - \varepsilon^2\left(\frac{1}{30}a_2^2r_1^2 + \frac{29}{120}a_2^2r_2^2 + e^{-2\tau}\frac{1}{32}a_4^2r_1^2\right). \end{cases} \tag{9.78}$$

We do not find $O(\varepsilon^2)$ terms for the amplitudes, such terms arise only for the angles. For the combination angle $\chi$, we find

$$\begin{cases} \frac{d\chi}{dt} &= \varepsilon a_4 e^{-\tau}(-r_2 + \frac{r_1^2}{8r_2})\cos\chi + \\ & \varepsilon^2 a_2^2\left(\frac{1}{30}r_1^2 + \frac{29}{120}r_2^2\right) + \varepsilon^2 a_4^2 e^{-2\tau}\left(-\frac{1}{32}(9r_1^2+4r_2^2) + \frac{1}{32}r_1^2\right). \end{cases} \tag{9.79}$$

We find, using Eq. (9.77), that in the resonance manifold we have that $\dot{\chi} \neq 0$. So the solutions with constant amplitude do not persist at second order:

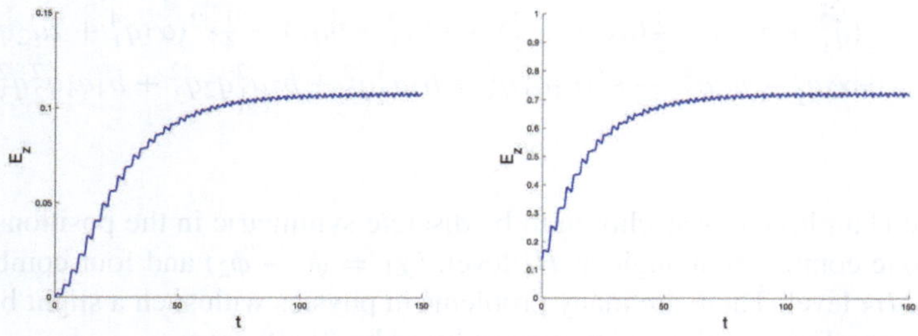


**Fig. 9.14** The behaviour of action $E_z(t)$ of system (9.69) in two cases with $a_2 = 1, a_4 = 2, \varepsilon = 0.05$. Starting near the $x$ normal mode (left) with $x(0) = 1, \dot{x}(0) = 0, z = 0.01, \dot{z}(0) = 0$. And right near an unstable solution in the resonance manifold with $x(0) = -\sqrt{2}, \dot{x}(0) = 0, z(0) = 0, \dot{z}(0) = 0.5$. In both cases, the dynamics has changed considerably by the evolution to mirror symmetry

1. During an interval of time of order $1/\varepsilon$, the integral (adiabatic invariant) (9.76) is active, and the system is still dominated by the asymmetric $a_4$ term.
2. When time goes on, the system (9.69) is dominated by the $a_2$ term. In [131] it is shown that for this system corresponding with a time-independent Hamiltonian, depending on the value of $a_2$, two resonance manifolds can exist on the energy manifold (see also Chap. 7).
3. An interesting consequence arises when considering orbits that change stability during the process of evolution to mirror symmetry. The instability of some periodic solutions with constant amplitude and some normal modes persists at second order but leads to stability in the final stage of mirror symmetry after a time longer than $1/\varepsilon$. The implication is that evolution to symmetry will generally produce a drastic change of the distribution of the actions. Another qualitative consequence in this example is that the epicyclic $x$-normal mode does not exist in the first stage (time order $1/\varepsilon$), but this normal mode exists in the final stage of mirror symmetry; again we can expect action changes. As discussed above, similar considerations hold for the unstable out-phase periodic solutions derived before, see Fig. 9.14.

## 9.10 Low and Higher Order 2 : 2 : 3 Resonance

Considering three or more dof may present a mix of dominant (low order) resonance and higher order resonance. The Hamiltonian 2 : 2 : 3 resonance contains the primary resonance 1 : 1 and twice the higher order resonance 2 : 3. We present some aspects of this case based on [154]. Consider for the 2 : 2 : 3 resonance the Hamiltonian $H_2 + \varepsilon^2 H_4 + \varepsilon^3 H_5$:

$$\begin{cases} H = & \frac{1}{2}(\dot{q}_1^2 + 4q_1^2) + \frac{1}{2}(\dot{q}_2^2 + 4q_2^2) + \frac{1}{2}(\dot{q}_3^2 + 9q_3^2) - \frac{1}{4}\varepsilon^2(\alpha_1 q_1^4 + 2\alpha_2 q_1^2 q_2^2 \\ & +\alpha_3 q_2^4 + \alpha_4 q_3^4) - \varepsilon^3(b_1 q_1^3 q_3^2 + b_3 q_2^3 q_3^2 + b_3 q_1^2 q_2 q_3^2 + b_4 q_1 q_2^2 q_3^2). \end{cases} \tag{9.80}$$

$H_4$ in the Hamiltonian was chosen to be discrete symmetric in the positions and to contain one combination angle at $H_4$ level: ($\chi_1 = \phi_1 - \phi_2$) and four combination angles at $H_5$ level. There are many problems in physics with such a slight breaking of symmetry. The equations of motion induced by (9.80) are

$$\begin{cases} \ddot{q}_1 + 4q_1 = & \varepsilon^2(\alpha_1 q_1^3 + \alpha_2 q_1 q_2^2) + \varepsilon^3(3b_1 q_1^2 q_3^2 + 2b_3 q_1 q_2 q_3^2 + b_4 q_2^2 q_3^2), \\ \ddot{q}_2 + 4q_2 = & \varepsilon^2(\alpha_2 q_1^2 q_2 + \alpha_3 q_2^3) + \varepsilon^3(3b_2 q_2^2 q_3^2 + b_3 q_1^2 q_3^2 + 2b_4 q_1 q_2 q_3^2), \\ \ddot{q}_3 + 9q_3 = & \varepsilon^2 \alpha_4 q_3^3 + \varepsilon^3(2b_1 q_1^3 q_3 + 2b_2 q_2^3 q_3 + 2b_3 q_1^2 q_2 q_3 + 2b_4 q_1 q_2^2 q_3). \end{cases} \tag{9.81}$$

The figures we show in this section have been obtained using system (9.81). To make some explicit numerical calculations, we have in most cases chosen:

$$\alpha_1 = 0.4, \alpha_2 = 1, \alpha_3 = 0.6, \alpha_4 = 4, b_1 = 1, b_2 = -1.5, b_3 = 1, b_4 = -1. \tag{9.82}$$

The choice of $\alpha_1, \ldots, \alpha_4$ excludes non-generic behaviour. Because of the resonances in $H_2$ and the resulting frequency ratios, we expect resonant interaction between the first two modes $q_1, q_2$ and a small interaction with the third mode.

**The primary resonance zones**

At $H_5$ four combination angles are possibly active:

$$\chi_2 = 3\phi_1 - 2\phi_3,\ \chi_3 = 3\phi_2 - 2\phi_3,\ \chi_4 = 2\phi_1 + \phi_2 - 2\phi_3,\ \chi_5 = \phi_1 + 2\phi_2 - 2\phi_3.$$

We will use amplitude-phase coordinates $r, \phi$ from (1.6). The transformation results in

$$H_2 = 2r_1^2 + 2r_2^2 + \frac{9}{2}r_3^2.$$

We find after averaging to $O(\varepsilon^2)$

$$\begin{cases} \dot{r}_1 = & -\varepsilon^2 \frac{1}{8} r_1 r_2^2 \sin 2\chi_1,\ \dot{\phi}_1 = -\varepsilon^2 \frac{1}{8}(\frac{6}{5}r_1^2 + 2r_2^2 + r_2^2 \cos 2\chi_1), \\ \dot{r}_2 = & +\varepsilon^2 \frac{1}{8} r_1^2 r_2 \sin 2\chi_1,\ \dot{\phi}_2 = -\varepsilon^2 \frac{1}{8}(2r_1^2 + r_1^2 \cos 2\chi_1 + \frac{9}{5}r_2^2), \\ \dot{r}_3 = & 0,\ \dot{\phi}_3 = -\varepsilon^2 \frac{3}{8} r_3^2. \end{cases} \tag{9.83}$$

The combination angles $\chi_2, \ldots, \chi_5$ will arise at $O(\varepsilon^3)$. For the averaging-normal form (9.83), we have $2r_1^2 + 2r_2^2 = E_1 + O(\varepsilon)$ with constant $E_1 \geq 0$ and $r_3 - r_3(0) =$

$O(\varepsilon)$ on the timescale $1/\varepsilon^2$. The normal form to $O(\varepsilon^2)$ is integrable with integrals $r_3 = r_3(0)$, $2r_1^2 + 2r_2^2 = E_1$ and the normalised Hamiltonian to quartic terms.

The equation for $\chi_1$ depends on $r_1, r_2$ and $\chi_1$. If the averaging-normal form (9.83) of the equations has a few solutions with $\sin 2\chi_1 = 0$, $\dot{\chi}_1 = 0$, they give approximate short-periodic solutions in general position (separate from the normal modes).

A *primary resonance zone* $M$ is defined as a neighbourhood of the solutions with $\sin 2\chi_1 = 0$, $\dot{\chi}_1 = 0$. In the 2 : 2 : 3 problem, such a resonance zone has for a chosen value $E_0$ with $H_2 = E_0$, $E_0 > 0$ the free parameter $r_3(0)$. We find with system (9.83):

$$\dot{\chi}_1 = \frac{1}{8}\varepsilon^2 \left( \frac{4}{5} r_1^2 - \frac{1}{5} r_2^2 + (r_1^2 - r_2^2) \cos 2\chi_1 \right). \tag{9.84}$$

Putting the right-hand side of Eq. (9.84) zero, we find two primary resonance zones $M_1, M_2$:

$$\begin{cases} M_1 : \; 2\chi_1 = & 0, 2\pi, r_1^2 = \frac{2}{3} r_2^2, \\ M_2 : \; 2\chi_1 = & \pi, 3\pi, r_1^2 = 4 r_2^2. \end{cases} \tag{9.85}$$

The periodic solutions of the two primary resonance zones are located on the energy manifold, which is topologically the five-dimensional sphere $S^5$ and its intersections with the elliptical cylinder $\frac{1}{2}(\dot{q}_1^2 + 4q_1^2) + \frac{1}{2}(\dot{q}_2^2 + 4q_2^2) = E_1$ and the hyperbolic cylinder $\dot{q}_1^2 + 4q_1^2 = 4r_1^2 = \mu(\dot{q}_2^2 + 4q_2^2)$ with respectively $\mu = \frac{2}{3}$ or 4; this defines two three-dimensional manifolds.

To study the complexity of the primary resonance zones, we use the recurrence Theorem 9.1. In Fig. 9.15 we start outside the resonance zones to observe the resonant interaction of the first two modes; the variations of the third mode are small in accordance with the normal form (9.83). We will use the Euclidean distance $d(t)$ for three dof of Eq. (9.5). Choosing $q_3(0) = \dot{q}_3(0) = 0$, the third mode remains zero, and the recurrence is fairly strong and regular; see Fig. 9.15 right. Starting with non-zero initial values of the third mode in Fig. 9.15 left, we find behaviour that suggests quasi-trapping in one or more resonance zones (see [163]). This suggests that we have to study the resonance zones in more detail (see Figs. 9.15 and 9.16).

We will extend the theory of higher order resonance, see [123], to three dof. Higher order (or secondary) resonance may take place in a primary resonance zone. Consider as an example the primary zone $M_1$ with $2\chi_1 = 0, 2\pi, r_1^2 = \frac{2}{3} r_2^2$. To $O(\varepsilon^2)$ the higher order combination, angles satisfy the equations:

$$\begin{cases} \dot{\chi}_2 = & -\frac{3}{4}\varepsilon^2 (\frac{3}{5} r_1^2 + \frac{3}{2} r_2^2 - r_3^2), \\ \dot{\chi}_3 = & -\frac{3}{4}\varepsilon^2 (\frac{3}{2} r_1^2 + \frac{9}{10} r_2^2 - r_3^2), \\ \dot{\chi}_4 = & -\frac{3}{4}\varepsilon^2 (\frac{9}{10} r_1^2 + \frac{13}{10} r_2^2 - r_3^2), \\ \dot{\chi}_5 = & -\frac{3}{4}\varepsilon^2 (\frac{6}{5} r_1^2 + \frac{11}{10} r_2^2 - r_3^2). \end{cases} \tag{9.86}$$

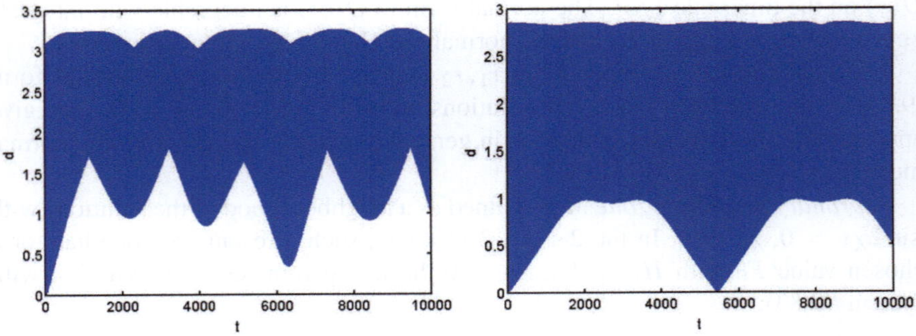


**Fig. 9.15** The 2 : 2 : 3 resonance. Left the Euclidean distance $d$ using Eq. (9.5) in 10,000 timesteps of the orbits to their initial conditions in phase space outside the primary resonance zones and parameter values Eq. (9.82); recurrence takes more than 10 000 timesteps. Right the Euclidean distance $d$ for the same Hamiltonian (9.80) but starting at $q_1(0) = 0.3, q_2(0) = 1.2, q_3(0) = 0$ and velocities zero; we have now $q_3(t) = \dot{q}_3(t) = 0, t \geq 0$. The recurrence is different and faster in this case of pure $q_1, q_2$ interaction. Courtesy Int. J. Bif. Chaos [154]

Outside the primary resonance zone we have terms in system (9.83) from the first two modes that dominate at $O(\varepsilon^2)$. In the zone $M_1$, this is not the case and other combination angles may play a part. If the derivative of a combination angle is sign-definite, the angle is timelike and we can average over the angle. A combination angle is not timelike if the right-hand side of the equation contains zeroes. In such a case, a secondary resonance will be found in the primary resonance zone. In $M_1$, we find for the 1 : 1 periodic solutions

$$2r_1^2 + 2r_2^2 = E_1,\ r_1^2 = \frac{2}{3}r_2^2 \ \rightarrow\ r_1^2 = \frac{1}{5}E_1, r_2^2 = \frac{3}{10}E_1.$$

The combination angles in (9.86) admit a zero (critical value) if in $M_1$:

$$r_3^2 = \frac{57}{100}E_1. \tag{9.87}$$

We repeat the calculation for primary resonance zone $M_2$. We find for the 1 : 1 periodic solutions in $M_2$

$$2r_1^2 + 2r_2^2 = E_1,\ r_1^2 = 4r_2^2 \ \rightarrow\ r_1^2 = \frac{2}{5}E_1, r_2^2 = \frac{1}{10}E_1.$$

The combination angles in (9.86) (we omit the details) admit a zero (critical value) if in $M_2$:

$$r_3^2 = \frac{29}{100}E_1. \tag{9.88}$$

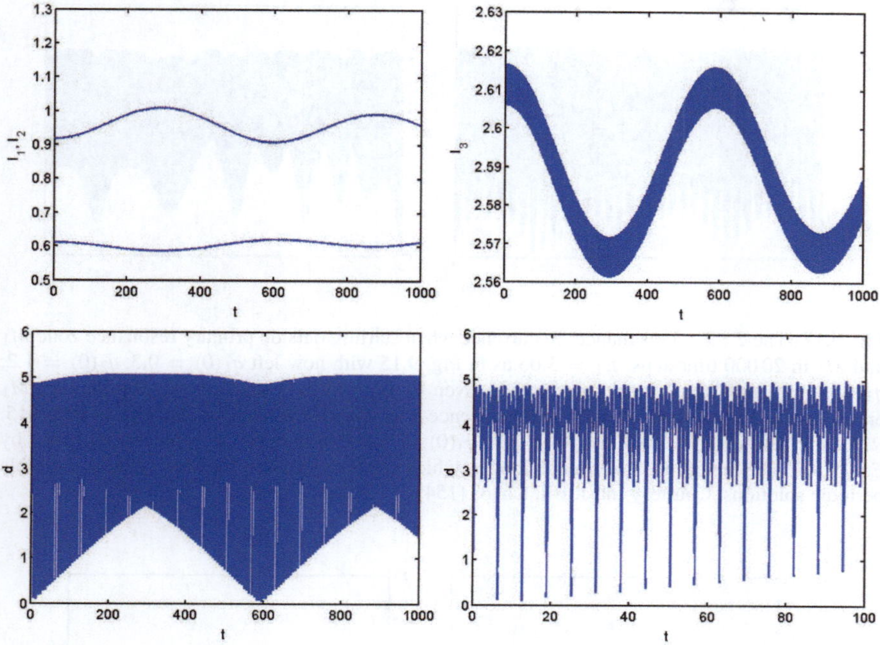


**Fig. 9.16** Recurrence when starting in the primary resonance zone $M_1$ in 1000 timesteps. $E_1 = 3.06$ as in Fig. 9.15 with now $q_1(0) = 0.7823$, $q_2(0) = 0.9581$; $q_3(0) = 1.32068$, the value given by Eq. (9.87), which puts the orbits in the secondary resonance zone. Initial velocities are zero. The top figure left shows the variations of the actions $I_1$, $I_2$, and the top figure right shows $I_3$ with small variations as predicted. The recurrence distance $d$ (below left) is strong and regular as the orbits are caught in the resonance zone $M_1$ with near fairly stable dynamics; the picture for 1000 timesteps shows some white segments but hides the fine structure of recurrence transitions shown when enlarged for 100 timesteps (below right). Courtesy Int. J. Bif. Chaos [154]

To improve our insight in the dynamics, we produce the actions corresponding with the two cases of Fig. 9.17. In Fig. 9.18 we present left $I_1$ and $I_2$ starting outside the primary resonance zones exchanging energy according to the 1 : 1 resonance; $I_3$ (right) varies within the error estimates. Next we give the actions in the unstable case of Fig. 9.17 (right). It is interesting that $I_3(t)$ changes strongly (between 0.65 and 1.33) when outside the time interval of validity, $1/\varepsilon^2$, of the normal form. Conservation of energy influences the exchanges of $I_1$ and $I_2$ at the same time (see Figs. 9.18 and 9.19).

To obtain a better understanding of the dynamics in the primary resonance zones, we have to extend the averaging-normal forms to $O(\varepsilon^3)$.

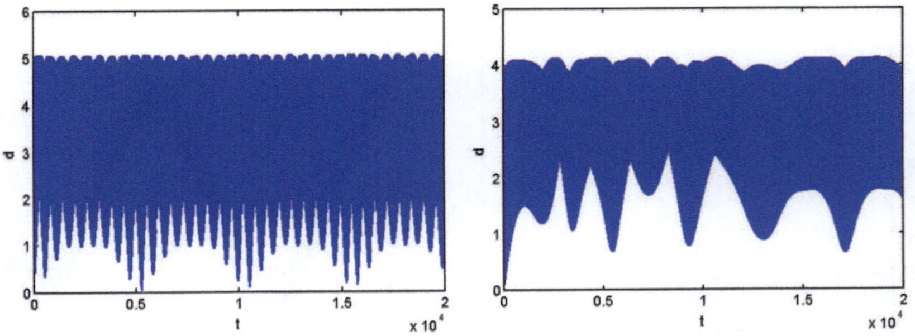


**Fig. 9.17** The 2 : 2 : 3 resonance. Recurrence when starting outside primary resonance zone $M_1$ and $M_2$ in 20 000 timesteps. $E_1 = 3.06$ as in Fig. 9.15 with now left $q_1(0) = 0.3, q_2(0) = 1.2$; $q_3(0) = 1.32068$, the critical value in $M_1$ given by Eq. (9.87). The orbits passing through $M_1$ encounter the critical case of $q_3(0)$, and recurrence is slow and different from the case of Fig. 9.15. Right the case $q_1(0) = 0.3, q_2(0) = 1.2$; $q_3(0) = 0.9420$, the critical value in $M_2$ given by Eq. (9.88). The recurrence takes much longer, which suggests that this location contains unstable periodic solutions. Courtesy Int. J. Bif. Chaos [154]

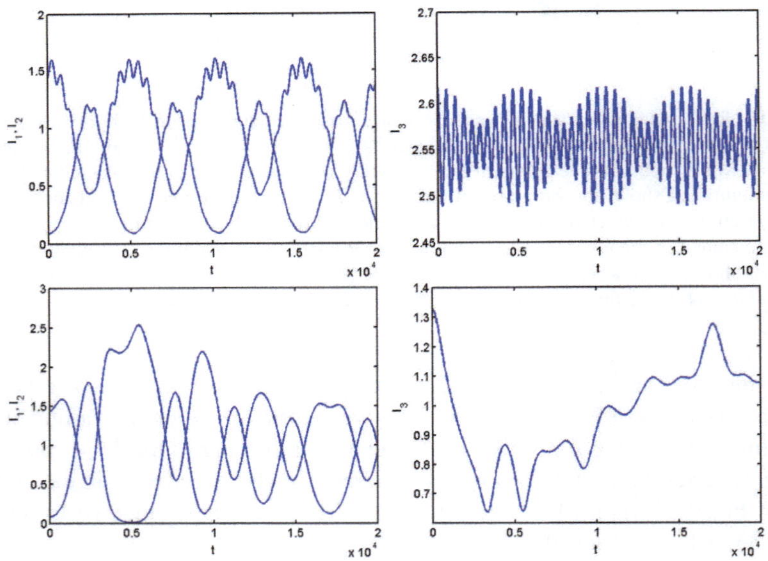


**Fig. 9.18** The 2 : 2 : 3 resonance. The actions when starting outside primary resonance zone $M_1$ and $M_2$ in 20 000 timesteps. Each orbit starts outside the primary resonance zones with $q_1(0) = 0.3$, $q_2(0) = 1.2$ and all velocities zero. In the first two figures (top), we took $q_3(0) = 1.32068$, the critical value in $M_1$ given by Eq. (9.87). $I_3(t)$ varies with magnitude 0.06 in agreement with the error estimate. The instability of the normal modes forces considerable exchange of energy of the first two modes. The next two figures show the instability of $M_2$. When passing this primary resonance zone, the critical value of $q_3(0) = 0.9420$ taken from Eq. (9.88) plays a part. Courtesy Int. J. Bif. Chaos [154]

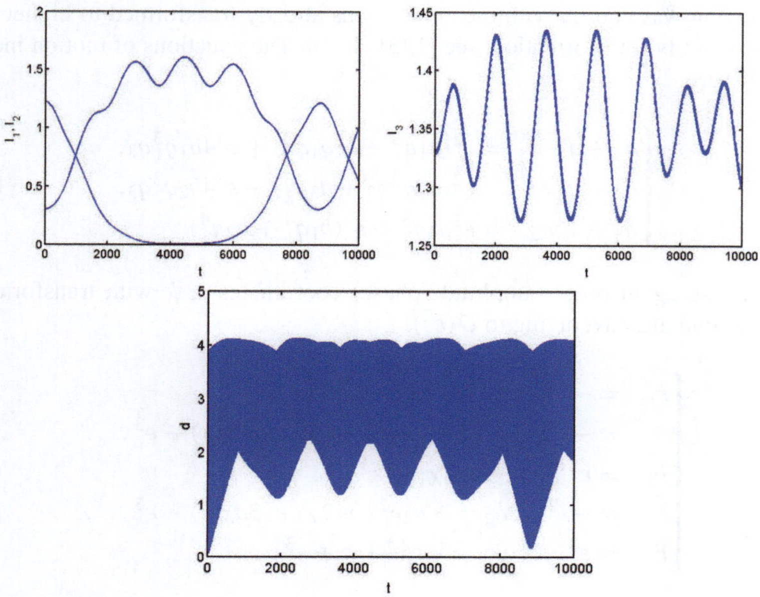


**Fig. 9.19** The 2 : 2 : 3 resonance. Dynamics when starting in the primary resonance zone $M_2$ in (because of the instability) 10,000 timesteps. $E_1 = 3.06$ as in Fig. 9.15 with now $q_1(0) = 1.1063, q_2(0) = 0.5532; q_3(0) = 0.94202$, values putting the orbits in $M_2$. Initial velocities are zero. The top figure left shows strong variations of the actions $I_1$, $I_2$, and the top figure right shows $I_3$; the variations are strong as we start near an unstable secondary resonance. The orbits are leaving the unstable resonance zone $M_2$. Below the corresponding recurrence $d$ from Eq. (9.5). Courtesy Int. J. Bif. Chaos [154]

## 9.11 Low and Higher Order 1 : 1 : 4 Resonance

Embedded double resonances can be discussed in the framework of interaction between lower and higher order Hamiltonian resonances of type $1 : 1 : \omega$. After the resonance 2 : 2 : 3, we present another example here showing interesting stability aspects. We illustrate the dynamics of the 1 : 1 : 4 resonance for the Hamiltonian based on [154]:

$$\begin{cases} H = & \frac{1}{2}(\dot{q}_1^2 + q_1^2) + \frac{1}{2}(\dot{q}_2^2 + q_2^2) + \frac{1}{2}(\dot{q}_3^2 + 16q_3^2) - \frac{1}{4}\varepsilon^2(a_1q_1^4 + 2a_2q_1^2q_2^2 \\ & +a_3q_2^4 + a_4q_3^4) - \varepsilon^3(b_1q_1^4q_3 + b_2q_2^4q_3). \end{cases} \tag{9.89}$$

Note that $H_3$ is discrete (mirror) symmetric in the three positions, and $H_4$ is discrete symmetric in $q_1, q_2$. Such symmetries are often found in physics applications. We define the actions $I_i = \frac{1}{2}(\dot{q}_i^2 + q_i^2), i = 1, 2$ and $I_3 = \frac{1}{2}(\dot{q}_3^2 + 16q_3^2)$. The

Hamiltonian was chosen with the cubic terms already transformed to higher order by near-identity transformation, see [123] ch. 10. The equations of motion induced by (9.89) are

$$\begin{cases} \ddot{q}_1 + q_1 & = \varepsilon^2(a_1 q_1^3 + a_2 q_1 q_2^2) + \varepsilon^3 4 b_1 q_1^3 q_3, \\ \ddot{q}_2 + q_2 & = \varepsilon^2(a_2 q_1^2 q_2 + a_3 q_2^3) + \varepsilon^3 4 b_2 q_2^3 q_3, \\ \ddot{q}_3 + 16 q_3 & = \varepsilon^2 a_4 q_3^3 + \varepsilon^3(b_1 q_1^4 + b_2 q_2^4). \end{cases} \tag{9.90}$$

We will use again polar (amplitude-phase) coordinates $r, \psi$ with transformation (1.6); we find after averaging to $O(\varepsilon^2)$

$$\begin{cases} \dot{r}_1 & = -\varepsilon^2 \frac{a_2}{8} r_1 r_2^2 \sin 2\chi_1 + \varepsilon^3 \ldots, \ \dot{\psi}_1 \\ & = -\varepsilon^2 \frac{1}{8}(3a_1 r_1^2 + 2a_2 r_2^2 + a_2 r_2^2 \cos 2\chi_1) + \varepsilon^3 \ldots, \\ \dot{r}_2 & = \varepsilon^2 \frac{a_2}{8} r_1^2 r_2 \sin 2\chi_1 + \varepsilon^3 \ldots, \ \dot{\psi}_2 \\ & = -\varepsilon^2 \frac{1}{8}(2a_2 r_1^2 + a_2 r_1^2 \cos 2\chi_1 + 3a_3 r_2^2) + \varepsilon^3 \ldots, \\ \dot{r}_3 & = \varepsilon^3 \ldots, \ \dot{\psi}_3 = -\varepsilon^2 \frac{3a_4}{32} r_3^2 + \varepsilon^3 \ldots, \end{cases} \tag{9.91}$$

with $\chi_1 = \psi_1 - \psi_2$. The dots represent higher order terms containing the amplitudes and, as we shall see, the combination angles $\chi_2 = 4\phi_1 - \phi_3$, $\chi_3 = 4\phi_2 - \phi_3$. For the averaging-normal form (9.91), we have the integrals:

$$r_1^2 + r_2^2 = 2E_1, \ r_3 = r(0)$$

($E_1$ constant) with both integrals $O(\varepsilon)$ valid on the timescale $1/\varepsilon^2$.

As before, a primary resonance zone $M$ is defined as a neighbourhood of the solutions of $\sin 2\chi_1 = 0$, $\dot{\chi}_1 = 0$. We find with system (9.91)

$$\dot{\chi}_1 = -\frac{1}{8}\varepsilon^2 \left((3a_1 - 2a_2)r_1^2 + (2a_2 - 3a_3)r_2^2 + a_2(r_2^2 - r_1^2)\cos 2\chi_1\right). \tag{9.92}$$

Putting the right-hand side of Eq. (9.92) zero, we find the primary resonance zones $M_1, M_2$:

$$\begin{cases} M_1 : \ r_1^2 = \frac{a_2 - a_3}{a_2 - a_1} r_2^2, \\ M_2 : \ r_1^2 = \frac{a_2 - 3a_3}{a_2 - 3a_1} r_2^2. \end{cases} \tag{9.93}$$

We require the right-hand sides of $r_1^2$ in (9.93) to be finite and positive. It is easy to check analytically and numerically that outside the primary resonance zones defined by (9.93) we find 1 : 1 resonant interaction of the first two modes and small modulation of the third mode.

**Secondary Resonance in the Primary Resonance Zones**

For the combination angles $\chi_2, \chi_3$, we find with $\sin 2\chi_1 = 0$ in the primary resonance zones

$$\begin{cases} \dot{\chi}_2 &= -\frac{1}{8}\varepsilon^2(12a_1r_1^2 + 8a_2r_2^2 + 4a_2r_2^2\cos 2\chi_1 - \frac{3}{4}a_4r_3^2), \\ \dot{\chi}_3 &= -\frac{1}{8}\varepsilon^2(8a_2r_1^2 + 12a_3r_2^2 + 4a_2r_1^2\cos 2\chi_1 - \frac{3}{4}a_4r_3^2). \end{cases} \tag{9.94}$$

Secondary resonances involving the combination angles in $M_1$ and $M_2$ may arise where the right-hand side of system (9.94) vanishes; in such a case the combination angles $\chi_2, \chi_3$ are not timelike. To study the stability within the resonance zones, we compute $\dot{\phi}_3$ to $O(\varepsilon^3)$. We find

$$\dot{\phi}_3 = -\varepsilon^2\frac{3\alpha_4}{32}r_3^2 - \varepsilon^3\frac{69}{128^2}\alpha_4^2r_3^4 - \frac{\varepsilon^3}{64r_3}A(r_1, r_2, \chi_2, \chi_3)$$

with

$$A(r_1, r_2, \chi_2, \ldots, \chi_6) = b_1r_1^4\cos\chi_2 + b_2r_2^4\cos\chi_3.$$

We conclude that unless the coefficients $b_1, b_2, \alpha_4$ vanish, $\phi_3$ may move out of resonance in the primary resonance zones on a timescale longer than $1/\varepsilon^2$. The higher order resonances included in $M_1, \ldots, M_4$ may generate instabilities, see Fig. 9.20. If secondary resonances arise from zeroes of $\dot{\chi}_2, \dot{\chi}_3$, we find stable and unstable periodic solutions from the normal forms in the resonance zones.

For illustrations we postponed the choice of the constants $\alpha_1, \ldots, \alpha_4$ to be able to exclude non-generic cases like $\alpha_1 = \alpha_3$. We choose now

$$\alpha_1 = 0.4,\ \alpha_2 = 1,\ \alpha_3 = 0.6,\ \alpha_4 = 4. \tag{9.95}$$

This choice results in

$$\dot{\chi}_1 = -\frac{1}{8}\varepsilon^2(-0.8r_1^2 + 0.2r_2^2 + (r_2^2 - r_1^2)\cos 2\chi_1). \tag{9.96}$$

Eq. (9.96) results in resonance zones generated by $\chi_1$, i.e. small neighbourhoods of

$$M_1 : 3r_1^2 - 2r_2^2 = 0,\ (2\chi_1 = 0, 2\pi),\ \ M_2 : r_1^2 - 4r_2^2 = 0,\ (2\chi_1 = \pi, 3\pi). \tag{9.97}$$

The combination angles $\chi_2, \chi_3$ from Eq. (9.94) are not timelike in a subset of $M_1$ if we can find values of the amplitudes (or actions) such that $\dot{\chi}_2 = 0$ or $\dot{\chi}_3 = 0$.

The recurrency properties of the phase flow are strongly dependent on the initial conditions. In Fig. 9.20 we show the Euclidean distance to the initial conditions in two cases. Left we start away from stable periodic solutions (outside $M_1$), and the orbits have a recurrence delayed by quasi-trapping in $M_1$ (see [163]). Right we start with the same Hamiltonian in the resonance zone $M_1$ but near a location where the

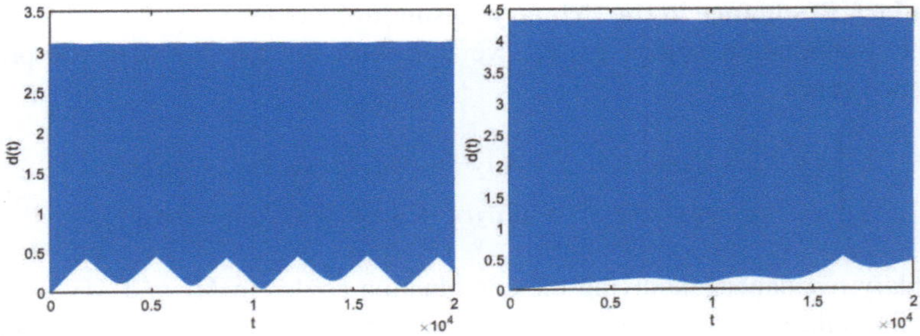


**Fig. 9.20** The 1 : 1 : 4 resonance in 20 000 timesteps, initial velocities zero, $b_1 = 1, b_2 = -1.5, \varepsilon = 0.1$. Left the Euclidean distance $d(t)$ of the orbits to their initial conditions in the case of initial conditions outside the resonance zones: $q_1(0) = 0.3, q_2(0) = 0.5, q_3(0) = 0.7$. Recurrence takes many timesteps because of passage through the resonance zones. Right the Euclidean distance $d(t)$ for the same Hamiltonian (9.89) but inside primary resonance zone $M_1$, starting at a location of higher order resonances $q_1(0) = 0.3674, q_2(0) = 0.45, q_3(0) = 1.0129$ ($\cos 2\chi_1 = 1$). First, the recurrence is stronger until the orbit drifts outside the higher order resonance location in $M_1$. $I_3$ (and $r_3$) experiences 0.01 variation. Courtesy Int. J. Bif. Chaos [154]

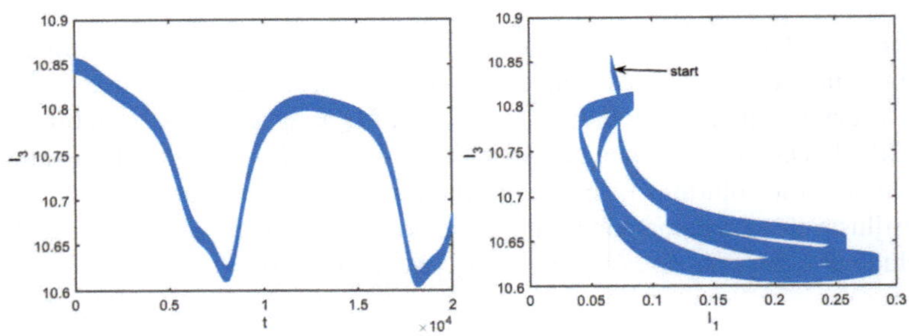


**Fig. 9.21** The 1 : 1 : 4 resonance in 20 000 timesteps, initial velocities zero, $b_1 = 1, b_2 = -1.5, \varepsilon = 0.1$. Dynamics in the stable part of the primary resonance zone $M_1$, initial values $q_1(0) = 0.3674, q_2(0) = 0.45, q_3(0) = 1.65$ and initial velocities zero. Left the time series $I_3(t)$, right the diagram $I_1, I_3$ dependent on $t$. The fluctuations are $O(\varepsilon)$ although the calculations go far beyond $1/\varepsilon^2$. Courtesy Int. J. Bif. Chaos [154]

higher order combination angles are not timelike; the recurrence is stronger, but then the variation of $\phi_3$ produces a drift-off of the solutions in $M_1$ and worse recurrence. To illustrate the stable and unstable behaviour of the solutions starting in $M_1$, we present in Fig. 9.21 left the case where $q_3(0) = 0.3$ ($\chi_2, \chi_3$ are timelike) and right the unstable case $q_3(0) = 1.0129$. The scales have been adjusted to the variations of $I_1(t), I_3(t)$.

## 9.12 Chaotic Hamiltonian Dynamics

In 1964 appeared the surprising paper by Hénon and Heiles [75]. In a simple looking Hamiltonian system with two dof but with quadratic non-linearities, they found numerically smooth and integrable looking dynamics near stable equilibrium. Increasing the energy with respect to equilibrium a little bit, some of the regular curves broke up and increasing the energy even more the phase space looked completely chaotic.

This result was not only very interesting but also confusing. Because of the regularity of the dynamics close to equilibrium, people thought that there existed a threshold of the energy for systems to become chaotic. This idea was natural but not correct. As we have indicated in Sect. 9.1, for two-dof Hamiltonian systems, the normal form is integrable with implication that for small values of the energy, where the normal form is valid, chaotic behaviour has an exponentially small measure. To demonstrate this presence of chaos numerically is difficult.

One of the methods to show the presence of chaos is to show that a suitable Poincaré map exists in the system that contains a horseshoe map; see, for instance, [66]. However, this requires a detailed technical knowledge that is difficult to obtain.

Showing non-integrability of a Hamiltonian system gives a strong indication for complexity of the dynamics or even chaotic behaviour, see for instance [42, 43] and [159], but in addition one still needs a description of the actual dynamics. We describe a few successful cases.

### 9.12.1 Šilnikov Bifurcation

Certain special solutions turn out be a source of remarkable dynamics. Poincaré identified in [110] the so-called homoclinic solutions. Suppose we have a Hamiltonian system, or more in general a dynamical system, with an invariant set $M$ like an equilibrium, a periodic solution, or a more complicated structure. If there exists a solution that starts arbitrarily close to M and that after leaving a neighbourhood of $M$ returns after some time arbitrarily close to $M$, we call this solution homoclinic to $M$.

If we have two invariant sets $M_1$ and $M_2$ and there exists a solution that connects in such a way the two invariant sets, Poincaré called this solution heteroclinic to $M_1$ and $M_2$.

In Fig. 9.22 we show schematically special saddle behaviour. The flow expands near a critical point (in the plane) along and near an isolated homoclinic solution while rotating around the homoclinic. This produces a horseshoe map in a transversal of the homoclinic with an infinite number of periodic solutions and chaos. This is a general Šilnikov scenario for chaos; Devaney showed in [47] how to apply this to Hamiltonian systems.

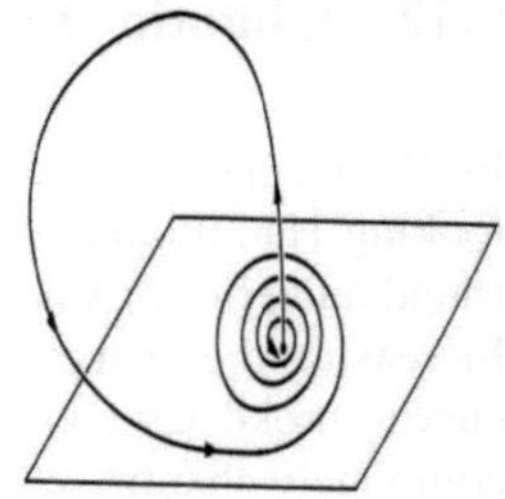

**Fig. 9.22** A critical point in a plane has locally an expanding and rotational flow around an homoclinic solution

In [159] a survey is given of methods to prove non-integrability. An application involving various methods is given to the discrete symmetric 1 : 2 : 1 resonance; both normal forms and numerics suggested by the Šilnikov scenario are part of the analysis. This case is complicated by the fact that no normal mode of the symmetric 1 : 2 : 1 resonance is complex unstable so one has to find other complex unstable periodic solutions in the normal form $H_2 + \bar{H}_3$.

## *9.12.2 The* 1 : 2 : 3 *Resonance*

For the three-dof 1 : 2 : 3 resonance, this was applied to the normal form in [78], see also the survey 9.6.1. An additional difficulty there was that there exists an infinite number of homoclinics in the first order normal form. This problem is resolved by computing a second order approximation.

It was shown in [78] that the normal form (9.53) is non-integrable for open sets of the parameters. The method is based on [47] which is an extension of the Šilnikov scenario to Hamiltonian systems. One has to start with a complex unstable periodic solution on the energy manifold. Next, one has to identify a transversal homoclinic orbit to this solution with inward and outward spiraling orbits. The dynamics near the homoclinic fits into the theory of Šilnikov showing that in such a case the dynamics contains locally a horseshoe map. The system induced by the normal form $H_2 + \bar{H}_3$ Hamiltonian (9.53) contains a complex unstable periodic solution associated with the $\tau_2$ normal mode, two heteroclinic orbits, and a continuous set of homoclinics. The dynamics based on Hamiltonian (9.53) is summarised in Fig. 9.23 (the $\tau_i$ are replaced by $x_i$).

It complicates the application of Devaney's theory of [47]. In [78] this is solved by extending the normal form with $\bar{H}^4$; it is shown by precise numerics that the infinite set of homoclinics breaks up to produce a transversal homoclinic orbit. This implies the presence of Šilnikov dynamics corresponding with chaotic behaviour. A Poincaré section in [78] illustrates the dynamics.

It should be noted that a horseshoe map is a structurally stable phenomenon in the context of symplectic maps, so the non-integrability also extends to the original Hamiltonian.

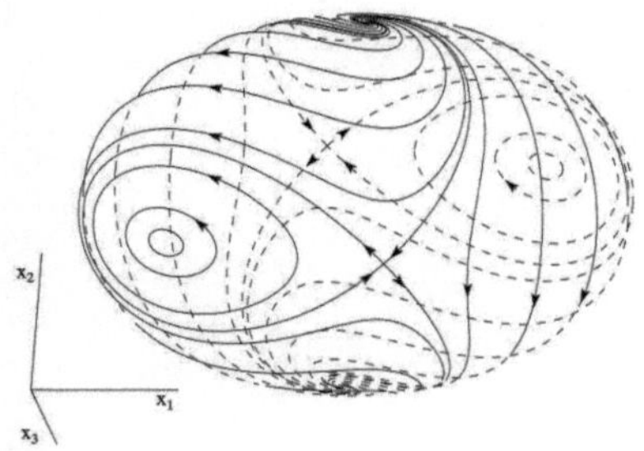


**Fig. 9.23** Summary of the dynamics of the 1 : 2 : 3 resonance based on normal form $H_2 + \bar{H}_3$ of (9.53). The complex unstable normal mode is at the top of the energy manifold. Homoclinic solutions emerge from the normal mode and return at the back of the manifold. A few stable periodic solutions are pictured. Figure courtesy Igor Hoveijn

Interestingly, when using Ziglin-Morales-Ramis theory and special solutions of the equations of motion the non-integrability of the normal form $H_2 + \bar{H}_3$ for open sets of the parameters was shown in [43] by studying the normal form (9.53) without need of $H_4$. On the other hand, the use of Šilnikov bifurcation provides more explicit knowledge of the dynamics. The energy simplex in Fig. 9.8 (left) shows the case of the $\tau_2$ normal mode HH unstable.

## 9.12.3 Laskar's Frequency Analysis

According to KAM theory, we will find in Hamiltonian systems near stable equilibrium an infinite number of tori (invariant manifolds), see [27] and [28]. The theory predicts the presence of chaos between the tori but as mentioned before the smallness of the chaotic sets makes the numerics difficult. Laskar [97] developed a method, originally for celestial mechanics, to find evidence of chaos. On a set of tori, a few have been pictured in Fig. 9.6, one can look for the dominant frequencies of periodic and quasi-periodic solutions present on the torus. A number of them will be prominent. Moving with small steps from one torus to another in the infinite set, the prominent frequencies are expected to change smoothly unless, in very small gaps between the tori, certain jumps in the frequencies are found. These jumps indicate small sets of chaos between the tori, and they will arise in Cantor sets of the frequencies.

It is interesting that infinite sets of tori can also be found in dissipative systems if certain symmetries in these systems are present. In these cases, Laskar's theory works as well, see for an example inspired by chemical physics [10] and Sect. 12.8.

# Chapter 10
# Fermi-Pasta-Ulam Chains

**Abstract** The original Fermi-Pasta-Ulam (FPU) problem considers a chain of N so-called monoatomic elements, i.e. all masses equal. The idea of the model was to demonstrate the statistical mechanics phenomenon of equipartition of energy. However, the system does show recurrence phenomena, even for a large number of particles. From averaging-normalisation we find generically in FPU chains invariant submanifolds that have a remarkable impact on the dynamics. Also it is possible to prove an embedding theorem that relates the behaviour of an FPU chain with n particles with a chain that contains an arbitrary multiple of n particles. FPU chains with unequal masses can produce very different resonance problems. Interesting cases are chains with alternating masses where even significant interactions between small and large masses are possible.

In this chapter the hybrid approach of using analysis, in particular, by averaging and inspired by computing plays an important part. Computing alone is sometimes unguided and may miss important phenomena. Mathematical analysis alone gives basic information but in its generality can ignore the role of timescales and new subtle phenomena; combination of analytic approximations with computing can be very inspiring as the history of FPU chains clearly shows. This hybrid approach will be present throughout the chapter.

In the original Fermi-Pasta-Ulam (FPU) problem, one considered a chain of $N$ so-called mono-atomic elements, i.e. all masses equal. We call this the classical FPU chain and will put for the masses $m_1 = m_2 = \cdots = m_N = 1$. The idea of the authors was to demonstrate numerically the statistical mechanics phenomenon of equipartition of energy. Starting with initial values close to one mode of the system, and although collisions are excluded, one expected the energy to spread permanently in the system. FPU chains are Hamiltonian, so this contradicts the recurrence theorem at low energy states, but it might be possible that at chaotic, larger energy states, the recurrence would take extremely long times. However, the authors soon found recurrence numerically for 16 and 32 particles. For the historical background see [59]. The formulation of the FPU chain started off an enormous



F. Verhulst, *Averaging for Nonlinear Dynamics with Applications and Numerical Bifurcations*, Applied Mathematical Sciences 223,
https://doi.org/10.1007/978-3-032-12745-7_10

amount of papers. It was also an inspiration to study many more types of chains and networks.

As we shall see one can consider FPU chains with unequal masses. It turns out that such systems can be far more interesting than the classical cases of equal masses. We will include three spectacular results.

1. First it will be shown that the classical FPU chain near stable equilibrium will have in most cases an integrable normal form; the implication is that chaos in the classical system near stability is a small affair, even for many dof.
2. A second result is that we have generically in FPU chains invariant submanifolds that have an enormous impact on the dynamics; the formulation of this result is summarised in the embedding Theorem 10.1. There is a relation with the results of Chechin et al. in [36] with applications to FPU chains in [37] and [38]; the idea is to identify so-called bushes that can be found in classical FPU chains. The embedding theorem will be applied to FPU chains with unequal masses; the theorem does not depend on 'small energy assumptions'. Equipartition of energy cannot be expected.
3. A third result is that we will find interaction between low- and high-frequency modes. This is of importance for large-scale chains, networks and wave equations, where after spectral analysis, often high frequency parts of the modes are truncated.

## 10.1 The Classical Fermi-Pasta-Ulam Chain

The formulation of the FPU chain runs as follows: We consider an $N$ dof Hamiltonian near stable equilibrium $p = q = 0$, and use a potential $V$ of the $\varepsilon$-rescaled form

$$V(z) = \frac{1}{2}z^2 + \varepsilon\frac{1}{3}\alpha z^3 + \varepsilon^2\frac{1}{4}\beta z^4.$$

So, as usual, $\varepsilon$ indicates an estimate of the distance to the origin. We speak of an $\alpha$-chain if $\alpha \neq 0, \beta = 0$ and of a $\beta$-chain if $\alpha = 0, \beta \neq 0$. Apart from $\alpha$- or $\beta$-chains we have two possible models: the FPU chain with fixed endpoints; see Fig. 10.1 and the periodic chain where the first mass and the last mass are identified. We have for the Hamiltonian so-called nearest-neighbour interaction

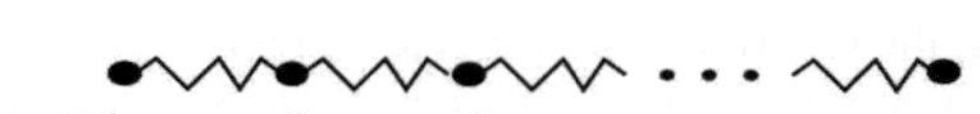


**Fig. 10.1** A Fermi-Pasta-Ulam chain of oscillators with fixed endpoints

$$H = \sum_{i=1}^{N} \left( \frac{1}{2} p_i^2 + V(q_{i+1} - q_i) \right). \tag{10.1}$$

The interaction is essentially different for $\alpha$- and $\beta$-chains, as for the former the interaction is asymmetric, and for the latter it is symmetric with respect to the particle. These physical differences show up in the results of the mathematical analysis.

**The Fixed Endpoints FPU Chain with Three Masses**

It is instructive to consider FPU chains with a few masses. In [59] the case with 3 masses is discussed for fixed endpoints and $\alpha$-chains. The Hamiltonian is (notation [59])

$$H = \sum_{i=0}^{2} \left( \frac{1}{2}(p_i^2 + (q_{i+1} - q_i)^2) + \varepsilon \frac{1}{3} \alpha (q_{i+1} - q_i)^3 \right). \tag{10.2}$$

The linearised system is not in quasi-harmonic form like for instance system (9.20) (coupled harmonic equations). The eigenvalues produce the frequencies (3, 3, 0). By a linear transformation we can obtain quasi-harmonic equations leads in [59] to the equivalent Hamiltonian:

$$H = \frac{1}{2}(P_1^2 + P_2^2 + P_3^2 + 3Q_2^2 + 3Q_3^2) + \varepsilon \frac{3\alpha}{\sqrt{2}} (Q_2 Q_3^2 - \frac{1}{3} Q_3^3). \tag{10.3}$$

The $P, Q$ coordinates were obtained by the diagonalising transformation. As mentioned in [59] Hamiltonian (10.3) is equivalent by rescaling coordinates to the Hénon-Heiles Hamiltonian presented in [75], also discussed in [144]. The equations of motion are then of the form

$$\ddot{x} + x = -\varepsilon 2xy, \; \ddot{y} + y = \varepsilon(-x^2 + y^2). \tag{10.4}$$

The Hénon-Heiles Hamiltonian was chosen in [75] to illustrate the motion of stars in an axi-symmetric galaxy near the plane of symmetry. This brings to mind what Henri Poincaré wrote in an essay: "Mathematics is the art of giving the same name to different things". As stated before the Hénon-Heiles Hamiltonian became famous as it was the first Hamiltonian system where the transition from regular to chaotic motion was demonstrated numerically. If $\varepsilon$ is small, say 0.05, we have dominating regular dynamics but on increasing the energy, say to 1/8 or near 1/6, chaos appears. It is remarkable that we have already such complex behaviour in the case of three masses. Note, however, that according to the introduction of Chap. 9 for two dof, we find no chaos in the normal form near stable equilibrium; chaos is a small phenomenon.

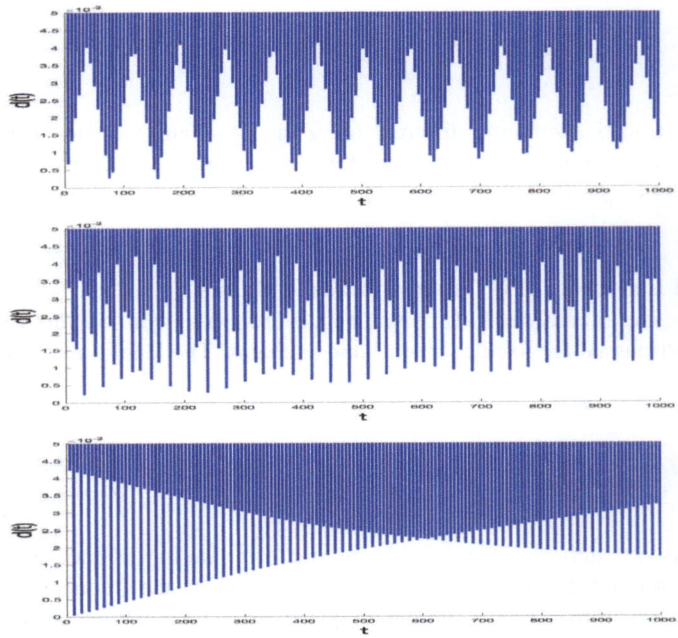

**Fig. 10.2** Recurrence of three initially close orbits of the Fermi-Pasta-Ulam chain of FPU-$\alpha$-coupled oscillators with fixed endpoints where system (10.4) was used on 1000 timesteps; relative and absolute precision $e^{-15}$. We have $\varepsilon = 0.05$ and plotted $0 \leq d(t) \leq 0.005$ (so 1 on the vertical axis means 0.001). We have initially for the three orbits $x(0) = 0.05$, $\dot{x}(0) = 0$, $\dot{y}(0) = 0$. Top figure $y(0) = 0.097$, middle 0.1 and below 0.101. Small differences in $y(0)$ produce large recurrence difference

In Fig. 10.2 we show at low energy, various recurrence behaviour for initially close orbits. We have in the three cases $\varepsilon = 0.05$ and initially $x(0) = 0.05, \dot{x}(0) = 0, \dot{y}(0) = 0$. For the fourth coordinate we have $y(0) = 0.97, 0.1, 0.101$. The complexity is caused by the presence of very small resonance zones between the large scale tori predicted by KAM theory (see section 15.6 of [146] or [163]); see Chap. 9 and [27] for the general theory. There will be many small resonance zones like this with qualitative different behaviour of orbits.

### *10.1.1 The Classical Periodic FPU Chain with Six Masses*

We will consider as a second example the periodic $\beta$-chain with six particles, but as the analysis is rather unwieldy, although mathematically not difficult, we will summarise the results.

We choose $N = 6$. The basic frequencies follow from the linearised part of the equations of motion (leaving out the cubic terms):

$$\begin{cases} \ddot{q}_1 + 2q_1 - q_2 - q_6 &= 0, \\ \ddot{q}_2 + 2q_2 - q_3 - q_1 &= 0, \\ \ddot{q}_3 + 2q_3 - q_4 - q_2 &= 0, \\ \ddot{q}_4 + 2q_4 - q_5 - q_3 &= 0, \\ \ddot{q}_5 + 2q_5 - q_6 - q_4 &= 0, \\ \ddot{q}_6 + 2q_6 - q_1 - q_5 &= 0. \end{cases} \tag{10.5}$$

The linear part of the equations is again not in quasi-harmonic form, and we use a diagonalising transformation by MATHEMATICA. Indicate the transformation by $(p, q) \mapsto \bar{p}, \bar{q}$. One can check that the complete non-linear equations of motion of the periodic chains have the first (momentum) integral:

$$\dot{q}_1 + \dot{q}_2 + \dot{q}_3 + \dot{q}_4 + \dot{q}_5 + \dot{q}_6 = \text{constant.} \tag{10.6}$$

This can directly be generalised to $N$ particles, the integral will produce a zero for one of the eigenvalues of the trivial solution. For the transformation to modes we need the eigenvalues and the frequencies $\omega_1, \omega_2, \ldots, \omega_6$ produced by the linear part. The frequencies are:

$$1,\ \sqrt{3},\ 2,\ \sqrt{3},\ 1,\ 0. \tag{10.7}$$

Frequency 0 corresponds with the momentum integral that reduces by transformation the equations of motion to 5. The transformation to a diagonal form mixes also the non-linear terms (this is the unwieldy part that we leave out; see [114]). We introduce the actions:

$$I_i = \frac{1}{2\omega_i}(\bar{p}^2 + \omega_i^2 \bar{q}_i^2),\ i = 1, \ldots, 5. \tag{10.8}$$

It is remarkable that the frequencies contain two $1 : 1$ resonances ($1, 1$ and $\sqrt{3}, \sqrt{3}$) and a three-dof resonance $(1, 2, 1)$. Applying averaging-normalisation we find the normalised Hamiltonian $\tilde{H}$ in action-angle $I_i, \phi_i, i = 1, \ldots, 5$ variables with

$$\begin{cases} \tilde{H} = I_1 + \sqrt{3}I_2 + 2I_3 + \sqrt{3}I_4 + I_5 + \\ \quad \varepsilon^2\beta[\frac{1}{4}\sqrt{3}\sqrt{I_1 I_2 I_4 I_5}(\cos(\phi_1+\phi_2-\phi_4-\phi_5) + \cos(\phi_1-\phi_2+\phi_4-\phi_5)) \\ \quad +\frac{1}{16}I_1 I_5(\cos(2\phi_1 - 2\phi_5) - 1) + \frac{3}{16}I_2 I_4(\cos(2\phi_2 - 2\phi_4) - 1) \\ \quad +\frac{1}{8}\sqrt{3}(I_1 I_2 + I_4 I_5 + 3I_1 I_4 + 3I_2 I_5) + \frac{1}{2}I_3(I_1 + \sqrt{3}I_2 + \sqrt{3}I_4 + I_5) \\ \quad +\frac{1}{4}I_3^2 + \frac{3}{32}(I_1 + I_5)^2 + \frac{9}{32}(I_2 + I_4)^2]. \end{cases} \tag{10.9}$$

The quartic terms of the $\beta$-chain produce the $O(\varepsilon^2)$ terms. From the Hamitonian $\tilde{H}$ we can write down the equations of motion in actions and angles $I, \phi$. As we can see from the Hamiltonian the two $1 : 1$ resonances play a part, but the $1 : 2 : 1$ resonance does not. This shows there is a hidden symmetry in the periodic FPU chain. It will become clear that $N = 6$ with these features is typical for classical periodic FPU chains.

In [114] many other aspects are discussed. We mention that there exist five independent first integrals of the equations of motion derived from Hamiltonian (10.9). Also, a number of invariant manifolds and periodic solutions for fixed values of the energy can be found.
The recurrence of the periodic FPU $\beta$-chain with six masses leading to integrable averaged equations of motion is obvious. This behaviour is typical for $N$ dof periodic FPU chains.

### *10.1.2 Classical Periodic FPU Chains with N Particles*

A remarkable paper [115] shows that because of symmetries and resonances, the periodic FPU chain with arbitrary number of $N$ masses shows similar behaviour as for the case $N = 6$. Using the eigenvalues of the trivial solution and transforming to a diagonal system, we find that for FPU systems near stable equilibrium, the averaged-normalised Hamiltonian is in many cases integrable with consequence that strong recurrence is expected in such systems near stable equilibrium.

The results can be generalised to FPU chains with fixed endpoints. We have from [115] that the classical FPU chain has discrete symmetries and other symmetries that remove the $1 : 2 : 1$ resonance from $\bar{H}_3$ as for the case $N = 6$. Even stronger, no resonance is active in $\bar{H}_3$.

1. Assume $\alpha \neq 0$ and $N$ is odd; then $H_2 + \bar{H}_4$ is integrable.
2. If $\alpha \neq 0$ and $N$ is even, then $H_2 + \bar{H}_4$ has at least $(3N - 4)/4$ quadratic integrals if $N$ divides 4, and $(3N - 2)/4$ quadratic integrals if $N$ does not divide 4.
3. For the $\beta$-chain, $H_2 + \bar{H}_4$ is integrable.

These results solve the classical FPU problems at low energy level in most cases.

## 10.2 FPU Chains with Unequal Masses

A largely unexplored field is the case of a periodic FPU chain with unequal masses. One of our conclusions will be that the dynamics in this more general case is qualitatively very different from the classical FPU chain; the classical chain is a non-generic case of FPU chains. As first-order resonances may come in, the normal form can be non-integrable at $\bar{H}_3$. With arbitrary non-zero mass values the Hamiltonian will be

$$H = \sum_{i=1}^{N} \left( \frac{1}{2} \frac{p_i^2}{m_j} + V(q_{i+1} - q_i) \right). \tag{10.10}$$

with $V(z)$ as before:

$$V(z) = \frac{1}{2} z^2 + \varepsilon \frac{1}{3} \alpha z^3 + \varepsilon^2 \frac{1}{4} \beta z^4.$$

The equations of motion will have (apart from the energy) the momentum integral

$$m_1 \dot{q}_1 + m_2 \dot{q}_2 + \ldots + m_N \dot{q}_N = \text{constant}. \tag{10.11}$$

To obtain more insight we will consider first the cases of a four-particle $\alpha$-chain, so $N = 4$. A more general analysis of the case of $N$ alternating masses will follow. The equations of motion for four particles are

$$\begin{cases} m_1 \ddot{q}_1 + 2q_1 - q_2 - q_4 &= -\varepsilon\alpha[(q_1 - q_4)^2 - (q_2 - q_1)^2], \\ m_2 \ddot{q}_2 + 2q_2 - q_3 - q_1 &= -\varepsilon\alpha[(q_2 - q_1)^2 - (q_3 - q_2)^2], \\ m_3 \ddot{q}_3 + 2q_3 - q_4 - q_2 &= -\varepsilon\alpha[(q_3 - q_2)^2 - (q_4 - q_3)^2], \\ m_4 \ddot{q}_4 + 2q_4 - q_1 - q_3 &= -\varepsilon\alpha[(q_4 - q_3)^2 - (q_1 - q_4)^2]. \end{cases} \tag{10.12}$$

The basic frequencies follow from the linearised part of the equations of motion with the four parameters $a_1, a_2, a_3, a_4$. Because of the momentum integral (10.11) we have one eigenvalue zero. It turns out that the frequency ratios $1 : 1 : 1$ and $1 : 2 : 2$ do not occur for any mass distribution with four particles. From the rather long computation in [30] we have sets of mass distributions that produce the first-order three dof resonances $1 : 2 : 1$ (organised in four dots in parameter-space), $1 : 2 : 3$ and $1 : 2 : 4$ (both organised in open curves); see also [73]. Ten second-order resonances are listed in [30]. We will consider the FPU $1 : 2 : 3$ and $1 : 2 : 4$ resonances.

### *10.2.1 Unequal Masses in* 1 : 2 : 3 *Resonance,* $N = 4$

Because of the momentum integral (10.11) the system can be reduced to three dof. The parameter sets that correspond with this resonance produces two different cases, the case where $m_1 = m_3$ and the case that all masses are different.

**The Case** $m_1 = m_3$

To be explicit we choose the parameter case $m_1 = m_3 = 28$, $m_2 = 7.8862$, $m_4 = 3.3138$. After rescaling time and transforming $p, q \rightarrow \dot{x}, x$, putting $\alpha = 1$ we find the equations of motion:

$$\begin{cases} \ddot{x}_1 + 9x_1 &= -\varepsilon 14(2d_3x_1x_3, +d_6x_2x_3) \\ \ddot{x}_2 + 4x_2 &= -\varepsilon 14(2d_{10}x_2x_3 + d_6x_1x_3), \\ \ddot{x}_3 + x_3 &= -\varepsilon 14(d_3x_1^2 + d_{10}x_2^2 + d_6x_1x_2). \end{cases} \tag{10.13}$$

For the parameters we have from [30] $d_3 = -9\sqrt{21}/490$, $d_6 = -3\sqrt{14}/490$, $d_{10} = 2\sqrt{21}/245$. We note that if $x_1(t)$ and $x_2(t)$ vanish we have the $x_3$ normal mode consisting of harmonic solutions. Averaging-normalisation in amplitude-angle variables $r, \phi$ shows that to first order only the $d_6$ terms survive. Transforming to action-phase variables by (1.8) we find with combination angle $\chi = \phi_1 = -\phi_1 - \phi_2 - \phi_3$ :

$$\begin{cases} \dot{r}_1 &= \varepsilon \frac{7}{6} d_6 r_2 r_3 \sin\chi, \\ \dot{r}_2 &= -\varepsilon \frac{7}{4} d_6 r_1 r_3 \sin\chi, \\ \dot{r}_3 &= -\varepsilon \frac{7}{2} d_6 r_1 r_2 \sin\chi, \\ \dot{\chi} &= \varepsilon \frac{7}{2} d_6 \frac{\cos\chi}{r_1 r_2 r_3} \left( \frac{r_2^2 r_3^2}{3} - \frac{r_1^2 r_3^2}{2} - \frac{r_1^2 r_2^2}{1} \right). \end{cases} \tag{10.14}$$

We have again the $H_2$ integral:

$$9r_1^2 + 4r_2^2 + r_3^2 = 2E_0. \tag{10.15}$$

Moreover we have the integrals:

$$3r_1^2 + 2r_2^2 = C_1,\ 2r_2^2 - r_3^2 = C_2, \tag{10.16}$$

with constants $C_1, C_2$ determined by the initial conditions. So the normal form of (10.13) is integrable. In [30] periodic solutions in general position, 3 normal modes and periodic solutions in the coordinate planes have been derived.

**General Position Periodic Solutions**

Looking for solutions with constant amplitude we put $\sin\chi = 0$ leading to $\chi = 0, \pi$. The right-hand side of the equation for $\dot{\chi}$ has to be zero; using an integral to eliminate $r_1$ we have the condition

$$2r_2^2 r_3^2 + \frac{4}{3} r_2^4 + \frac{1}{6} r_3^4 = \frac{1}{3} E_0 (2r_2^2 + r_3^2) = 0.$$

This is a quadratic form in the squares of the amplitudes. We find positive solutions if

$$0 < r_2 < \sqrt{\frac{E_0}{2}},\ 0 < r_3 < \sqrt{2E_0}. \tag{10.17}$$

This is a remarkable result. The two families of general position periodic solutions are not standard; they are embedded in 2-tori on the energy manifold. Their period is modulated by their position on the tori; they connect the $x_2$ and $x_3$ normal modes.

As described by Poincaré in [110] vol. 1, ch. 4, such families of periodic solutions on the energy manifold point at the existence of extra first integrals.

**Normal Modes**

The averaged system (10.14) shows that normal mode solutions (and solutions in coordinate planes) show singularities and should be studied by transformation (1.17):

$$\begin{cases} x_1 &= y_1 \cos 3t + \frac{1}{3} y_2 \sin 3t,\ \dot{x}_1 = -3y_1 \sin 3t + y_2 \cos 3t, \\ x_2 &= z_1 \cos 2t + \frac{1}{2} z_2 \sin 2t,\ \dot{x}_2 = -2z_1 \sin 2t + z_2 \cos 2t, \\ x_3 &= u_1 \cos t + u_2 \sin t,\ \dot{x}_1 = -u_1 \sin t + u_2 \cos t. \end{cases} \tag{10.18}$$

Transforming and averaging over time we have from [30]

$$\begin{cases} \dot{y}_1 &= \varepsilon \frac{7}{6} d_6 (z_1 u_2 + \frac{1}{2} z_2 u_1),\ \dot{y}_2 = -\varepsilon \frac{7}{2} d_6 (z_1 u_1 - \frac{1}{2} z_2 u_2), \\ \dot{z}_1 &= \varepsilon \frac{7}{4} d_6 (-y_1 u_2 + \frac{1}{3} y_2 u_1),\ \dot{z}_2 = -\varepsilon \frac{7}{2} d_6 (y_1 u_1 + \frac{1}{3} y_2 u_2), \\ \dot{u}_1 &= \varepsilon \frac{7}{2} d_6 (-\frac{1}{2} y_1 z_2 + \frac{1}{3} y_2 z_1),\ \dot{u}_2 = -\varepsilon \frac{7}{2} d_6 (y_1 z_1 + \frac{1}{6} y_2 z_2). \end{cases} \tag{10.19}$$

It is clear from the averaged system that the three normal modes exist and are harmonic functions. From [30] we have that the $x_1$ normal mode is unstable (type HH), and the $x_3$ normal mode is stable (type EE). As an example we study the $x_2$ normal mode. We have

$$x_2(t) = A \cos 2t + B \sin 2t, \tag{10.20}$$

with $A, B$ constants and energy $E_0$ fixed; we introduce small perturbations $(w_1, w_2, w_3)$ for the solution $(0, x_2(t), 0)$. Substituting the perturbed solutions into system (10.19) and linearising we have

$$\begin{cases} \dot{y}_1 &= -\varepsilon \frac{7}{6} d_6 (B u_1 + A u_2),\ \dot{y}_2 = \varepsilon \frac{7}{2} d_6 (A u_1 - B u_2), \\ \dot{u}_1 &= \varepsilon \frac{7}{2} d_6 (B y_1 - \frac{1}{3} A y_2),\ \dot{u}_2 = \varepsilon \frac{7}{2} d_6 (A y_1 + \frac{1}{3} B y_2). \end{cases} \tag{10.21}$$

After linearisation we have $w(t)$ is harmonic. We find that the eigenvalues of the corresponding matrix with entries $A, B$ are purely imaginary and coincident. This opens the possibility of a Hamilton-Hopf bifurcation when introducing small perturbations of the vector field (10.13).

**The Case of Four Masses Different**

We choose now a case with less symmetry which is more generic. In [30] various cases of different masses are considered as the corresponding dynamics shows many different aspects. This will be clear when using the momentum integral producing the transformed Hamiltonian:

$$H_2 = \frac{1}{2}\left(\dot{x}_1^2 + \frac{9}{14}x_1^2 + \dot{x}_2^2 + \frac{4}{14}x_2^2 + \dot{x}_3^2 + \frac{1}{14}x_3^2\right). \tag{10.22}$$

Reducing canonically to three dof, we have ten parameters for $H_3$; see [30]. In the first-order normal form $\bar{H}_3$ only two parameters, $d_6$ and $d_9$ survive. We leave out the eight parameters that are relevant only at higher order. After rescaling time and transforming $p, q \to \dot{x}, x$, putting $\alpha = 1$ we find the equations of motion valid to first order in $\varepsilon$:

$$\begin{cases} \ddot{x}_1 + 9x_1 &= -\varepsilon 14 d_6 x_2 x_3 \\ \ddot{x}_2 + 4x_2 &= -\varepsilon 14(d_6 x_1 x_3 + d_9 x_3^2), \\ \ddot{x}_3 + x_3 &= -\varepsilon 14(d_6 x_1 x_2 + 2d_9 x_2 x_3). \end{cases} \tag{10.23}$$

In [30] a systematic analysis of general position periodic solutions, solutions in the coordinate planes and the normal modes are given. The analytic techniques are the same as for the case with two equal masses in 1 : 2 : 3 resonance, but an essential difference is that an integral of the normalised system as (10.16) does not exist. As an example we consider the $x_2$ normal mode that was shown to be dynamically interesting in the case with two equal masses.

Note that this normal mode is a solution of system (10.23). We study the stability question. Consider a perturbation of the $x_2$ normal mode:

$$x_1(t) = w_1,\ x_2(t) = A\cos 2t + B\sin 2t + w_2,\ x(t) = w_3.$$

Substitution into system (10.23) and linearisation yields two coupled Mathieu-equations:

$$\begin{cases} \ddot{w}_1 + w_1 &= -\varepsilon 14 d_6(A\cos 2t + B\sin 2t)w_3, \\ \ddot{w}_3 + w_3 &= -\varepsilon 14[d_6(A\cos 2t + B\sin 2t)w_1 + 2d_9(A\cos 2t + B\sin 2t)w_3], \end{cases} \tag{10.24}$$

and a harmonic equation for $w_2$. We average the linear system (10.24) to obtain a four-dimensional matrix with entries dependent on $A, B$. The eigenvalues are given by

$$\lambda^2 = -(A^2 + B^2)\left[\left(\frac{1}{3}d_6^2 - 2d_9^2\right) \pm 2d_9\sqrt{d_9^2 - \frac{1}{3}d_6^2}\,\right]. \tag{10.25}$$

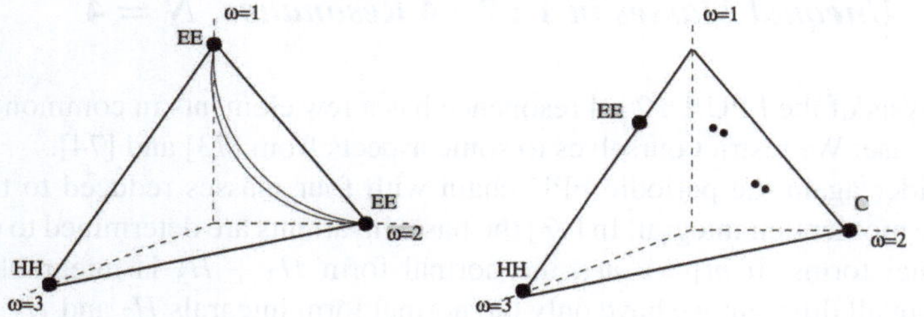


**Fig. 10.3** Energy simplices for the 1 : 2 : 3 resonant FPU chain with the case of two equal masses including a sketch of the tori consisting of families of periodic solutions linking 2 normal modes on the left. Dots indicate periodic solutions. The general asymmetric case of the simplex with complex unstable $x_2$ normal mode on the right. Courtesy Acta Appl. Math

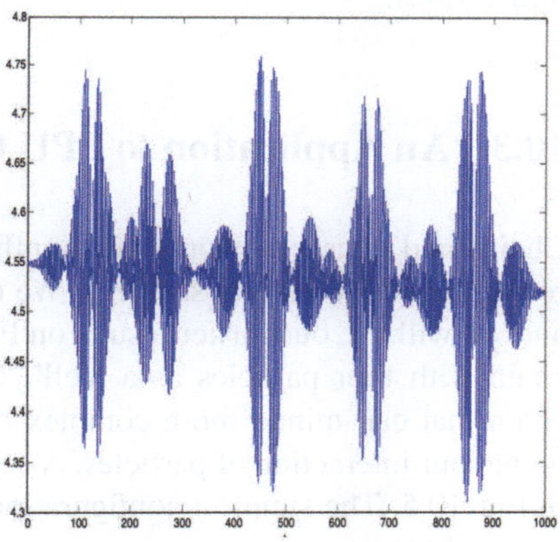

**Fig. 10.4** Time series $H_2(t)$ in 1000 timesteps for the 1 : 2 : 3 resonant FPU chain starting near the complex unstable normal mode $x_2$. We have $x_1(0) = 0.1, x_2(0) = 1.5, x_3(0) = 0.1, \varepsilon = 0.5$, $H_2(0) = 4.52$ and the case $m_1 = 195, 9662, m_2 = 8.5277, m_3 = 11.7095, m_4 = 3.4220$. Courtesy Acta Appl. Math.

In the case of two equal masses we found coincident purely eigenvalues and so Hamilton-Hopf bifurcation in the case $d_9 = 0$. It is clear that the same bifurcation arises for values of $d_9$ in a neighbourhood of 0. See Fig. 10.3.

In [30] two numerical illustrations are presented. We consider the rather unbalanced case:

$$m_1 = 195.9662,\ m_2 = 8.5277,\ m_3 = 11.7095,\ m_4 = 3.4220.$$

For the parameters we have from [30] $d_6 = -0.0306229, d_9 = -0.0089438$. In Fig. 10.4 we show the fluctuations of $H_2$ near the complex unstable normal mode $x_2$. The parameter $\varepsilon = 0, 5$ to enhance the instability effect.

### *10.2.2 Unequal Masses in* 1 : 2 : 4 *Resonance, $N = 4$*

The analysis of the FPU 1 : 2 : 4 resonance has a few elements in common with the 1 : 2 : 3 case. We restrict ourselves to some aspects from [73] and [74].

Consider again the periodic FPU chain with four masses reduced to three dof using the momentum integral. In [73] the basic invariants are determined to compute the normal forms. If $m_1 = m_3$, the normal form $H_2 + \bar{H}_3$ is integrable. If the masses are all different we have only the normal form integrals $H_2$ and $H_2 + \bar{H}_3$ (or just $\bar{H}_3$). Both for the $\alpha$- and for the $\beta$-chain we find besides periodic solutions in general position three normal mode periodic solutions. Small detuning of the three frequencies has been included in the analysis.

The analysis is continued in [74] to describe the rich collection of bifurcations in the 1 : 2 : 4 FPU chain. Results are given on Hamiltonian flip bifurcations, centre-saddle bifurcations and Krein collisions associated with Hamilton-Hopf bifurcations.

## 10.3 An Application to FPU Cell Chains

Chains and neural networks are significant tools for disciplines as varied as brain research and the natural sciences. We discuss briefly a formulation based on [153] and we will use our earlier results on FPU chains. Suppose we treat a periodic FPU chain with four particles as a 'cell'. Connecting such FPU cells produces a cell chain that can mimic more complex or at least different interaction than nearest-neighbour interaction of particles. A cell chain consisting of three cells is illustrated in Fig. 10.5. The simplest configuration has two cells. In this case the equations of motion become

$$\begin{cases} \dot{q}_1 = v_1, \; m_1\dot{v}_1 = -2q_1 + q_2 + q_4 - \varepsilon((q_1-q_4)^2 - (q_2 - q_1)^2), \\ \dot{q}_2 = v_2, \; m_2\dot{v}_2 = -2q_2 + q_3 + q_1 - \varepsilon c_1(q_2-q_6) - \varepsilon((q_2 - q_1)^2 - (q_3 - q_2)^2), \\ \dot{q}_3 = v_3, \; m_3\dot{v}_3 = -2q_3 + q_4 + q_2 - \varepsilon((q_3-q_2)^2 - (q_4 - q_3)^2), \\ \dot{q}_4 = v_4, \; m_4\dot{v}_4 = -2q_4 + q_1 + q_3 - \varepsilon((q_4-q_3)^2 - (q_1 - q_4)^2), \\ \dot{q}_5 = v_5, \; m_1\dot{v}_5 = -2q_5 + q_6 + q_8 - \varepsilon((q_5-q_8)^2 - (q_6 - q_5)^2), \\ \dot{q}_6 = v_6, \; m_2\dot{v}_6 = -2q_6 + q_7 + q_5 + \varepsilon c_1(q_2-q_6) - \varepsilon((q_6 - q_5)^2 - (q_7 - q_6)^2), \\ \dot{q}_7 = v_7, \; m_3\dot{v}_7 = -2q_7 + q_8 + q_6 - \varepsilon((q_7-q_6)^2 - (q_8 - q_7)^2), \\ \dot{q}_8 = v_8, \; m_4\dot{v}_8 = -2q_8 + q_5 + q_7 - \varepsilon((q_8-q_7)^2 - (q_5 - q_8)^2). \end{cases} \tag{10.26}$$

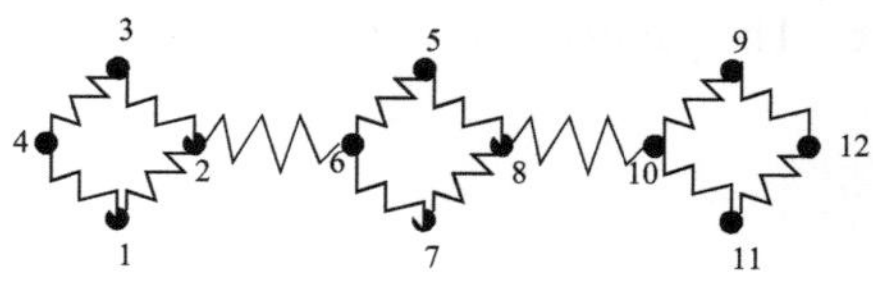


**Fig. 10.5** An example of three periodic FPU chains connected to form a cell chain

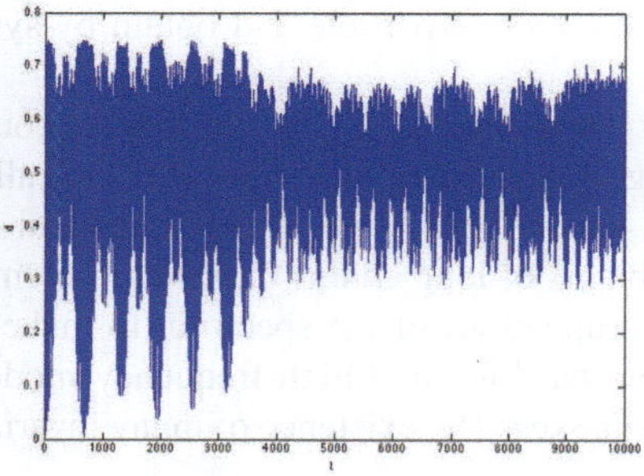


**Fig. 10.6** Time series of the Euclidean distance $d$ in the case of two cells described by system (10.26). The dynamics starts near the unstable (C) normal mode of the $1 : 2 : 3$ resonance. After around 3000 timesteps the orbit wanders off, recurrence takes longer than 10 000 timesteps. Courtesy Int. J. Bif. Chaos [153]

The coefficient $c_1$ indicates the strength of the connection between cells 1 and 2. If $c_1$ is small, it acts as a detuning. The momentum integral for this system is

$$\sum_{1}^{8} m_i v_i = \text{constant}. \tag{10.27}$$

The choice for initial conditions is to excite particles or excite normal modes; the second one is based on the canonical transformation to quasi-harmonic form based on the momentum integral. In [153] the normal modes are chosen. As an illustration we show the interaction starting in the first cell near the unstable (C) normal mode of the $1 : 2 : 3$ resonance, the transfer of energy to the second cell takes place. In Fig. 10.6 we show the Euclidean distance $d$ for the system in the case where the second cell starts with zero initial conditions. Recurrence will take a long time.

## 10.4 FPU Chains with $N$ Alternating Masses

An interesting variation of the FPU chain was studied in [61]. The authors studied a chain with fixed endpoints and two different alternating masses; their analysis is focused on obtaining insight in equipartition of energy and uses mainly numerical tools. We will consider periodic FPU chains with Hamiltonian (10.10) and mass distribution:

$$1, m, 1, m, \ldots, 1, m.$$

If the total number of particles $N$ is even, put $N = 2n, n = 1, 2, \ldots$ The mass of the first particle is normalised to 1 and $m > 1$. Note that choosing $m = 1$ we have the classical FPU chain. The case $0 < m < 1$ is also included as in that case we can normalise the mass of the second particle and obtain by symmetry considerations similar results.

We will show that choosing $m \gg 1$ we will encounter a spectrum with $O(1)$ and $O(1/m)$ eigenvalues; the first type will be called the optical part of the spectrum and the second eigenvalues the acoustical part of the spectrum. Usually, in related problems of long chains or wave problems in partial differential equations, one applies truncations of the spectrum to make the problem treatable. However, we will show that low- and high-frequency modes can interact in FPU chains with alternating masses. The existence of many invariant manifolds obstructs equipartition of energy.

We formulate some basic results from [32]. Consider the chain

$$m_j \ddot{q}_j = -2q_j + q_{j+1} + q_{j-1} + \dots, \tag{10.28}$$

for $1 \leq j \leq 2n$ modulo $2n$. The masses for $j$ even are 1, for $j$ odd $m > 1$. The dots stand for quadratic $\alpha$ terms and cubic $\beta$ terms. The system is of the form $\ddot{q} = -A_{2n}C_{2n}q$ with $A_{2n}C_{2n}$, a square $2n$-dimensional matrix. The eigenvalues of matrix $A_{2n}C_{2n}$ are

$$1 + a \pm \sqrt{1 + 2a\cos(2\pi j/n) + a^2}, \tag{10.29}$$

If the number of particles is a fourfold ($N = 4n$), we have among the eigenvalues 2 and $2a$. with $a = 1/m$. If $m \gg 1$ (or $0 < a \ll 1$), we have two sets of eigenvalues for $a \to 0$:

$$2 + O(a),\ a\left(1 - \cos\frac{2\pi j}{n}\right) + O(a^2). \tag{10.30}$$

The modes with eigenvalues near 2 are called the optical group, with size $O(a)$ the acoustical group. We show examples below. It may be surprising to have interactions between the two groups.

The *embedding theorem* proved in [32] is as follows:

**Theorem 10.1** *Consider the equations of motion induced by Hamiltonian* (10.10) *for periodic $\alpha$ and $\beta$ chains and also if $\alpha\beta \neq 0$, with alternating masses $1, m$ and $2n$ (even) particles. Suppose $k$ is a multiple of $n$ and consider the equations of motion induced by Hamiltonian* (10.10) *with $2kn$ dof and identical $\alpha, \beta, m$; then there exists an invariant submanifold of this larger Hamiltonian system that is equivalent to the first system with $2n$ dof.*

The theorem implies that equipartition of energy is excluded. Another implication is that studying systems with a low number of dof like 4 or 6 is useful as they will emerge in systems with a multiple of dof number like $4n$ or $6n$. Suppose we consider systems with $2pn$ dof, $p$ a natural number. If $p$ is even, we can demonstrate interactions between acoustic and optical eigenmodes by considering four and eight dof. For $p$ odd this is not so simple; in the case of $p$ prime we have an infinite number of separate possibilities in the dynamics. We will consider some details in the next sections.

As this is a rich topic with many cases, part of the asymptotics will be left to the reader.

## 10.5 Alternating Masses with 4n and 8n Dof

In [32] the cases with the number of dof a 4-fold or 8-fold are considered with interesting chains if $m(= 1/a)$ is large, so $a$ is small. For eight dof there are also a few cases with interesting $O(1)$ values of $m$. We note that the case of four particles is relevant for systems with $4n$ dof, a chain with eight particles is relevant for $8n$ dof, $n = 2, 3, \ldots$

### *10.5.1 Four Particles with Alternating Mass*

Following [31], consider first the case of four alternating masses. The linear part of system (10.12) in the alternating case has the eigenvalues

$$\lambda_1 = 2(1 + a),\ \lambda_2 = 2,\ \lambda_3 = 2a,\ \lambda_4 = 0. \tag{10.31}$$

The eigenvalue zero is produced by the presence of the momentum integral (10.11). The acoustical group has one component, the optical group 2.

The cases $a = 1/m = 1/2, 3/4$ represent first-order resonances.

**The $\alpha$-Chain with Four Particles**

Transformation to quasi-harmonic form produces for the $\alpha$-chain a symmetric form:

$$\begin{cases} \ddot{x}_1 + 2(1 + a)x_1 &= \gamma x_2 x_3, \\ \ddot{x}_2 + 2x_2 &= \gamma x_1 x_3, \\ \ddot{x}_3 + 2ax_3 &= \gamma x_1 x_2, \end{cases} \tag{10.32}$$

with $\gamma = 2\sqrt{2a(1 + a)}$. It is reminiscent of system (10.14) where $m_1 = m_3$ and the other masses are different. The system has three exact harmonic normal mode solutions. If $a$ is an $O(1)$ quantity, we find from [31] in a restriction to two dof two first-order resonances ($a = 1/4, 1/3$) and two second-order resonances ($a = 1/8, 1/9$) and in these cases integrability of the normal forms.

### 10.5.2 Alternating Masses, Four Particles, *m* = 2

If $a = 1/m = 1/2$, we have the $3 : 2 : 1$ resonance, $\gamma = 2\sqrt{3/2}$. With the scaling near stable equilibrium $x \to \varepsilon x/\gamma$ we average outside the normal modes in amplitude-phases (1.8) to find

$$\begin{cases} \dot{r}_1 &= -\frac{\varepsilon}{12} r_2 r_3 \sin\chi\,,\ \dot{\phi}_1 = -\frac{\varepsilon}{12}\frac{r_2 r_3}{r_1}\cos\chi\,, \\ \dot{r}_2 &= \frac{\varepsilon}{8} r_1 r_3 \sin\chi\,,\ \dot{\phi}_2 = -\frac{\varepsilon}{8}\frac{r_1 r_3}{r_2}\cos\chi\,, \\ \dot{r}_3 &= \frac{\varepsilon}{4} r_1 r_2 \sin\chi\,,\ \dot{\phi}_2 = -\frac{\varepsilon}{4}\frac{r_1 r_2}{r_3}\cos\chi\,, \end{cases} \tag{10.33}$$

with $\chi = \phi_1 - \phi_2 - \phi_3$. See also section (9.6.1) for the general case of the $3 : 2 : 1$ resonance. Periodic solutions with constant amplitude are found if $\chi = 0, \pi$. The system has the integral

$$9r_1^2 + 4r_2^2 + r_3^3 = 2E_0,$$

with $E_0$ a positive constant; the left-hand side is $H_2$ in the terminology of Chap. 9. In addition one has the quadratic integrals $3r_1^2 + 2r_2^2 =$ constant and $2r_2^2 - r_3^2 =$ constant. One can use the integral $\bar{H}_3 = r_1 r_2 r_3 \cos\chi =$ constant. For periodic solutions in general position to be possible, the equation is as in Sect. 10.2.1:

$$\dot{\chi} = -\varepsilon\left(\frac{1}{12}\frac{r_2 r_3}{r_1} - \frac{1}{8}\frac{r_1 r_3}{r_2} - \frac{1}{4}\frac{r_1 r_2}{r_3}\right)\cos\chi = 0,\ \chi = 0, \pi.$$

The case of the resonance for $a = 3/4$ is less interesting.

### 10.5.3 Alternating Masses, Four Particles *m* Large

Interesting is the case of large mass ($a$ small). With $\gamma$ in Eq. (10.32) as a small parameter we can average in amplitude-phase variables (1.8).

**The $\alpha$-Chain with Four Particles**
We find to first order in $\gamma$ integrability of the normal form and no interaction between acoustical ($x_3$) and optical ($x_1, x_2$) group. However averaging to second order we find instabilities and interaction; see for details [31]. In Fig. 10.7 we display $I_3(t) = 0.5(\dot{x}_3^2 + 2ax_3^2)$ showing irregular second-order interactions.

**The $\beta$-Chain with Four Particles**
The transformation to quasi-harmonic form becomes

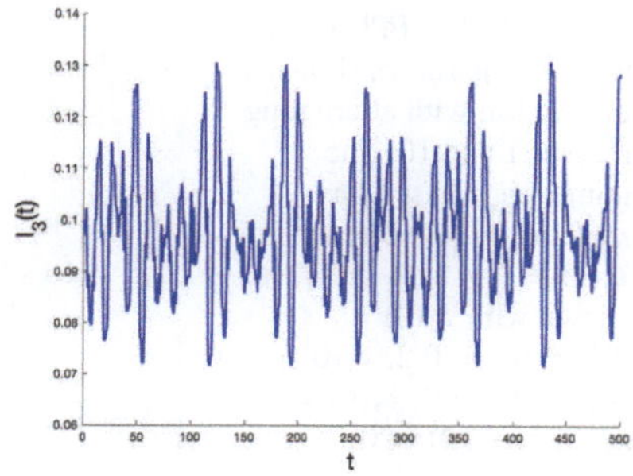


**Fig. 10.7** An $\alpha$ FPU chain with four particles showing interaction with alternating masses 1 and 10. The numerics is based on system (10.32). The action $I_3(t) = 0.5(\dot{x}_3^2 + 2ax_3^2)$ is shown with $I_3(0) = 0.5, x_1(0) = 0.2, x_2(0) = 0.4, x_3(0) = 0, \dot{x}_3(0) = 1$ and $\dot{x}_1(0) = \dot{x}_2(0) = 0$

$$\begin{cases} \ddot{x}_1 + 2(1+a)x_1 & = -(1+a)^2 x_1^3 - 3(1+a)x_1(x_2^2 + ax_3^2), \\ \ddot{x}_2 + 2x_2 & = -x_2^3 - 3ax_2x_3^2 - 3(1+a)x_1^2 x_2, \\ \ddot{x}_3 + 2ax_3 & = -3ax_2^2 x_3 - a^2 x_3^3 - 3a(1+a)x_1^2 x_3. \end{cases} \tag{10.34}$$

The normal modes are again exact solutions but elliptic functions. In the case of large second mass $m$ we put in a first approximation $a = 0$ and so obtain the $1:1$ resonance between the $x_1$ and $x_2$ modes, whereas $x_3(t)$ will be harmonic with a long period. This suggests that interactions between the acoustical and the optical group needs a second-order approximation as for the $\alpha$ chain with four particles.

One can average over elliptic functions, but this is technically more complicated. Various rescalings are possible. In [31] the scaling is $x_1, x_2 \to a^{1/4}x_1, a^{1/4}x_2$ and $x_3 \to (2a)^{-1/4}x_3$.

We will use a different scaling here that uses that the $x_1, x_2$ components remain to first order small when initially started small. We scale $x_1, x_2 \to a^{1/2}x_1, a^{1/2}x_2$ and leave $x_3$ without scaling. We find the system

$$\begin{cases} \ddot{x}_1 + 2(1+a)x_1 & = -ax_1^3 - 3ax_1x_2^2 + O(a^2), \\ \ddot{x}_2 + 2x_2 & = -ax_2^3 - 3ax_2x_3^2 - 3ax_1^2x_2 + O(a^2), \\ \ddot{x}_3 + 2ax_3 & = -3a^2x_2^2x_3 - a^2x_3^3 - 3a^2x_1^2x_3 + O(a^3). \end{cases} \tag{10.35}$$

It is clear that a first-order approximation in $a$ does not produce interaction between acoustical and optical group. In a second-order approximation the slow $x_3$ mode is forced by the $1:1$ resonance of the $x_1, x_2$ modes; see Fig. 10.8 for a numerical illustration (the asymptotics of the second-order calculation is left to the reader).

### 10.5.4 Alternating Masses, Eight Particles m Large

The eight eigenvalues of the linearised system are

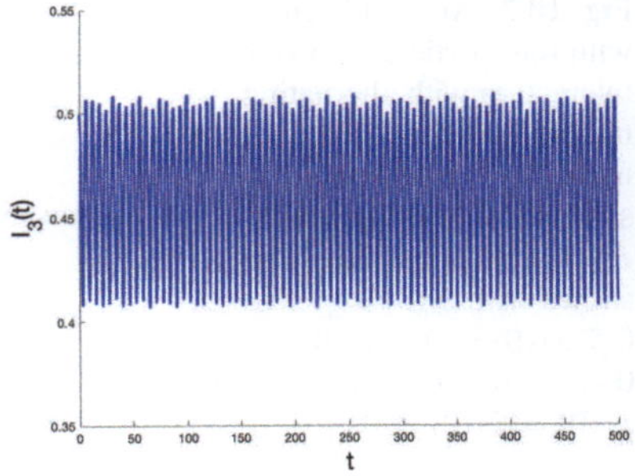


**Fig. 10.8** A $\beta$ FPU chain with four particles showing interaction with alternating masses 1 and 10. The numerics is based on system (10.34). The action $I_3(t) = 0.5(\dot{x}_3^2 + 2ax_3^2)$ is shown with $I_3(0) = 0.5, x_1(0) = 0.2, x_2(0) = 0.4, x_3(0) = 0, \dot{x}_3(0) = 1$ and $\dot{x}_1(0) = \dot{x}_2(0) = 0$

$$\lambda_1 = 2(1+a),\ \lambda_{2,3} = 1 + a + \sqrt{1+a^2},\ \lambda_4 = 2,\ \lambda_5 = 2a,$$
$$\lambda_{6,7} = 1 + a - \sqrt{1+a^2},\ \lambda_8 = 0, \tag{10.36}$$

with two double eigenvalues. Transforming to eigenmodes (quasi-harmonic form) we have for $H_3$ and $H_4$:

$$\begin{aligned} H_3 &= (d_{225}x_2^2x_5 + d_{335}x_3^2x_5) + (x_{566}x_6^2x_5 + d_{577}x_7^2x_5) + (d_{127}x_1x_2x_7 + d_{136}x_1x_3x_6) \\ &+ d_{234}x_2x_3x_4 + d_{145}x_1x_4x_5 + (d_{256}x_2x_5x_6 + d_{357}x_3x_5x_7) \\ &+ (d_{346}x_3x_4x_6 + d_{247}x_2x_4x_7) + d_{467}x_4x_6x_7\,, \\ H_4 &= e_{1111}x_1^4 + e_{1122}x_1^2x_2^2 + e_{1133}x_1^2x_3^2 + e_{1224}x_1x_2^2x_4 + e_{1144}x_1^2x_4^2 + e_{1235}x_1x_2x_3x_5 \\ &+ e_{1155}x_1^2x_5^2 + e_{1126}x_1^2x_2x_6 + e_{1246}x_1x_2x_4x_6 + e_{1166}x_1^2x_6^2 + e_{1137}x_1^2x_3x_7 \\ &+ e_{1257}x_1x_2x_5x_7 + e_{1177}x_1^2x_7^2. \end{aligned}$$

The coefficients $d_{ijk}$ and $e_{ijkl}$ are algebraic functions of the parameter $a$; the expressions are given in [32]. The equations of motion show the existence of three invariant manifolds consisting of three modes; we indicate them by $M_{kij}$ with the index for the mode numbers. $M_{145}$ is composed of the $x_1, x_4, x_5$ modes and corresponds with the four-particle case. Such an invariant manifold was predicted by the embedding theorem (10.1). The other two invariant manifolds are $M_{256}$ and $M_{357}$; the three manifolds have the mode $x_5$ in common.

In [32] many cases with various parameter values are studied with respect to periodic solutions, integrability and recurrence. We will discuss the $\alpha$-chain with eight particles. A conclusion in [32] is that for eight-particle $\beta$-chains, no significant interaction between optical and acoustical modes takes place in the three invariant manifolds.

**The $\alpha$-Chain with Eight Particles and $m$ Large**

With $a$ small (large mass $m$) the eigenvalues are to $O(a)$:

$$\lambda_1 = 2(1+a),\ \lambda_{2,3} = 2+a,\ \lambda_4 = 2,\ \lambda_5 = 2a,\ \lambda_{6,7} = a,\ \lambda_8 = 0.$$

We have four modes in the optical, and three modes in the acoustical group. We analyse the case with $m$ large, put $\alpha = 1$ and rescale $x_i \to \sqrt{a}x_i, i = 1, \ldots, 7$. We find for the optical group

$$\begin{cases} \ddot{x}_1 + 2x_1 &= \alpha\sqrt{a}(2x_4x_5 + \sqrt{2}x_3x_6 + \sqrt{2}x_2x_7) + O(a^{\frac{3}{2}}), \\ \ddot{x}_2 + 2x_2 &= \alpha\sqrt{a}(\sqrt{2}x_7x_4 + x_2x_5 + \sqrt{2}x_1x_7) + O(a^{\frac{3}{2}}), \\ \ddot{x}_3 + 2x_3 &= \alpha\sqrt{a}(-\sqrt{2}x_6x_4 - x_3x_5 + \sqrt{2}x_1x_6) + O(a^{\frac{3}{2}}), \\ \ddot{x}_4 + 2x_4 &= \alpha\sqrt{a}(-\sqrt{2}x_6x_3 + 2x_1x_5 + \sqrt{2}x_2x_7) + O(a^{\frac{3}{2}}). \end{cases} \tag{10.37}$$

In the same way the acoustical group with non-linear terms to $O(\sqrt{a})$ becomes

$$\begin{cases} \ddot{x}_5 + 2ax_5 &= \alpha\sqrt{a}(x_2^2 - x_3^2 + 2x_1x_4) + O(a^{\frac{3}{2}}), \\ \ddot{x}_6 + ax_6 &= \alpha\sqrt{2a}(x_1x_3 - x_4x_3) + O(a^{\frac{3}{2}}), \\ \ddot{x}_7 + ax_7 &= \alpha\sqrt{2a}(x_1x_2 + x_4x_2) + O(a^{\frac{3}{2}}). \end{cases} \tag{10.38}$$

As suggested by the terms of the Hamiltonian $H_3$ there are terms leading to interaction of the optical and acoustical group. Also we have the invariant manifolds $M_{145}, M_{256}, M_{357}$ with a key role played by mode 5. The complexity of the dynamics is directly connected with the resonance zones in the system. In [163] quasi-trapping and so delay of recurrence is discussed. We illustrate this delay and the interaction by choosing non-zero initial values in the optical group, zero initial values in the acoustical group, and vice versa; see Fig. 10.9. We use the distances to the initial values of the optical and acoustical groups:

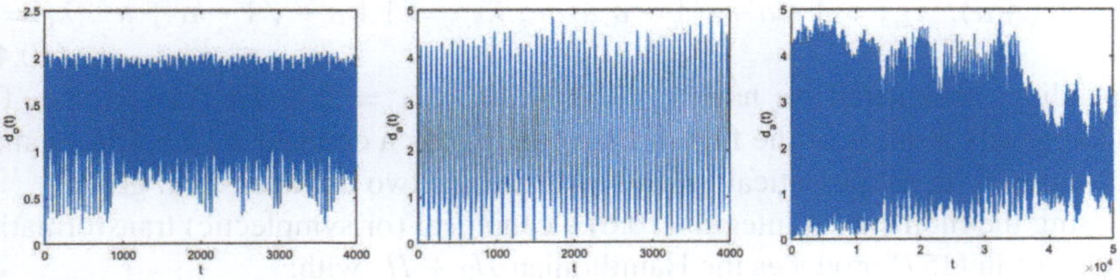

**Fig. 10.9** We illustrate for $a = 0.01, \alpha = 1$ the interactions of the optical and acoustical groups of systems (10.37) and (10.38) (derived for $\alpha$-chains with eight particles with large mass). Initial positions $x_j(0) = 0.5, j = 1, \ldots, 4,, x_j(0) = 0, j = 5, 6, 7$ and initial velocities zero. Left: $d_o(t)$ for 4000 timesteps. The acoustical group is mainly excited by $x_5(t)$ (middle figure shows $d_a(t)$ on 4000 timesteps), as an illustration $d_a(t)$ is also shown for 50 000 timesteps. Courtesy J. Nonlinear Science [32]

$$d_o(t) = \sqrt{\sum_{j=1}^{4}((x_j(t) - x_j(0))^2 + (\dot{x}_j(t) - \dot{x}_j(0))^2)}, \tag{10.39}$$

and

$$d_a(t) = \sqrt{\sum_{j=5}^{7}((x_j(t) - x_j(0))^2 + (\dot{x}_j(t) - \dot{x}_j(0))^2)}. \tag{10.40}$$

On a time interval of 1000 steps the interaction is dominated by $x_5(t)$; for a more complete picture we also give in the case of interactions produced by the acoustical group $d_a(t)$ for 50 000 timesteps in Fig. 10.9.

### 10.5.5 Explorations for Six and Ten Particles and Large Mass

With four and eight particles we have studied some important aspects of the cases with large mass. The case of $2pn$ particles, $p$ odd, is technically slightly different. Such cases have many more possibilities; each case with $p$ prime has to be studied separately until general theorems are available. This is of special interest if we meet with new phenomena. We sketch some of the results near stable equilibrium based on [157] and [141].

**Six Particles,** $p = 3$
The eigenvalues are

$$\lambda_1 = 2(1+a),\ \lambda_{2,3} = 1+a+\sqrt{1-a+a^2},\ \lambda_{4,5} = 1+a-\sqrt{1-a+a^2},\ \lambda_6 = 0. \tag{10.41}$$

Expanding for $a$ small we have $\lambda_1 = 2 + 2a$, $\lambda_{2,3} = 2 + \frac{1}{2}a$, $\lambda_{4,5} = \frac{3}{2}a$. The optical group consists of the first three modes and is a detuned $1 : 1 : 1$ resonance. A question is: does the optical group react with the two dof acoustical group.

Using the momentum integral (10.6) a canonical (or symplectic) transformation to five dof in [157] produces the Hamiltonian $H_2 + H_3$ with:

$$H_3 = \sum_{i,j,k} d_{ijk} x_i x_j x_k.$$

The coefficients $d_{ijk}$ are given in the appendix of [157]. We put $\sqrt{a} = \varepsilon$; for the interaction analysis the terms $O(\sqrt{a})$ and $O(a)$ are sufficient. The equations of motion for the optical group $x_1, x_2, x_3$ and the acoustical group $x_4, x_5$ after rescaling $x_i = \sqrt{\varepsilon}\bar{x}_i$, $i = 1, 2, 3$ and leaving out the bars (dots for higher-order terms) become

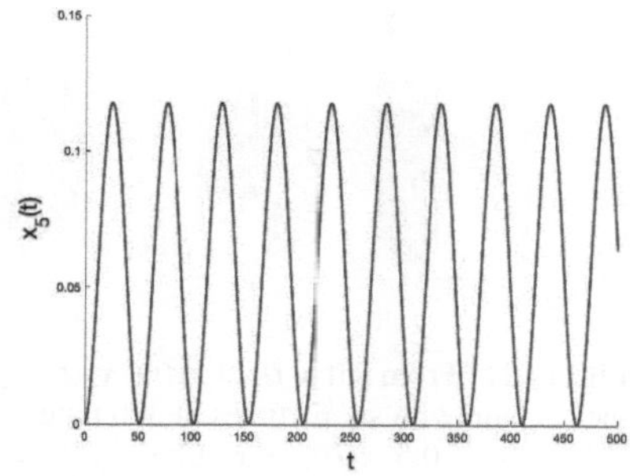


**Fig. 10.10** We show the excitation of mode $x_5$ in an $\alpha$ FPU chain with six particles showing in 500 timesteps interaction with alternating masses 1 and 100 ($\sqrt{a} = 0.1$). The mode $x_5$ is excited with in system (10.43) the initial values $x_3(0) = 0.5, \dot{x}_3(0) = x_5(0) = \dot{x}_5(0) = 0$

$$\begin{cases} \ddot{x}_1 + 2x_1 & = \varepsilon(-2x_2x_5 - 2x_3x_4) + \dots, \\ \ddot{x}_2 + 2x_2 & = \varepsilon(-2x_1x_5 - \sqrt{2}x_2x_5 + \sqrt{2}x_3x_4) + \dots, \\ \ddot{x}_3 + 2x_3 & = \varepsilon(-2x_1x_4 + \sqrt{2}x_2x_4 + \sqrt{2}x_3x_5) + \dots, \\ \ddot{x}_4 + \frac{3}{2}\varepsilon^2 x_4 & = \varepsilon^2(-2x_1x_3 + \sqrt{2}x_2x_3) + \dots, \\ \ddot{x}_5 + \frac{3}{2}\varepsilon^2 x_5 & = \varepsilon^2(-2x_1x_2 - \frac{1}{2}\sqrt{2}x_2^2 + \frac{1}{2}\sqrt{2}x_3^2) + \dots. \end{cases} \tag{10.42}$$

In system (10.42) we observe quadratic forcing terms for the acoustical modes 4 and 5, so we expect interaction of acoustical and optical modes. Fortunately we can identify a two dof invariant manifold that is easier to study. In system (10.42) putting $x_1 = x_2 = x_4 = 0$ we have an invariant manifold consisting of the modes 3 and 5:

$$\ddot{x}_3 + 2x_3 = \varepsilon\sqrt{2}x_3x_5,\ \ddot{x}_5 + \frac{3}{2}\varepsilon^2 x_5 = \frac{\varepsilon^2}{2}\sqrt{2}x_3^2. \tag{10.43}$$

As the right-hand side of the equation for $x_5$ is positive if $x_3(0) \neq 0$, we observe that mode $x_5$ is forced by mode $x_3$. In Fig. 10.10 the acoustical mode $x_5$ is excited by optical mode $x_3$.

Another invariant manifold, $M_{125}$, consisting of the modes 1, 2 and 5 contains two optical modes and one acoustical; assume that modes $x_3, x_4$ have zero initial values and vanish from the system. System (10.42) is a system with widely different frequencies and it merits a deeper analysis, part of which is given in [157] and [141]. We illustrate the interactions numerically in Fig. 10.11 starting with zero energy in the acoustical modes.

1. For the $\beta$-chain with six particles there is little evidence for significant interactions between the optical and the acoustical group.
2. A few remarks on the case of 18 particles ($p = 9$) are made in [141]. The dynamics of the case $p = 3$ is included; the optical group contains more $1 : \ldots : 1$ resonances.
3. There exist a remarkable circularity for the optical-acoustical interactions. For instance the acoustical mode $x_1$ excites the optical mode $x_6$, $x_6$ excites acoustical $x_7$, $x_7$ excites optical $x_2$, $x_2$ excites acoustical $x_5$, $x_5$ excites optical $x_8$, etc.

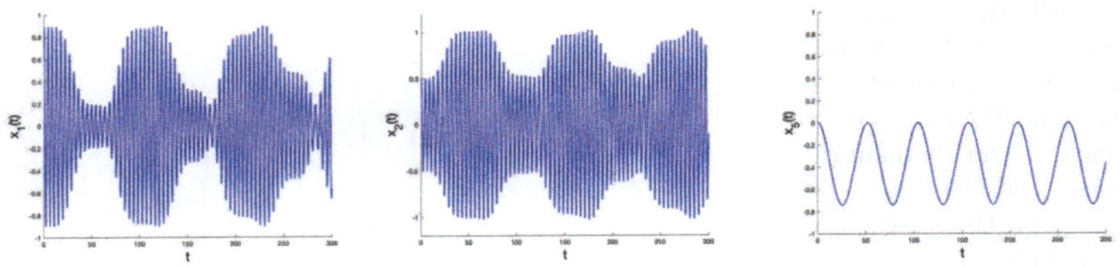

**Fig. 10.11** From left to right $x_1(t), x_2(t), x_5(t)$ for modes 1, 2, 5 of the invariant manifold $M_{125}$ of the $\alpha$-chain with six particles in 300 timesteps; see system (10.42). The initial values are $x_1(0) = 0.9, x_2(0) = 0.5, x_3(0) = x_4(0) = x_5(0) = 0$ and all initial velocities zero. We observe forcing of the acoustical mode $x_5$. Courtesy Int. J. Nonl. Mech. [141]

It would be interesting to study whether this circularity persists for more than 18 particles.

**Ten Particles, $p = 5$ and More**

Considering the original FPU formulation with alternating masses derived from Hamiltonian (10.10) we find two equivalent four dof subsystems. Transforming to quasi-harmonic equations we note that for $\alpha$-chains again quadratic terms on the right-hand side produce considerable interactions between the optical and acoustical groups. A few remarks are made in [141] for the multiples 15 and 25 (30 and 50 particles) of $p = 5$.

We summarise a few conclusions:

1. A direct consequence of the embedding theorem (10.1) is that for $\alpha$- and $\beta$-chains, equipartition of energy by exciting a finite number of modes is not possible.
2. In general one can expect interactions between optical and acoustical groups because of the quadratic terms in the $\alpha$ FPU systems transformed to eigenmode equations. For $\beta$-chains the interactions seem to be smaller.
3. If $N = 2n$ ($n$ even) is fourfold, we have from the expression for the eigenvalues (10.29) the eigenvalues $2, 2(1 + a), 2a, 0$ and $2n - 4$ double eigenvalues. If $a \ll 1$, we have a detuned $1 : 1 : \ldots : 1$ resonance for the $2n - 4$ eigenmodes.
4. If $N = 2n$, $n$ is odd, $n \geq 3$, we have from (10.29) the eigenvalues $2(1+a), 0$ and $2n - 2$ double eigenvalues. If $a \ll 1$, we have a detuned $1 : 1 : \ldots : 1$ resonance for the $2n - 2$ eigenmodes.
5. A conclusion is that for $N = 2n, n \geq 2$ and $n \leq 47$, most phenomena can be explained by the embedding Theorem 10.1. The bound 47 comes from the technical possibility to compute canonical transformations, decompositions and circularity phenomena by MATHEMATICA if the dimensions are very high.

# Chapter 11
# Parametric and Autoparametric Oscillations

**Abstract** The presence of parametric excitation is one of the possible excitation mechanisms with as iconic example the Mathieu-equation. It is often used for describing a pendulum with oscillating support, but the equation comes up in many applications. A basic element is the calculation of instability tongues in force of excitation-frequency space. More complex applications are concerned with nonlinear interaction with other oscillators and the study of flow-induced vibrations, see also Chap. 12. Autoparametric systems are using parametric oscillators as energy absorbers. An example is quenching of Rayleigh self-excited oscillations.

Excitation of an oscillator can take different forms; we have paid attention to external, parametric and self-excited excitation. In this chapter we will focus on parametric excitation where forces are acting to produce a varying frequency in the system. The theory of linear equations with periodic coefficients is extensively described in [162]. The classical example is a pendulum with oscillating suspension point.

We discussed the Mathieu-equation in the introduction, Chap. 1; it is natural to add a non-linear term and study again the stability of the trivial solution.

## 11.1 The Non-linear Mathieu-Equation

Consider the Mathieu-equation of Chap. 1 with a cubic term added:

$$\ddot{x} + (1 + \varepsilon \cos 2t)x + \varepsilon a x^3 = 0,\ a \neq 0.$$

From amplitude-phase transformation (1.6) we find for the variational system

$$\dot{r} = \varepsilon \sin(t + \psi)[\cos 2t\, r \cos(t + \psi) + a r^3 \cos^3(t + \psi)],$$



F. Verhulst, *Averaging for Nonlinear Dynamics with Applications and Numerical Bifurcations*, Applied Mathematical Sciences 223,
https://doi.org/10.1007/978-3-032-12745-7_11

$$\dot{\psi} = \frac{\varepsilon}{r}\cos(t+\psi)[\cos 2t\, r\cos(t+\psi) + ar^3\cos^3(t+\psi)].$$

Averaging produces, excluding a neighbourhood of the origin

$$\dot{r} = \frac{1}{4}\varepsilon r\sin 2\psi,\ \dot{\psi} = \frac{1}{4}\varepsilon(\cos 2\psi + \frac{3}{2}ar^2). \tag{11.1}$$

This system is based on polar coordinates and is not convenient to analyse near $r = 0$, but for positive radius we find stationary $r_0, \psi_0$ if $\psi_0 = 0, \pi/2, \pi, 3\pi/2$. Two periodic solutions are found for $a > 0$ if $\psi_0 = \pi/2, 3\pi/2$, for $a < 0$ if $\psi_0 = 0, \pi$. The Jacobian of the right-hand sides of system (11.1) is, omitting the scale factor $\frac{1}{4}\varepsilon$

$$J = \begin{pmatrix} \sin 2\psi_0 & 2r_0\cos 2\psi_0 \\ 3ar_0 & -2\sin 2\psi_0 \end{pmatrix}.$$

For the determinant we have

$$det(J) = -2\sin^2 2\psi_0 - 6ar_0^2\cos 2\psi_0.$$

In both cases of the sign of $a$ we have $det(J) \neq 0$.

If $a > 0$, we find two periodic solutions near $r_0 = \sqrt{2a/3}$; if $a < 0$, we have two periodic solutions near $r_0 = \sqrt{-2a/3}$. To study the behaviour near the trivial solution we use for the variational equations transformation (1.16). We find

$$\dot{y}_1 = \varepsilon\sin t(\cos 2t(y_1\cos t + y_2\sin t) + a(y_1\cos t + y_2\sin t)^3),$$
$$\dot{y}_2 = \varepsilon\cos([\cos 2t(y_1\cos t + y_2\sin t) + a(y_1\cos t + y_2\sin t)^3).$$

Averaging produces

$$\dot{y}_1 = \frac{1}{4}\varepsilon(-y_2 + \frac{3a}{2}y_1^2y_2 + \frac{3a}{2}y_2^3),\ \dot{y}_2 = \frac{1}{4}\varepsilon(y_1 + \frac{3a}{2}y_1^3 + \frac{3a}{2}y_1y_2^2).$$

Linearisation produces near $y_1 = y_2 = 0$ a saddle with eigenvalues $\pm\frac{1}{4}\varepsilon$, so the trivial solutions of both the linear and non-linear Mathieu-equation are unstable.

## 11.2 The Mathieu-Equation with Damping

Consider the Mathieu-equation with damping in regions including the trivial solution

$$\ddot{x} + \varepsilon\kappa\dot{x} + (\omega^2 + \varepsilon\cos 2t)x = 0, \tag{11.2}$$

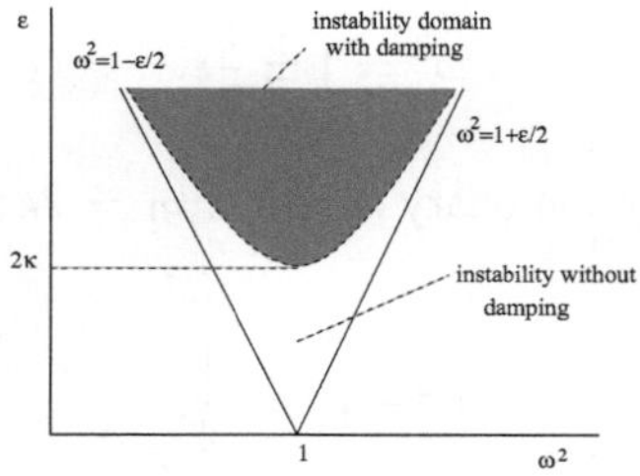


**Fig. 11.1** The $(\omega^2, \varepsilon)$ parameter-space of the damped Mathieu-equation for $\varepsilon$ small (the scaling of the axes is different). The approximate solutions are described by system (11.3). The instability domain emanating from the horizontal axis at $\omega^2 = 1$ in the case without damping is called a Floquet tongue. With damping the tongue is lifted from the horizontal axis

with $\varepsilon\kappa > 0$ the damping coefficient. We need more information about the stability and instability regions. The behaviour of the solutions depends strongly on the parameters; as an example we choose the frequency $\omega$ near 1. This is expressed by putting $\omega^2 = 1+\varepsilon\beta$ with $\beta$ a constant independent of $\varepsilon$. The equation is transformed to amplitude-phase variables by transformation Eq. (1.6) and becomes

$$\begin{aligned}\dot{r} &= \varepsilon(-\kappa r \sin^2(t+\psi) + \beta r \sin(t+\psi)\cos(t+\psi)\\ &\qquad + r\sin(t+\psi)\cos 2t \cos(t+\psi)),\\ \dot{\psi} &= \varepsilon(-\kappa \sin(t+\psi)\cos(t+\psi) + \beta\cos^2(t+\psi) + \cos^2(t+\psi)\cos 2t).\end{aligned}$$

Again the singularity at $r = 0$ is removed. Averaging over time $t$ produces

$$\dot{r} = -\frac{\varepsilon}{2} r(\kappa - \frac{1}{2}\sin 2\psi),\ \dot{\psi} = \frac{\varepsilon}{2}(\beta + \frac{1}{2}\cos 2\psi). \tag{11.3}$$

In the $\omega, \varepsilon$-parameter-space the stable and unstable solutions are located in different domains; see Fig. 11.1. The domains are separated in parameter-space by curves, and as these curves separate stability and instability we infer that the separating curves correspond with the parameters of the periodic solutions.

To find the separating curves we will look for periodic solutions using the equilibria of averaged system (11.3). We assume that such simple and prominent types of periodic solutions are characterised by constant amplitude $r$ and constant phase $\psi$. The equilibria follow from the equations

$$2\kappa = \sin 2\psi,\ 2\beta = -\cos 2\psi.$$

The results from Eq. (11.3) are as follows.

Without damping $\kappa = 0$ choose $\beta = \pm 0.5$. We find $\sin 2\psi = 0$ or $\cos 2\psi = \pm 1$, producing two straight lines in parameter-space.:

$$\omega^2 = 1 \pm \frac{1}{2}\varepsilon.$$

With damping we have for stationary $r_0$: $\sin 2\psi_0 = 2\kappa$; eliminating the phase $\psi_0$ we find

$$\kappa^2 + \beta^2 = \frac{1}{4},$$

and by eliminating $\beta$:

$$\omega^2 = 1 \pm \frac{\varepsilon}{2}\sqrt{1 - 4\kappa^2},\ \kappa^2 \leq \frac{1}{4}. \tag{11.4}$$

In the case of damping the curve separating stability and instability in parameter-space is lifted off the horizontal frequency-axis and stabilises the trivial solution. See Fig. 11.1.

## 11.3 Instability Tongues by the Poincaré-Lindstedt Method

We found that in parameter-space the stable and unstable solutions are separated by curves forming a boundary of the so-called instability tongues. On the separating curves the solutions are periodic. The Poincaré-Lindstedt method should be able to describe these curves. We will use again the variational equations.

In the Introduction, Example 1.4, we discussed the equation by averaging with $\omega = 1$ and excitation frequency 2. We reformulate as

$$\ddot{x} + (\omega^2 + \varepsilon \cos 2t)x = 0.$$

We assume that $\omega$ is $\varepsilon$-close to 1 and that we have a convergent expansion of the frequency by

$$\omega^2 = 1 - \beta_0\varepsilon + O(\varepsilon^2)$$

The period $\pi$ is present by the excitation in the equation. If $\varepsilon = 0$, we give the solution as

$$x|_{\varepsilon=0}(t) = y_1(0)\cos t + y_2(0)\sin t,\ \dot{x}|_{\varepsilon=0}(t) = -y_1(0)\sin t + y_2(0)\cos t$$

Using the variational system with transformation (1.16) we have to first order:

$$\dot{y}_1 = -\varepsilon(\beta_0 - \cos 2t)(y_1(t)\cos t + y_2(t)\sin t)\sin t,$$
$$\dot{y}_2 = -\varepsilon(\beta_0 - \cos 2t)(y_1(t)\cos t + y_2(t)\sin t)\cos t.$$

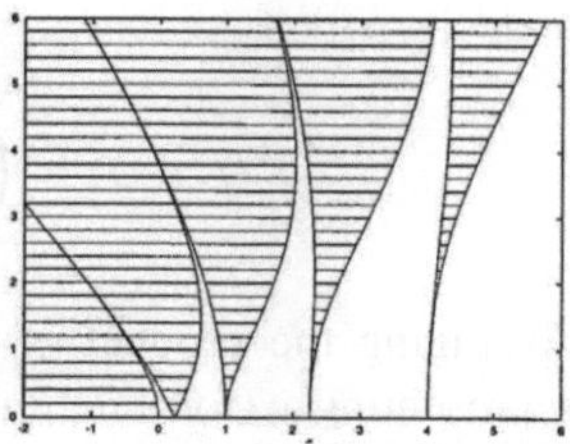

**Fig. 11.2** The instability tongues of the Mathieu-equation $\ddot{x} + (a + \varepsilon \cos t)x = 0$. The vertical left scales $\varepsilon$ from 0 to 6

The equivalent integral equations for $y_1(t)$, $y_2(t)$ produce the periodicity conditions:

$$\begin{cases} \int_0^{2\pi} (\beta_0 - \cos 2s(y_1(s)\cos s + y_2(s)\sin s)\sin s ds &= 0, \\ \int_0^{2\pi} (\beta_0 - \cos 2s(y_1(s)\cos s + y_2(s)\sin s)\cos s ds &= 0. \end{cases} \tag{11.5}$$

The periodicity conditions to first order are

$$y_2(0)\left(\beta_0 + \frac{1}{2}\right) = 0,\ y_1(0)\left(\beta_0 - \frac{1}{2}\right) = 0.$$

We have the boundaries of the instability tongue $\beta_0 = \pm\frac{1}{2}$ as we found by averaging in section 11.2. See Fig. 11.2 for the instability tongues for $\varepsilon$ near zero and to value 6; the white fields correspond with stable parameter values.

A few additional examples can be found in the monographs [130] and [148].

## 11.4 Instability Tongues for Other Excitation Frequencies

Consider the Mathieu-equation with a different excitation frequency:

$$\ddot{x} + (1 + 2\varepsilon \cos \omega t)x = 0,\ \ \omega = 1. \tag{11.6}$$

In slowly varying phase-amplitude variables we find with the transformation (1.6]

$$\begin{cases} \dot{r} &= \frac{\varepsilon}{2} r[\sin(3t + 2\phi) + \sin(t + 2\phi)], \\ \dot{\phi} &= \varepsilon[\cos t + \frac{1}{2}\cos(3t + 2\phi) + \frac{1}{2}\cos(t + 2\phi)]. \end{cases} \tag{11.7}$$

In general averaging over the quasi-periodic right-hand side associated with Eq. (11.6) for which $\omega$ is not close to 2 produces zero. We construct a second-order approximation to look for less trivial results. Using the notation of Theorem 5.1 we have for $Df, u^1, f_1^0$

$$Df(t, r, \phi) = \begin{pmatrix} \frac{1}{2}[\sin(3t + 2\phi) + \sin(t + 2\phi)] & r[\cos(3t + 2\phi) + \cos(t + 2\phi)] \\ 0 & -\sin(3t + 2\phi) - \sin(t + 2\phi) \end{pmatrix},$$

and

$$u^1(t, r, \phi) = \begin{pmatrix} -r[\frac{1}{6}\cos(3t + 2\phi) + \frac{1}{2}\cos(t + 2\phi)] \\ \sin t + \frac{1}{6}\sin(3t + 2\phi) + \frac{1}{2}\sin(t + 2\phi) \end{pmatrix}.$$

Averaging the product of matrix $Df$ and vector $u^1$ produces a system for $\tilde{r}, \tilde{\phi}$ representing $y$ (formulation Theorem 6.1) in expression (6.2):

$$f_1^0(\tilde{r}, \tilde{\phi}) = \begin{pmatrix} -\frac{\tilde{r}}{2}\sin 2\tilde{\phi} \\ -\frac{1}{3} - \frac{1}{2}\cos 2\tilde{\phi} \end{pmatrix}.$$

An $O(\varepsilon)$ approximation valid on the timescale $1/\varepsilon^2$ can be obtained from the system

$$\frac{d\tilde{r}}{dt} = -\varepsilon^2 \frac{\tilde{r}}{2}\sin 2\tilde{\phi}, \ \frac{d\tilde{\phi}}{dt} = \varepsilon^2 (\frac{1}{3} + \frac{1}{2}\cos 2\tilde{\phi}). \tag{11.8}$$

Solving system (11.8) this approximation is the transpose of

$$(\tilde{r}, \tilde{\phi}) + \varepsilon u^1(t, \tilde{r}, \tilde{\phi}). \tag{11.9}$$

**Remark 11.1** The formulation to slowly varying variables by transformation (1.6) excludes a neighbourhood of $r = 0$, but the solutions of system (11.8) suggest instability. Using transformation (1.16) we can show that also in the case of the Mathieu-equation with $\omega = 1$, the origin is unstable.

## 11.5 Interaction with Parametric Excitation

A simplified case of a system of coupled oscillators involving parametric excitation studied in [6] is the system

$$\begin{cases} \ddot{x} + \varepsilon\kappa_1\dot{x} + (1 + \varepsilon\cos 2t)x + \varepsilon(axy + \gamma x^3) & = 0, \\ \ddot{y} + \varepsilon\kappa_2\dot{y} + 4(1 + \varepsilon)y - \varepsilon b x^2 & = 0. \end{cases} \tag{11.10}$$

This is a system with parametric excitation of the non-linear $x$ oscillator and non-linear coupling to a damped $y$ oscillator; $\kappa_1, \kappa_2 \geq 0$ damping coefficients independent of $\varepsilon$. In addition the oscillators contain the 1 : 2 resonance but with small detuning.

If $\kappa_1, \kappa_2 > 0$, the trivial solution is stable, and the $(y, \dot{y})$ coordinate plane contains a damped oscillator. Away from the coordinate planes we may use amplitude-phase variables by transformation (1.6). First-order averaging produces the approximating system

$$\begin{cases} \dot{r}_1 &= \varepsilon r_1 \left( a\frac{r_2}{4} \sin(2\phi_1 - \phi_2) + \frac{1}{4} \sin 2\phi_1 - \frac{1}{2}\kappa_1 \right), \\ \dot{\phi}_1 &= \varepsilon \left( a\frac{r_2}{4} \cos(2\phi_1 - \phi_2) + \frac{1}{4} \cos 2\phi_1 + \frac{3}{8}\gamma r_1^2 \right), \\ \dot{r}_2 &= \varepsilon \frac{r_2}{2} \left( b\frac{r_1^2}{4r_2} \sin(2\phi_1 - \phi_2) - \kappa_2 \right), \\ \dot{\phi}_2 &= \frac{\varepsilon}{2} \left( -b\frac{r_1^2}{4r_2} \cos(2\phi_1 - \phi_2) + 2 \right). \end{cases} \tag{11.11}$$

To study periodic solutions near the coordinate planes (normal modes) we have to use a different coordinate transformation, but to locate a torus containing quasi-periodic solutions we are more interested in general position periodic orbits. In the averaged system we obtained the combination angle $\chi = 2\phi_1 - \phi_2$.

Putting the right-hand sides equal to zero produces various solutions, but as we have five free parameters the linear algebra becomes already complicated. A systematic study in [6] produces stable and unstable periodic solutions.

A simple but rather special case is the choice:

$$a = b = 1, \gamma = -\frac{1}{6}, \kappa_1 = \kappa_2 = 0, \sin\chi = \sin 2\phi_1 = 0,$$

$$\cos\chi = \cos 2\phi_1 = 1, r_1 = 2\sqrt{2}, r_2 = 1.$$

Linearising near this critical point of the vector field of system (11.11) we find the eigenvalues

$$\lambda_{1,2} = \lambda_{3,4} = \pm 0.7071i,$$

This symmetric result is caused by the absence of damping; the flow in the extended phase-space is measure-preserving.

Consider the periodic solution obtained by the choice $a = 1, b = 4, \kappa_1 = \gamma = 0, \kappa_2 = 1$, leading to the critical point given by

$$\sin\chi = -\sin 2\phi_1 = \frac{1}{\sqrt{5}}, \cos\chi = -\cos 2\phi_1 = \frac{2}{\sqrt{5}}, r_1^2 = \sqrt{5}, r_2 = 1.$$

Linearising near this critical point of the vector field of system (11.11) we find the eigenvalues

$$\lambda_{1,2} = 0.09444 \pm 0.9384i, \lambda_3 = -0.0354, \lambda_4 = 0.5615,$$

pointing at complex instability in two dimensions and attraction in two other dimensions. These special cases are instructive but rather laborious, so it is surprising that in [6] the averaging result of system (11.11) can be used to find explicitly cases of periodic solutions with two purely imaginary eigenvalues. Systematically using MATCONT [102] and increasing the coefficients $\kappa_1, \kappa_2$ from small positive values, we find that for critical values of the parameters the real parts of two eigenvalues vanish. The critical point itself corresponds with a periodic solution but the critical imaginary values corresponds with a Hopf bifurcation of the periodic solution . This in its turn corresponds with a torus in the original equations (in $x$ and $y$), which is a Neimark-Sacker bifurcation. For more details see [6] and for this bifurcation Chap. 12.

## 11.6 Flow-Induced Vibrations

In a long series of books (partly published by the Bechovice Research Institute) and articles, A. Tondl, a Czech engineer, constructed models describing non-linear phenomena in engineering problems. One of his many research topics was concerned with the interaction of self-excited and parametric vibrations; see [129]. In a private communication A. Tondl gave in 2002 a model, see Fig. 11.3, to quench undesirable flow-induced vibrations $y_2(t)$. The idea is to quench the flow-induced vibrations of mass $m_2$ by non-linear coupling to two parametrically excited oscillators with masses $m_1$, $m_3$ and deflections $y_1$, $y_3$. The system for the deflections $y_1, y_2, y_3$ is

$$\begin{cases} m_1\ddot{y}_1 + b_0\dot{y}_1 + k_0(1+\varepsilon\cos\omega t)y_1 - k_1(y_2 - y_1) &= 0, \\ m_2\ddot{y}_2 + \beta U^2(1-\gamma\dot{y}_2^2)\dot{y}_2 + 2k_1y_2 - k_1(y_1+y_3) &= 0, \\ m_3\ddot{y}_3 + b_0\dot{y}_3 + k_0(1+\varepsilon\cos\omega t)y_3 - k_1(y_2 - y_3) &= 0. \end{cases} \tag{11.12}$$

All parameters are assumed to be positive, and the flow-induced vibrations of mass $m_2$ with flow strength $U$ are caused by a so-called Raleigh self-excitation term.

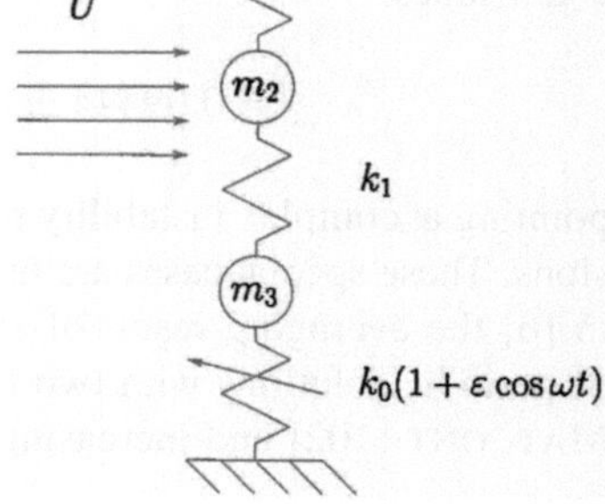


**Fig. 11.3** Vibrations of mass $m_2$ caused by flow $U$ are quenched by two parametric energy absorbers connected to mass $m_2$ with springs and masses $m_1, m_3$. The dynamics is described by system (11.12)

Because of the many parameters the analysis is complicated; it uses MATHEMATICA and MATCONT leading to periodic solutions that by Neimark-Sacker bifurcation evolve to tori. By slowly changing the parameters in this three-mass system one finds break-up of the tori and evolution to strange attractors. This is basically an instructive example of the Ruelle-Takens scenario [118] that sketches the origin of turbulence by such a sequence of bifurcations from unstable periodic solutions to tori to strange attractors. For extensive details of this flow-induced vibration model see [9]. Omitted in this study are details of the case $m_1 = m_3$, but it is shown in [9] that in this special case the averaged system of (11.12) decouples and can be treated as a two-degree of freedom (dof) system.

Instead of giving more details of the complicated but very interesting system (11.12) we consider in the next section a related but simplified system that contains the same elements but has lower dimension.

## 11.7 Interaction of Self-Excited and Parametric Excitation

Consider a system with two oscillators where we intend to quench the second one:

$$\begin{cases} \ddot{x} + \varepsilon\kappa\dot{x} + (1 + \varepsilon\cos 2t)x + \varepsilon c y &= 0, \\ \ddot{y} - \varepsilon\mu(1 - y^2)\dot{y} + y + 3ay^3 + \varepsilon c x &= 0. \end{cases} \tag{11.13}$$

It is discussed in [11] for $a = 0$ and in [12]. We have weak linear interaction between the two oscillators governed by parameter $c$, with parameters $a$ and $\kappa, \mu \geq 0$. The parameter $a$ shifts the phase $\phi_2$ and period of the $y$-oscillator; it will cause phenomena qualitatively different from the case $a = 0$.

General position periodic solutions are candidates to produce tori. Using transformation (1.6) we find variational equations for amplitudes $r_1, r_2$ and phases $\phi_1, \phi_2$. We leave out the variational equations. After averaging over time we find with $\chi = \phi_1 - \phi_2$

$$\begin{cases} \dot{r}_1 &= \frac{\varepsilon}{2}(-\kappa r_1 + \frac{1}{2} r_1 \sin 2\phi_1 + c r_2 \sin\chi), \\ \dot{\phi}_1 &= \frac{\varepsilon}{2r_1}(\frac{1}{2} r_1 \cos 2\phi_1 + c r_2 \cos\chi), \\ \dot{r}_2 &= \frac{\varepsilon}{2}(\mu r_2 (1 - \frac{1}{4} r_2^2) - c r_1 \sin\chi), \\ \dot{\phi}_2 &= \varepsilon\frac{3}{8} a r_2^2 + \frac{\varepsilon}{2} c \frac{r_1}{r_2} \cos\chi. \end{cases} \tag{11.14}$$

We have to discuss neighbourhoods of the normal modes $r_1 = 0$ and $r_2 = 0$ separately.

**We Start with the Case $a = 0$**

The averaged system has a general position critical point from the equation for $\phi_2$ if $\cos\chi = 0$ or $\sin\chi = \pm 1$ from the equation for $\phi_1$. The other conditions for critical points become in this case

$$-\kappa r_1 + \frac{1}{2} r_1 \sin 2\phi_1 + c r_2 \sin\chi = 0,\ \mu r_2 (1 - \frac{1}{4} r_2^2) - c r_1 \sin\chi = 0,\ \cos 2\phi_1 = 0. \tag{11.15}$$

We have for the critical values

$$r_1 = \frac{-c\sin\chi}{-\kappa + 0.5\sin 2\psi_1} r_2,\ r_2 = 2\sqrt{1 + \frac{c^2}{\mu(-\kappa + 0.5\sin 2\psi_1)}},\ \sin 2\psi_1 = \pm 1. \tag{11.16}$$

Non-trivial equilibria can be found for $\sin\chi = 1,\ c > 0,\ \sin 2\phi_1 = \pm 1,\ \kappa > 0$ and $\mu > 0$ large enough. Equation (11.16) becomes in the case $\sin 2\psi_1 = -1$

$$r_1 = \frac{c}{\kappa + 0.5} r_2,\ r_2 = 2\sqrt{1 - \frac{c^2}{\mu(\kappa + 0.5)}}. \tag{11.17}$$

The Jacobian matrix becomes at the non-trivial equilibrium:

$$\begin{pmatrix} -\kappa - \frac{1}{2} & 0 & -c & 0 \\ 0 & \frac{1}{2} - \kappa & 0 & \kappa + \frac{1}{2} \\ c & 0 & \frac{6c^2}{2\kappa - 1} - 2\mu & 0 \\ 0 & -\frac{2c^2}{2\kappa + 1} & 0 & \frac{2c^2}{2\kappa + 1} \end{pmatrix} \tag{11.18}$$

The eigenvalues at the equilibrium can in this case be computed by MATHEMATICA. We find from [11]

$$\lambda_1 = \frac{-\sqrt{-8c^2(4\kappa(\kappa+2)+3) + 16c^4 + \left(1 - 4\kappa^2\right)^2} + 4c^2 - 4\kappa^2 + 1}{8\kappa + 4}, \tag{11.19}$$

$$\lambda_2 = \frac{\sqrt{-8c^2(4\kappa(\kappa+2)+3) + 16c^4 + \left(1 - 4\kappa^2\right)^2} + 4c^2 - 4\kappa^2 + 1}{8\kappa + 4}, \tag{11.20}$$

$$\lambda_3 = \frac{-\sqrt{8c^2(2\kappa - 1)(2\kappa - 12\mu + 5) + 144c^4 + (1 - 2\kappa)^2(2\kappa - 4\mu + 1)^2} + R}{8\kappa - 4}, \tag{11.21}$$

$$\lambda_4 = \frac{\sqrt{8c^2(2\kappa - 1)(2\kappa - 12\mu + 5) + 144c^4 + (1 - 2\kappa)^2(2\kappa - 4\mu + 1)^2} + R}{8\kappa - 4} \tag{11.22}$$

$$R = 12c^2 - 4\kappa^2 - 8\kappa\mu + 4\mu + 1. \tag{11.23}$$

From equations (11.19) and (11.20) we conclude that the real part (i.e. the part outside the square root symbol) of $\lambda_1$ and $\lambda_2$ becomes zero in two cases:

$$c(\kappa) = 1/2\sqrt{-1+4\kappa^2},\ c(\kappa,\mu) = \frac{1}{2}\sqrt{(1+4\kappa+4\mu+4\kappa^2+8\upsilon\kappa\mu)/3},$$

with $\kappa \geq \frac{1}{2}$. The term under the square root in the first case is negative for $\kappa > \frac{1}{2}$ so that the parameter curve $c(\kappa) = 1/2\sqrt{-1+4\kappa^2}$ is a Hopf curve in the averaged system and consequently a Neimark-Sacker bifurcation in the original system, leading to torus dynamics. For this type of bifurcation see again Chap. 12. Along this curve the eigenvalues $\lambda_1$ and $\lambda_2$ are purely imaginary. Adding a small perturbation $\delta \ll 1$ to $c$ yields a transversal crossing of the imaginary axis and definitely a Hopf bifurcation. Remarkably this curve is independent of the self-excitation parameter $\mu$. An example and figures can be found in [11]. The use of the expression for $c(\kappa)$ leads to a stable torus; the use of the expression for $c(\kappa,\mu)$ also leads to a Neimark-Sacker curve in parameter-space but an unstable torus.

A second equilibrium arises if $\sin\chi = 1, \sin 2\phi_1 = 1$. We leave out the calculations of eigenvalues and Neimark-Sacker curve in this case as they lead to an unstable torus; see [11]. For reference we summarise some results in this case:

$$r_1 = \frac{c}{\kappa - 0.5} r_2,\ r_2 = 2\sqrt{1+\frac{c^2}{\mu(-\kappa+0.5)}}. \tag{11.24}$$

For the Neimark-Sacker curve in parameter-space we find

$$c(\kappa,\mu) = \frac{1}{2}\sqrt{(2\kappa-1)(2\kappa+4\mu-1)/3},$$

**The Case $a \neq 0$ and the Normal Modes**

This part is based on [12]. Involving the cubic term $y^3$ in the equation for $y$ produces richer dynamics. Although the analytic computation is more complicated we can repeat the calculation to obtain from system (11.14) the critical points that correspond with periodic solutions of the original equation. It is shown in [12] that taking the limit $a \to 0$ we obtain the same critical points as before. Also that transforming to co-varying variables (1.17) and averaging, the normal modes do not exist. Inspired by the averaging results one can start with numerical bifurcation analysis.

From system (11.14) we find two critical points corresponding with periodic solutions if

$$a = 2.4357, c = 1.61, \kappa = 0.25, \mu = 4.$$

There are many more solutions, but instead of considering the analytic expressions we use MATCONT by continuation to find the corresponding amplitude $r_1$ when

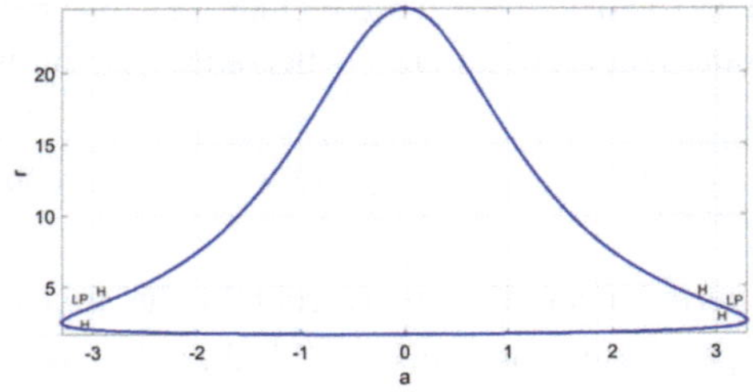


**Fig. 11.4** Bifurcation picture obtained by changing parameter $a$ to produce the corresponding amplitudes $r_1$ of system (11.14). For $-a_c < a < a_c$ we find two $r_1$ values; at the points $a = \pm a_c = \pm 3.2969$ the corresponding two periodic solutions collapse in a fold bifurcation indicated by LP and disappear. Courtesy Springer [12]

varying $a$. The result is shown in Fig. 11.4 where at critical points $\pm a_c$ the periodic solutions vanish. In the points H in Fig. 11.4 a Hopf bifurcation of the periodic solutions (Neimark-Sacker bifurcation) takes place corresponding with stable and unstable tori. As mentioned above these results can be obtained both analytically and by MATCONT numerically; hybrid combination is efficient. The critical value $a_c$ from averaging is

$$a_c = \frac{2c^2\mu}{3\sqrt{(4k^2-1)(-8c^2\kappa\mu + 4c^4 + (4k^2-1)\mu^2)}}. \tag{11.25}$$

In [12] the presence of Neimark-Sacker bifurcations (see Fig. 11.4) suggested to continue parameters starting at these points. This continuation resulted in finding stable $\mathbb{T}_2$ and $\mathbb{T}_3$ tori.

## 11.8 Autoparametric Energy Absorbers

In dynamical systems, in general in mathematics, one is interested in vibrational phenomena but in actual engineering problems, we may wish to diminish or even extinguish the vibration amplitudes of a system (called the Primary System). Think of two examples of Primary System, a sea-going vessel that is excited by the waves or machinery on a foundation that is slightly excited by neighbouring machinery. The dynamics is interesting but in real life undesirable.

A straightforward technique is using energy absorption by using dampers, but they have to be replaced often and sometimes ‘quenching of undesirable vibrations’ can be achieved by coupling the Primary System to a Secondary System that absorbs the energy. A systematic description of such quenching models using the theory of autoparametric systems can be found in [130]. As the calculations are long we present in what follows only a sketch of the results from [130]; this will give an impression of these interesting engineering problems.

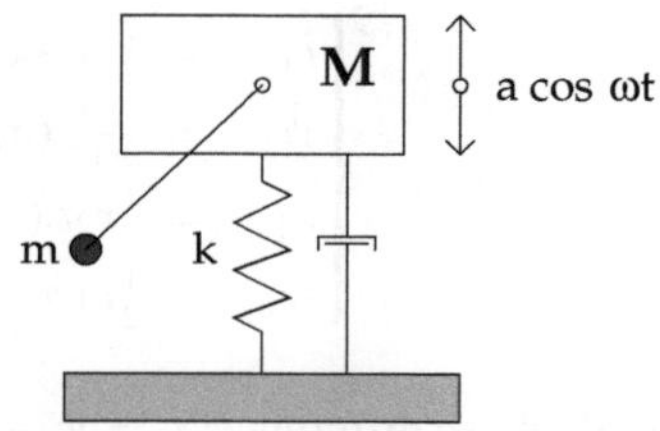


**Fig. 11.5** Two coupled oscillators with vertical oscillations as Primary System and parametric excitation of the coupled pendulum (Secondary System). Courtesy A. Tondl

## *11.8.1 Energy Absorption by an Attached Pendulum*

A typical example of autoparametric excitation is shown in Fig. 11.5. The Primary System that we would like to be stationary is the mass $M$ that rests on an elastic foundation and is forced vertically by oscillations with amplitude $a$ and frequency $\omega$. We attach to the mass $M$ a pendulum with mass $m$ as a Secondary System. The vertical (normal mode) vibrations of the Primary System with mass $M$ can (hopefully) act as parametric excitation of the (pendulum) Secondary System which will no longer remain at rest. Such parametric excitation will take place if the parameters of the vertical normal mode of the Primary System are in the instability (parameter) intervals of this normal mode solution in the full, coupled system.

The deflections of mass $M$ in the vertical direction will be indicated by $y$; for $y = 0$ the mass will be at rest. The pendulum deflections are indicated by the angle $\phi$ that the pendulum makes with the vertical axis. Following [130] with some rescaling we formulate the equations of motion for small deflections and assumptions on the frequencies (with small detunings $\sigma_1, \sigma_2$) of the vertical oscillations and the pendulum motion:

$$\begin{cases} \ddot{y} + y &= -\varepsilon(\kappa_1 \dot{y} + \sigma_1 y - \frac{1}{4}\phi^2 + \mu\dot{\phi}^2 + a\cos t), \\ \ddot{\phi} + \frac{1}{4}\phi &= -\varepsilon(\kappa_2 \dot{\phi} + \frac{1}{2}\sigma_2\phi - y\phi). \end{cases} \tag{11.26}$$

We have the positive damping coefficients $\kappa_{1,2}$, the detunings $\sigma_{1,2}$ of the basic frequencies 1, 1/2 and forcing amplitude $a > 0$, also parameter $\mu > 0$. This type of spring pendulum is a classical problem that was studied by many authors; see for references [130]. We have the normal mode solution $\phi = \dot{\phi} = 0$ corresponding with forced linear $y$-oscillations of the mass $M$. We can solve the equation to find a periodic normal mode solution:

$$y_n(t) = -a\frac{\sigma_1}{\kappa_1^2 + \sigma_1^2}\cos t - a\frac{\kappa_1}{\kappa_1^2 + \sigma_1^2}\sin t, \tag{11.27}$$

We introduce slowly varying variational variables by transformation (1.6) where $r_1$ is associated with the deflection $y$, $r_2$ with $\phi$. First-order averaging produces for general position orbits

$$\begin{cases} \dot{r}_1 & = \frac{1}{2}\varepsilon(-\kappa_1 r_1 - \frac{1}{8}(1+\mu)r_2^2 \sin\chi + a\sin\psi_1), \\ \dot{\psi}_1 & = \frac{1}{2r_1}\varepsilon(\sigma_1 r_1 - \frac{1}{8}(1+\mu)r_2^2 \cos\chi + a\cos\psi_1), \\ \dot{r}_2 & = \frac{1}{2}r_2\varepsilon(-\kappa_2 + r_1\sin\chi), \\ \dot{\psi}_2 & = \frac{1}{2}\varepsilon(\sigma_2 - r_1\cos\chi), \end{cases} \tag{11.28}$$

with phase difference $\chi = \psi_1 - 2\psi_2$. Putting $\phi = \dot{\phi} = 0$ in system (11.26) we obtain after averaging the first two equations of system (11.28) with $r_2 = 0$. Consider now system (11.28) with $r_2 > 0$. Looking for critical points we find from the last two equations

$$\sin\chi = \frac{\kappa_2}{r_1}, \cos\chi = \frac{\sigma_2}{r_1}, \; r_1 = \sqrt{\kappa_2^2 + \sigma_2^2}. \tag{11.29}$$

From the first two equations we find after some manipulations a quadratic equation for $r_2^2$:

$$(\kappa_1 r_1^2 + \frac{\kappa_2}{8}(1+\mu)r_2^2)^2 + (\sigma_1 r_1^2 - \frac{\sigma_2}{8}(1+\mu)r_2^2)^2 - a^2 r_1^2 = 0.$$

or

$$\frac{1}{64}(1+\mu)^2 r_2^4 + \frac{1}{4}(\kappa_1\kappa_2 - \sigma_1\sigma_2)(1+\mu)r_2^2 + (\kappa_1^2 + \sigma_1^2)(\kappa_2^2 + \sigma_2^2) - a^2 = 0.$$

One can easily deduce from this equation conditions on the parameters to have no, one or two positive solutions for $r_2$. For instance if $(\kappa_1^2+\sigma_1^2)(\kappa_2^2+\sigma_2^2) - a^2 > 0$ and $\kappa_1\kappa_2 - \sigma_1\sigma_2 > 0$, there are no positive solutions. For these details and the stability study we refer the reader to section 4.2 of [130].

### 11.8.2 Quenching Raleigh Self-Excitation

The Raleigh equation arises in system (11.30) below if $a = 0$; the equation is often used in models for flow-induced vibrations. The equation is equivalent to the Van der Pol equation which can be shown by differentiating the equation and replacing $\dot{x}$ by $z$. In system (11.30) without interaction ($a = 0$) the equation has a stable limit cycle with amplitude $2/\sqrt{3} + O(\varepsilon)$. Suppose we want to reduce the Raleigh amplitude. Consider the Rayleigh self-excited equation in $x$ coupled to a non-linear equation in $y$ with the purpose to absorb energy from the self-excited system; the interaction is derived from a potential. We have after rescaling the system near equilibrium

$$\begin{cases} \ddot{x} + x & = \varepsilon\mu(1 - \dot{x}^2)\dot{x} - \varepsilon a y^2, \\ \ddot{y} + q^2 y & = -\varepsilon\kappa\dot{y} - \varepsilon 2axy. \end{cases} \tag{11.30}$$

The parameters $\mu, \kappa$ are positive; $a$ is given; and $q$ and $b$ have to be chosen suitably. The analysis is partly based on [1]. Transforming to amplitude-phase variables by (1.8) $x = r_1 \cos(t + \phi_1)$, $y = r_2 \cos(qt + \phi_2)$, etc. we note that a first-order result in $\varepsilon$ will be non-trivial if the frequency ratios are close to 1 : 2. We choose $q^2 = 1/4 + \varepsilon\sigma$. Interesting detuning bifurcations arise when $\sigma \neq 0$; these phenomena are studied in [1]. Here we will consider exact resonance, $\sigma = 0$. Transforming and averaging we have

$$\begin{cases} \dot{r}_1 &= \varepsilon \frac{\mu}{2} r_1 (1 - \frac{3}{4} r_1^2) + \varepsilon \frac{a}{4} r_2^2 \sin\chi,\ \dot{\phi}_1 = \varepsilon \frac{a}{4} \frac{r_2^2}{r_1} \cos\chi, \\ \dot{r}_2 &= -\frac{\varepsilon}{2} r_2 (\kappa + 2 a r_1 \sin\chi),\ \dot{\phi}_2 = \varepsilon a r_1 \cos\chi. \end{cases} \tag{11.31}$$

with $\chi = \phi_1 - 2\phi_2$. We have

$$\dot{\chi} = \varepsilon a \left( \frac{r_2^2}{4 r_1} - r_1 \right) \cos\chi .$$

A general position orbit has constant combination angle $\chi_0$ if $r_2(t)^2 = 4 r_1(t)^2$, $t \geq 0$ or if $\chi = \pi/2, 3\pi/2$. Looking for constant amplitude periodic solutions we have to omit $\chi = \pi/2$; we find $r_1 = \kappa/2a$. Substitution into system (11.31) we find

$$r_2^2 = \mu \frac{\kappa}{a^2} \left( 1 - \frac{3\kappa^2}{16 a^2} \right). \tag{11.32}$$

The resulting dynamics for $a = 0$ and $a = 1$, $\chi(0) = 3\pi/2$ is illustrated in Fig. 11.6. Consider the case $r_2 = 2r_1$. From the equation for $r_2$ in system (11.31) we find $\sin\chi_0 = -\kappa/(2 a r_1)$. Substitution into the right-hand side of the equation for $r_2$ and looking for a critical point we find

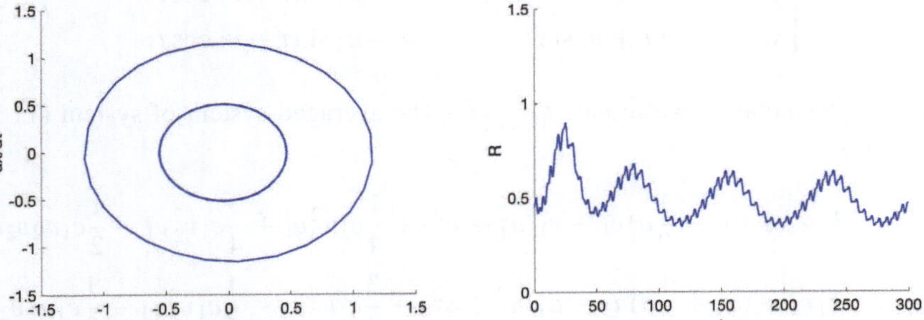


**Fig. 11.6** The $x, \dot{x}$ phase-plane of the Raleigh self-excited part of system (11.30). We have $\mu = \kappa = 1, \varepsilon = 0.1$. The large cycle for $a = 0$ is the Raleigh limit cycle with $r_1 = 2/\sqrt{3} + O(\varepsilon)$. For $a = 1$ we have autoparametric energy absorption leading to $r_1 = 0.5 + O(\varepsilon)$ (small cycle); the initial values are $x(0) = 0, \dot{x}(0) = 0.5, y(0) = \sqrt{13}/4, \dot{y}(0) = 0, \chi(0) = 3\pi/2$. Right: we show $R(t) = \sqrt{\dot{y}^2 + y^2/4}$ in the case of coupling ($a = 1$)

$$r_1 = \frac{2}{\sqrt{3}}\sqrt{1 - \frac{\kappa}{\mu}},\ r_2 = 2r_1,\ 0 < \frac{\kappa}{\mu} < 1. \tag{11.33}$$

The numerics of this case suggests instability. The dynamics of system (11.30) shows many interesting complications that are omitted here; see [1].

### 11.8.3 Autoparametric Excitation with Complex Dynamics

As an instructive but more complicated example we consider briefly the following autoparametric system studied in [54]:

$$\begin{aligned} x'' + x + \varepsilon(k_1 x' + \sigma_1 x + a\cos 2\tau x + \frac{4}{3}x^3 + c_1 y^2 x) &= 0 \\ y'' + y + \varepsilon(k_2 y' + \sigma_2 y + c_2 x^2 y + \frac{4}{3}y^3) &= 0 \end{aligned} \tag{11.34}$$

where $\sigma_1$ and $\sigma_2$ are the detunings from the $1:1$ resonance of the two oscillators. In this system $y(t) = y'(t) = 0$ corresponds with a normal mode of the $x$-oscillator (Primary System), and the question is whether non-trivial $y$-oscillation can reduce the oscillations of the Primary System. The system is rather artificial but contains many interesting phenomena that are typical for autoparametric interaction.

System (11.34) is invariant under $(x, y) \to (x, -y)$, $(x, y) \to (-x, y)$, and $(x, y) \to (-x, -y)$.

Using averaging we can investigate the stability of solutions of system (11.34) following [54] in more detail. Introduce the transformation leading to variational systems suitable for non-autonomous systems (1.16):

$$\begin{cases} x = u_1\cos t + v_1 \sin t \quad ; \quad x' = -u_1 \sin t + v_1 \cos t \\ y = u_2 \cos t + v_2 \sin t \quad ; \quad y' = -u_2 \sin t + v_2 \cos t. \end{cases} \tag{11.35}$$

For simpler notation we rescale $t = \frac{\varepsilon}{2}\tilde{t}$ . The averaged system of system (11.34) becomes

$$\begin{aligned} u_1' &= -k_1 u_1 + (\sigma_1 - \frac{1}{2}a)v_1 + v_1(u_1^2 + v_1^2) + \frac{1}{4}c_1 u_2^2 v_1 + \frac{3}{4}c_1 v_2^2 v_1 + \frac{1}{2}c_1 u_2 v_2 u_1 \\ v_1' &= -k_1 v_1 - (\sigma_1 + \frac{1}{2}a)u_1 - u_1(u_1^2 + v_1^2) - \frac{3}{4}c_1 u_2^2 u_1 - \frac{1}{4}c_1 v_2^2 u_1 - \frac{1}{2}c_1 u_2 v_2 v_1 \\ u_2' &= -k_2 u_2 + \sigma_2 v_2 + v_2(u_2^2 + v_2^2) + \frac{1}{4}c_2 u_1^2 v_2 + \frac{3}{4}c_2 v_1^2 v_2 + \frac{1}{2}c_2 u_1 v_1 u_2 \\ v_2' &= -k_2 v_2 - \sigma_2 u_2 - u_2(u_2^2 + v_2^2) - \frac{3}{4}c_2 u_1^2 u_2 - \frac{1}{4}c_2 v_1^2 u_2 - \frac{1}{2}c_2 u_1 v_1 v_2. \end{aligned} \tag{11.36}$$

Although there are many parameters involved this system can be analysed for critical points, periodic and quasi-periodic solutions, producing existence of periodic solutions and detailed stability diagrams in parameter-space; see [54]. A criterion for energy absorption by the $y$-oscillator is that the normal mode $x$-oscillator (the Primary System) is unstable. In system (11.36) this normal mode corresponds with $u_2 = v_2 = 0$. To find periodic solutions we have to compute critical points of the equations for $u_1, v_1$. It turns out that to find a critical point $u_1 = \alpha, v_1 = \beta$ we have the condition $a > 2k_1$. The stability of the normal mode $(u_1, v_1, u_2, v_2) = (\alpha, \beta, 0, 0)$ is obtained by computing the Jacobian of the right-hand side of system (11.36) and determining the eigenvalues in the critical point. It turns out there exist open sets of parameter values of damping, detuning and parametric excitation where the normal mode is unstable. This enables quenching of the $x$-oscillator by autoparametric coupling to the $y$-oscillator.

System (11.34) has also been analysed by the numerical bifurcation program CONTENT, a routine followed up later by MATCONT. Interestingly, it is shown in [54] that the system contains a sequence of period doubling bifurcations leading to chaotic solutions.

# Chapter 12
# Interactions, Bifurcations and Chaos

**Abstract** Bifurcations and chaos can be found in autonomous three- and four-dimensional systems. Concepts like hyperbolicity, normal hyperbolicity and fractal dimension of chaotic sets are important tools to study such systems. A relatively easy way to determine a fractal dimension is the Kaplan-Yorke calculation that uses Lyapunov exponents; we discuss the basic drawback of these calculations. It is remarkable that KAM theory can be extended to dissipative systems in cases with symmetries. We demonstrate this for two different systems of the Jafari-Sprott-Golpayegani project. Torus bifurcations and chaos can be demonstrated in several coupled second-order systems involving parametric and self-excited systems

In this chapter we will consider systematically three- and four-dimensional non-linear systems together with the necessary theoretical background. A reader with basic papers on chaotic attractors is [80]; it contains the theory as it was developed until around 1990 and numerous illustrative examples that inspired scientists in the starting period.

In many physical problems and in particular in engineering, non-linear coupled oscillations with dissipation of energy play an important role. The modelling of these phenomena leads often to complex dynamical systems where various parameters play a part. When studying dynamical systems like differential equations or maps, a first step is to locate and characterise the classical ingredients of such systems. These ingredients are critical points (equilibrium solutions), periodic solutions, invariant manifolds (in particular tori containing quasi-periodic solutions), homoclinics, heteroclinics, fractal manifolds arising in chaotic dynamics and in general stable and unstable manifolds emanating from special solutions. We will consider aspects of the theory needed for the applications; see also [66] and [146] for older introductions directed at physics and engineering. An excellent introduction in modern mathematical language is [28], very instructive with many examples is [109].



F. Verhulst, *Averaging for Nonlinear Dynamics with Applications and Numerical Bifurcations*, Applied Mathematical Sciences 223,
https://doi.org/10.1007/978-3-032-12745-7_12

**Complex Dynamics**
Two-dimensional ordinary differential equations (ODEs) with the usual smoothness assumptions produce a two-dimensional phase-plane that may be complicated but excludes tori, chaos and complex attracting manifolds. For complexity we have to go to at least three-dimensional autonomous ODEs or two-dimensional non-autonomous ODEs (forced or parametrically excited). The analysis and visualisation of such higher-dimensional problems is not easy. It helps if one can identify for such problems a map that characterises the dynamics at a lower dimension. An example is the Poincaré map; if one can construct a transversal of the flow of a system with a certain recurrence, the flow induces a map of the transversal into itself. A periodic solution will be a fixed point of the map; a torus surrounding a periodic solution will appear as a closed cycle. For a three-dimensional autonomous system the Poincaré map is two-dimensional.

This idea has produced an enormous amount of interesting papers on typical maps, in particular two-dimensional ones; for instance the Hénon, Rössler and Thom maps; see [28]. Studying such maps gives deep insight, but it is still a non-trivial matter to associate the maps with physical problems and actual ODEs.

This chapter starts with a few remarks on periods of two-dimensional autonomous ODEs (Sect. 12.1) as they are often ingredients of higher-dimensional systems. The chapter contains applications to three-dimensional problems with partly examples of the Jafari-Sprott-Golpayegani project (Sects. 12.5–12.12), followed by problems with more than three dimensions from Sect. 12.13 onwards, interspersed with necessary pieces of theory.

## 12.1 Periodic Solutions of Two-Dimensional ODEs

Reducing a perturbed ODE by putting $\varepsilon = 0$ often leads to finding families of periodic solutions in conservative two-dimensional equations of the form

$$\ddot{q} + f(q) = 0. \tag{12.1}$$

Suppose $f(0) = 0$ and that the origin of the phase-plane, $q(0) = \dot{q}(0) = 0$, is a Lyapunov-stable equilibrium. With

$$F(q) = \int_0^q f(s)ds, \ \frac{1}{2}\dot{q}^2 + F(q) = E,$$

we have the first integral; the constant $E$ is usually called 'energy'. Any periodic orbit crosses the $q$-axis precisely at two points, $q_1, q_2$. As the integral is even in $\dot{q}$ the period $T$ of this periodic solution is

$$T = 2\int_{q_1}^{q_2} \frac{dq}{\sqrt{2(E - F(q))}}. \tag{12.2}$$

If, for instance, $F(q)$ is a quadratic expression, the integral leads to elliptic functions as we saw in Chap. 1. We consider first two examples of Eq. (12.1). The equations are

$$\ddot{q} + q + aq^3 = 0, a = \pm 1.$$

We have $F(q) = \frac{1}{2}q^2 + \frac{a}{4}q^4$.

The cycles in the phase-plane corresponding with periodic solutions are shown in Fig. 12.1. Near the origin the period $T$ of the periodic solutions is close to $2\pi$ in both cases. In the case $a = 1$ (left) the origin is the only equilibrium. In this case increasing $q(0)$ (or $E$) shortens the period $T$. In the case $a = -1$ (right) we have two more equilibria, saddles, at $q(0) = \pm 1, \dot{q}(0) = 0$. Increasing $q(0)$ from zero to 1 increases the period without bound.

Information about the periods and monotonicity of solutions of Eq. (12.1) is important when studying interactions with other oscillators and regarding resonance in the case of forcing. A few remarks on the literature of these questions are given in [40]; one of their results is that if $f(q)$ in Eq. (12.1) is quadratic the period function (of parameter $q(0)$ or $E$) is always monotonic. If $f(q)$ is cubic, there are at most three equilibria, and it can be shown that the corresponding period functions (of parameter $q(0)$ or $E$) have at most three critical points that may change monotonicity.

**Example: Bifurcations of the Josephson Equation**

The Josephson effect is the macroscopic quantum phenomenon arising when two superconductors with a small barrier in between experience current and voltage across the weak link. The basic equation is

$$\beta\ddot{\phi} + (1 + \gamma \cos\phi)\dot{\phi} + \sin\phi = \alpha, \tag{12.3}$$

$0 \leq \phi \leq 2\pi$ with $\alpha, \beta, \gamma$ parameters; see [122]. A suitable rescaling produces the perturbed pendulum equation

$$\ddot{\phi} + \sin\phi = \varepsilon(a - (1 + \gamma \cos\phi)\dot{\phi}), \tag{12.4}$$

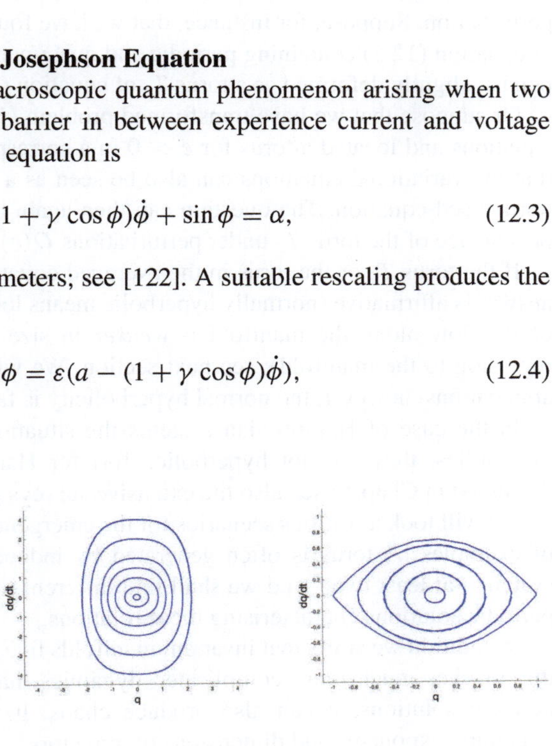

**Fig. 12.1** Phase-plane with periodic solutions of eq. $\ddot{q} + q + aq^3 = 0$. Left: the case $a = 1$. Right: the case $a = -1$ producing two saddle equilibria at $q = \pm 1, \dot{q} = 0$ with the two heteroclinic connections

with $a, \gamma$ parameters and a small forcing. One expects that the Josephson effect is connected with the limit cycles of Eq. (12.4). Using the solutions of the mathematical pendulum equation we can formulate variational equations as in Chap. 1. The averaging is then based on complete elliptic integrals, but this is technically complicated and less transparent. A special technique avoids this; see [122]; using the energy variations of the perturbation equation one analyses the integrals appearing in the averaged equation. This leads to saddle connections in the phase-plane and critical points of the averaged equations corresponding with limit cycles (stable and unstable periodic solutions) of Eq. (12.4). Monotonicity arguments for the periods play a part. Many phase-plane pictures were presented in [122].

## 12.2 Hyperbolic and Normally Hyperbolic Manifolds

Consider a perturbation problem like

$$\dot{x} = f(x) + \varepsilon R(t, x, \varepsilon) \tag{12.5}$$

where as usual $\varepsilon$ will indicate a small, positive parameter and $R$ represents a smooth perturbation. Suppose, for instance, that we have found an isolated torus $T_0$ if $\varepsilon = 0$ in equation (12.5) containing periodic and quasi-periodic solutions. Does this torus persist, slightly deformed as a torus $T_\varepsilon$ of equation (12.5)?

Or suppose that we have transformed problem (12.5) by introducing variational equations and located a torus for $\varepsilon > 0$ in a corresponding averaged system. Note that the variational equations can also be seen as a perturbation of an averaged or normalised equation. The question can then again be rephrased as the question of persistence of the torus $T_\varepsilon$ under perturbations $O(\varepsilon)$.

If the torus $T_0$ or the torus in the averaged equation is *normally hyperbolic*, the answer is affirmative (normally hyperbolic means loosely speaking that the strength of the flow along the manifold is *weaker* in size than the rates of attraction or repulsion to the manifold); see next section. We will discuss such cases. In many applications, however, the normal hyperbolicity is far from easy to establish.

In the case of Hamiltonian systems the situation is different as the tori arise in families, they are not hyperbolic. Tori for Hamiltonian problems have been discussed in Chap. 9; see also the extensive surveys [25, 27, 28, 72] and [163].

We will look at various scenarios for the emergence of tori and chaos in a number of examples. A torus is often generated by independent rotational motions in a system—at least two—and we shall find different timescales and quasi- or almost-periodic solutions characterising these rotations.

In addition we stress that invariant manifolds like tori can also show bifurcations; this invokes much more complicated dynamics than bifurcations of equilibria or periodic solutions, it can also produce chaos. It will be necessary to consider Lyapunov exponents and dimensions of attractors.

Consider the dynamical system in $\mathbb{R}^n$, described by the equation

$$\dot{x} = f(x)$$

and assume that the system contains a smooth ($C^r$) invariant manifold $M$ with as important example a torus. The smoothness enables us to define a *tangent* bundle $T(M)$ describing the flow along $M$ and a *normal* bundle $N(M)$ of $M$, describing the perpendicular flow to $M$. A typical situation in mechanics involves $N$ coupled two-dimensional oscillators containing a $m$-dimensional torus with $2 \leq m \leq N$ (the system is described by $N$ amplitudes and $N$ angles). In this case the phase-space dimension is $n = 2N$, the tangent bundle is $m$-dimensional and the normal bundle is $(2N - m)$-dimensional.

*Hyperbolicity* is introduced as follows: Assume that we can split the corresponding normal bundle of $M$ with respect to the flow generated by the dynamical system completely in an exponentially stable one $N(M)^s$ and an exponentially unstable one $N(M)^u$, with no other components.

In this case the manifold $M$ is called hyperbolic. If this hyperbolic splitting does not contain an unstable manifold $N(M)^u$, $M$ is stable. For a more detailed discussion of these classical matters see, for instance, Hirsch, Pugh and Shub [77].

Note that the smoothness of $M$ is needed in this description. In many cases the manifolds under consideration loose smoothness at certain bifurcation points when varying parameters. In most cases Lyapunov exponents can still be used to characterise the stability.

The manifold $M$ is moreover *normally hyperbolic* if, measured in the matrix- and vector norms in $\mathbb{R}^n$, $N(M)^u$ expands more sharply than the flow associated with $T(M)$ and $N(M)^s$ contracts more sharply than $T(M)$ under the flow.

Often in the literature the concept of normal hyperbolicity is used without explicit definition or even mentioning the term, but it will be implicitly present in the conditions.

In a number of applications the situation is simple because a small parameter is present which induces slow and fast dynamics in the dynamical system. Consider the system

$$\begin{cases} \dot{x} &= \varepsilon f(x, y), \quad x \in D \subset \mathbb{R}^n, t \geq 0 \\ \dot{y} &= g(x, y), \quad y \in G \subset \mathbb{R}^m \end{cases} \tag{12.6}$$

with $f$ and $g$ sufficiently smooth vector functions in $x, y$. Putting $\varepsilon = 0$ we have $x(t) = x(0) = x_0$ and from the second equation $\dot{y} = g(x_0, y)$. For $\varepsilon > 0$ and small we expect $x(t)$ to be slowly varying. Different things happen in the region where $g(x, y)$ vanishes or is $O(\varepsilon)$. An isolated root $y(x)$ of the equation $g(x, y) = 0$ corresponding with a compact manifold will be called a slow manifold. Fenichel (1971-79) has shown that if this root is hyperbolic with $x$ considered a parameter, it corresponds with a nearby hyperbolic invariant manifold of the full system. We can look for periodic and quasi- periodic solutions or tori on a slow manifold. For more explanation and references see [148].

In the analysis of system (12.6), the fact that if this root is hyperbolic, the corresponding manifold is also normally hyperbolic and is inherent in the problem formulation. The fibres of the slow manifold are ruled by the fast time variable $t$, whereas the dynamics of the drift along the slow manifold is ruled by the timelike variable $\varepsilon t$.

## 12.3 Bogoliubov-Mitropolsky-Hale Tori by Continuation

Periodic solutions are interesting and the emphasis on them in the literature is justified. However, quasi-periodic solutions that usually organise in tori are also a very general phenomenon in non-linear dynamics.

The branching off of tori is more complicated than the emergence of periodic solutions in dynamical system theory. For reference we mention briefly that emergence of tori was considered extensively by Bogoliubov and Mitropolsky [20], using basically continuation of quasi-periodic motion under perturbations. Another survey together with new results can be found in Hale [71] and [69]. A formulation in the more general context of bifurcation theory can be found in Chow and Hale [41].

Technically these results have a complicated formulation because of the aspiration to obtain general theorems. When distinguishing between perturbed and unperturbed parts of the problems it simplifies if one assumes constant frequency vectors in the unperturbed problem.

## 12.4 Tori Created by Neimark-Sacker Bifurcation

An important and practical scenario to create a torus arises from Neimark-Sacker bifurcation. As a periodic solution may be produced from a bifurcation of an equilibrium with two double imaginary eigenvalues, in a similar way a periodic solution with two double imaginary eigenvalues may produce a torus. This is the case of a Neimark-Sacker bifurcation, also called Hopf-Hopf bifurcation. For an instructive and detailed introduction see [94]).

Suppose that we have obtained an averaged equation $\dot{x} = \varepsilon f(x, a)$, with dimension 3 or higher, by variation of constants and subsequent averaging; $a$ is a parameter or a set of parameters. It is well known, see Chap. 3, that if this equation contains a critical point, the original equation may contain nearby a periodic solution. The first-order approximation of this periodic solution is characterised by the timelike variables $t$ and $\varepsilon t$.

Suppose now that by varying the parameter $a$ a pair of eigenvalues of the critical point becomes purely imaginary for $a = a_0$. The variation of $a$ through $a_0$ should be transversal. For this value of $a_0$ the averaged equation undergoes a Hopf bifurcation, producing a periodic solution of the averaged equation; the typical timelike variable of this periodic solution is $\varepsilon t$ and so the period will be $O(1/\varepsilon)$. As it branches off an existing periodic solution in the original equation it will produce a torus; it is associated with a Hopf bifurcation of the corresponding Poincaré map, and the bifurcation has a different name: Neimark-Sacker bifurcation. The result in the system will be a two-dimensional torus which contains two-frequency oscillations, one with a timescale of order 1 and the other a timescale $O(1/\varepsilon)$.

## 12.5 The Forced Pendulum, Conservative and Damped

A classical problem of mechanics is the mathematical pendulum with forcing described by the equation:

$$\ddot{\phi} + \kappa\dot{\phi} + \omega^2 \sin\phi = a\cos\Omega t. \tag{12.7}$$

The equation models the motion of a point mass in a vertical plane suspended by a string fixed at the top and making an angle $\phi$ with the vertical with an external forcing; $\kappa \geq 0$ is the damping coefficient, $\omega^2, a > 0$ and $\Omega$ is the frequency of the forcing. The model is an iconic example in physics and mathematics studied with many aspects. If $\kappa = 0$, the system is conservative and contains Hamiltonian chaos, near $\phi = \dot{\phi} = 0$ KAM theory applies. If $\kappa > 0$, topological theory can be used to show the presence of periodic solutions; see for instance [60] and [29]. Such qualitative results are important, but usually they give no information about stability and normal hyperbolicity of the solutions. In Fig. 12.2 we expect to display periodic solutions and tori, but proofs for such explicit examples have to be worked out in detail to obtain confirmation. The difference between the two figures is only $\kappa = 0$, left, and $\kappa = 0.1$, right; the qualitative and quantitative differences are striking.

Note that the parameter values admit averaging as we consider a neighbourhood of the origin. Taylor-expand $\sin\phi = \phi - \frac{1}{6}\phi^3 + O(\phi^5)$, rescale $\phi, \dot{\phi} \to \varepsilon\phi, \varepsilon\dot{\phi}$ and replace $\kappa, a$ by $\varepsilon\kappa, \varepsilon a$. We obtain the equation

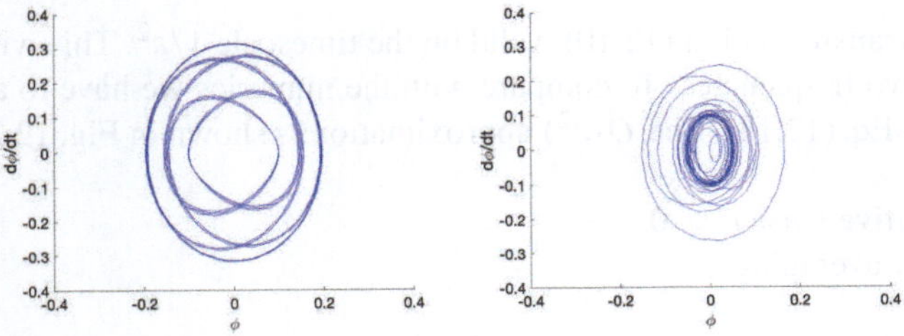


**Fig. 12.2** The mathematical pendulum with forcing described (numerically) by Eq. (12.7). We have initially $\phi(0) = 0.1$, $\dot{\phi}(0) = 0$ and $\omega = 1.43$, $a = 0.1$, $\Omega = 2$. Left: we have the conservative case $\kappa = 0$ and right: $\kappa = 0.1$ (the thin lines are transients)

$$\ddot{\phi} + \varepsilon\kappa\dot{\phi} + \omega^2(\phi - \frac{1}{6}\varepsilon^2\phi^3) + O(\varepsilon^4) = a\cos\Omega t. \tag{12.8}$$

The estimate $O(\varepsilon^4)$ is uniform on the interval $[0, 2\pi]$. If $\varepsilon = 0$, a particular solution of Eq. (12.8) is

$$\phi_p(t) = \frac{a}{\omega^2 - \Omega^2}\cos\Omega t,\ \omega \neq \Omega. \tag{12.9}$$

Following (1.28) of Chap. 1 we transform

$$\phi = r\cos(\omega t + \psi) + \phi_p(t), \dot{\phi} = -\omega r\sin(\omega t + \psi) + \dot{\phi}_p(t). \tag{12.10}$$

We exclude the singularity at the origin. The variational equations are to $O(\varepsilon^2)$:

$$\begin{cases} \dot{r} & = \varepsilon\frac{\kappa}{\omega}\sin(\omega t + \psi)(-\omega r\sin(\omega t + \psi) + \dot{\phi}_p(t)) \\ & \quad -\varepsilon^2\frac{\omega}{6}\sin(\omega t + \psi)(r\cos(\omega t + \psi) + \phi_p(t))^3, \\ \dot{\psi} & = \varepsilon\frac{\kappa}{\omega r}\cos(\omega t + \psi)(-\omega r\sin(\omega t + \psi) + \dot{\phi}_p(t)) \\ & \quad -\varepsilon^2\frac{\omega}{6r}\cos(\omega t + \psi)(r\cos(\omega t + \psi) + \phi_p(t))^3 \end{cases} \tag{12.11}$$

Assume we have no low-order commensurability of $\omega$ and $\Omega$ (not 1:1, 1:2, 1:3:).

**The Case Without Dissipation** $\kappa = 0$
We find after averaging

$$\dot{r} = 0, \dot{\psi} = -\varepsilon^2\frac{\omega}{8}\left(\frac{1}{2}r^2 + \frac{a^2}{(\omega^2 - \Omega^2)^2}\right). \tag{12.12}$$

Putting $r = r(0)$ we abbreviate $\dot{\psi} = -\varepsilon^2 C$. We have the approximation:

$$\phi(t) = r(0)\cos(\omega t - \varepsilon^2 Ct) + \frac{a}{\omega^2 - \Omega^2}\cos\Omega t + O(\varepsilon^2), \tag{12.13}$$

and $\dot{\phi}$ from transformation (12.10), valid on the timescale $1/\varepsilon^2$. This will produce a torus with two frequencies. To compare with the numerics we have to absorb $\varepsilon^2$ in $r^2$ and $a^2$ in Eq. (12.12). The $O(\varepsilon^2)$ approximation is shown in Fig. 12.3.

**The Dissipative Case** $\kappa > 0$
We find after averaging

$$\dot{r} = -\varepsilon\frac{\kappa}{2}r, \dot{\psi} = 0. \tag{12.14}$$

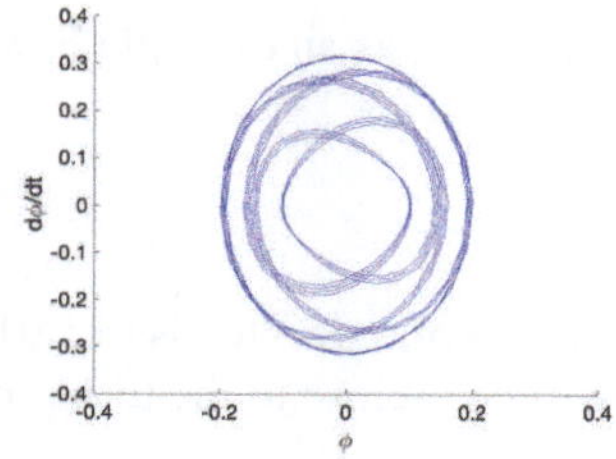


**Fig. 12.3** $O(\varepsilon^2)$ approximation (12.13) of the mathematical pendulum with forcing described by Eq. (12.7) in the conservative case $\kappa = 0$; compare Fig. 12.2

We have the approximation

$$\phi(t) = r(0)e^{-\varepsilon\frac{\kappa}{2}t}\cos(\omega t + \psi(0)) + \frac{a}{\omega^2 - \Omega^2}\cos\Omega t,\ \psi(t) = \psi(0), \tag{12.15}$$

and $\dot{\phi}$ from transformation (12.10), valid to $O(\varepsilon)$ for $t \geq 0$. The solution tends to the particular solution $\phi_p(t)$ as $t$ tends to $\infty$.

For related calculations see also Sect. 6.3.3.

## 12.6 Quasi-periodicity in the Forced Van der Pol Equation

Consider the forced Van der Pol equation

$$\ddot{x} + x = \varepsilon(1 - x^2)\dot{x} + a\cos\omega t$$

with $a$ and $\omega$ constants. This time we assume that the non-linearity is small and the forcing can be $O(1)$ as $\varepsilon \to 0$.

Assume $\omega$ is not $\varepsilon$-close to 1; for if $\omega$ is near to 1, the solutions move away from an $O(1)$ neighbourhood of the origin because of linear resonance. We introduce the standard transformation $x, \dot{x} \to r, \psi$ as in (1.27), Chap. 1:

$$x = r\cos(t + \psi) + \frac{a}{1 - \omega^2}\cos\omega t,\ \dot{x} = -r\sin(t + \psi) - \frac{a\omega}{1 - \omega^2}\sin\omega t.$$

The resulting slowly varying system can be averaged, producing periodic solutions in which various values of $\omega$ play a part. Returning to the corresponding expressions for $x$ and $\dot{x}$ we conclude to the presence of tori in the phase space extended with $t$.

The approximate behaviour is studied in Sect. 5.3.4. Averaging produces in this case

$$\dot{r} = \frac{\varepsilon}{2}r\left(1 - \frac{1}{4}r^2 - \frac{1}{2}\frac{a^2}{(1 - \omega^2)^2}\right) + \varepsilon^2\cdots,\ \dot{\psi} = \varepsilon^2\cdots, \tag{12.16}$$

Choosing as an example $a = 0.4$, $\omega = \sqrt{2}$ we have for the approximate $r$

$$\dot{r} = (0.92 - \frac{1}{4}r^2) + \varepsilon^2 \cdots$$

The amplitude $r$ tends to 1.918. In the case of $a = O(1)$ the corresponding solutions will be $\varepsilon$-close to the torus solutions in the extended $(x, \dot{x}, t)$-space described by

$$x = 1.918 \cos(t + \psi(0)) - \frac{2}{5} cos \sqrt{2}t, \dot{x} = -1.918 \sin(t + \psi(0)) + \frac{2\sqrt{2}}{5} \sin \sqrt{2}t.$$

## 12.7 The Jafari-Sprott-Golpayegani Project

In [125] a three-dimensional system was analysed; the system is of importance in chemical physics for molecular dynamics with a certain thermostatic control. As we shall see in Sect. 12.9 the control affects the energy exchanges between the particles. The system was an inspiration to formulate and analyse 17 systems in the so-called Jafari-Sprott-Golpayegani project, all of them containing chaos; see [81]. The system of [125] is called Sprott A or NE1; see Sect. 12.9, and the systems are numbered NE1, ..., NE17. They are three-dimensional, autonomous, dissipative ODEs with one free parameter and no equilibria, showing interesting chaotic behaviour, and as we shall see even in some cases KAM-like nested tori. The analysis of these 17 systems can be very instructive; they contain many phenomena arising in higher-dimensional systems. Some studies are inspired by these 17 systems, apart from the examples included in this chapter we mention [99] and many related papers of the authors.

First we will discuss relevant parts of the dynamics theory, including briefly the Lorenz attractor, and in addition variations of the Sprott A system. After the three-dimensional systems we will discuss coupled second-order ODEs.

## 12.8 KAM Theory, Fractal Dimension and Chaos

In Chap. 9 we have seen that for Hamiltonian systems that are close to being integrable, invariant manifolds in the form of nested tori may arise near stable equilibrium. There are an infinite number of these tori, but its measure in phase-space does generally not equal 1; between the tori irregular dynamics may take place. In Chap. 9 we discussed the theory of Laskar [97] enabling us to discover such irregularities between the tori.

### 12.8.1 KAM Theory in the Dissipative Case

Interestingly, in the presence of certain symmetries KAM theory can also be applicable to dissipative systems. General references are [44, 96] and [116]. The conditions are reversibility of the flow and that certain symmetries are present. However, the number of applications is still restricted. We will consider a few cases of so-called thermostatic control problems initiated by J.C. Sprott [125]. Some of these systems admit application of dissipative KAM theory.

### 12.8.2 The Kaplan-Yorke Dimension

When discovering irregular behaviour in dynamical systems the possibility of the presence of chaos arises. If in such a case an attracting object can be identified there are a number of techniques to determine the dimension of this object. We have to distinguish between a geometric dimension and a dynamical one; see for an introduction [146]. In [107] an introductory discussion of examples and fractal dimensions is given; the paper also contains examples of three-dimensional ODEs.

The Hausdorff dimension is geometrical; it can be obtained by a limit procedure of coverings of the object. To carry this out technically is not easy, but it is often shortened by using cubes for the covering; the dimension obtained in this way is expected to be an approximation and is called the *limit capacity*.

A useful tool is the Lyapunov exponent (or L-exponent) $\mu$. For a one-dimensional map $X \to X$ we take the limit (if it exists) starting at $x_0$:

$$\mu(x_0) = \lim_{n \to \infty} \frac{\ln |df^n(x_0)|}{n},$$

where $df^n(x_0)$ denotes the derivative of the $n$ times iterated map at $x_0$. The quantity $\mu(x_0)$ is called the Lyapunov exponent associated with $x_0$; it generalises the concept of eigenvalue obtained by linearisation near an object and, of course, it assumes that the limit exists. The L-exponent indicates local expansion or contraction.

We can generalise for $n$-dimensional maps to find $n$ L-exponents $\mu_1, \mu_2, \ldots, \mu_n$. A detailed description is given in [66] section 5.8; see also [53] and [64].

Kaplan and Yorke used this concept in [84] to formulate the Kaplan-Yorke dimension based on the $n$ L-exponents calculated when starting in a neighbourhood of a geometric object. Arrange the L-exponents $\mu_1, \mu_2, \ldots, \mu_n$ from largest to smallest. The index $j \leq n$ is the largest so that we have still $\sum_{i=1}^{j} \mu_j > 0$ so $\sum_{i=1}^{j+1} \mu_j < 0$; then the dimension of the attractor is according to Kaplan and Yorke

$$D_{KY} = j + \frac{\sum_{i=1}^{j} \mu_i}{|\mu_{j+1}|}. \tag{12.17}$$

If we find positive and negative L-exponents, the object can be a chaotic and fractal attractor; positive exponents denote chaos and two positive L-exponents hyperchaos. Two L-exponents zero indicate a torus.

A discussion of the use of L-exponents is given in [14]. A basic problem is that for a fractal object the L-exponents may not exist, but the Kaplan-Yorke dimension has still practical use by calculating the exponents for many initial values and averaging the results; see the remarks below on the Lorenz attractor and details in [14].

### *12.8.3 The Lorenz Attractor*

The first example of chaos in a dissipative system was discovered by Edward Lorenz who studied Rayleigh-Bénard convection. This is the instability that arises if we have a gas between a high temperature surface below and a low temperature surface above; it models the rising gas flow on a relative warm Earth surface meeting cooler regions higher up and after contracting going down again. Lorenz approximated the energy modes of the corresponding partial differential equation (PDE) by three ODEs:

$$\begin{cases} \dot{x} &= 10(-x + y), \\ \dot{y} &= -xz + 28x - y, \\ \dot{z} &= xy - \frac{8}{3}z. \end{cases} \tag{12.18}$$

This choice of coefficients produces irregular motion as depicted in the two-dimensional projection of Fig. 12.4. Started at (1, 1, 1) the solution moves aperiodically around the depicted manifold. Many papers have been published on the equations (12.18); the scientists Ruelle and Takens coined the name "strange attractor" for the phenomenon. Detailed analysis showed that the attractor is not a smooth object and has a fractal dimension. A Hausdorff dimension of 2.0627160 was computed in [160]; this is a geometric dimension based on a complicated analysis of Cantor sets, but it ignores the dynamics. The Kaplan-Yorke dimension

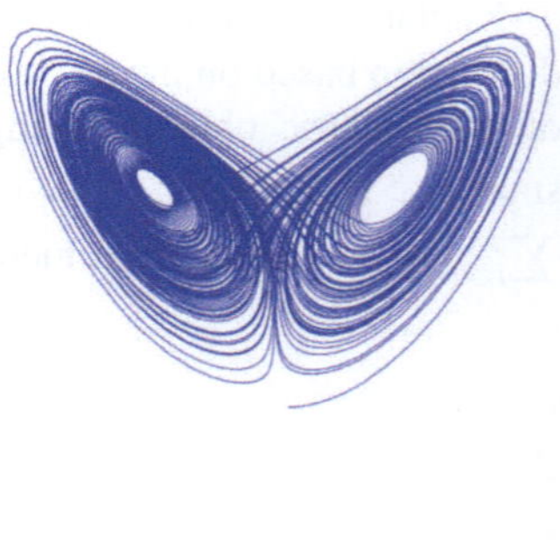

**Fig. 12.4** Projection of the Lorenz system described by Eq. (12.18)

gives additional information. As it takes the attraction and repelling into account we should have $D_{KY} \leq 2.0627160$. In the literature the L-exponents are given as $0.906, 0, -14.572$. Starting from 100 different initial conditions one finds in [14] very small fluctuations around these exponents and consequently a dimension with little fluctuation around $D_{KY} = 2.062$. The Lorenz attractor is nearly flat.

### *12.8.4 Period Doubling Leading to Chaos*

In a number of cases one can find upon changing a parameter a sequence of periodic solutions with repeatedly doubling of the period. Continuing the parameter one may reach a critical parameter value where chaos emerges. We will discuss such a case in the treatment of system NE9 in Sect. 12.10. See also the introductory chapter of [28] for period doubling (and other bifurcation phenomena) and also [109].

An inspiration came from the analysis of the logistic or quadratic difference equation:

$$x_{n+1} = ax_n(1 - x_n),\ n = 0, 1, \dots \tag{12.19}$$

with $0 \leq x_n \leq 1, a > 0$. It was started in [103] where the equation was studied as a simple looking model for population dynamics. The analysis of Eq. (12.19) gave many surprises.

It is easy to see that if $0 < a \leq 1$, the population $(x_n)$ tends to zero as $n \to \infty$. Another fixed point (equilibrium) is found by putting $x_0 = ax_0(1 - x_0)$. Solving the equation we find for $a > 1$ the fixed point $x_0 = (a - 1)/a$. We can also try to find a period two point by solving

$$x_1 = ax_0(1 - x_0),\ x_0 = ax_1(1 - x_1).$$

This involves solving a quartic algebraic equation; this is not difficult as we have already three fixed points: $x = 0, 1, (a - 1)/a$. We can go on and look for a period three point, a period four point, etc. It turns out that we find solutions for an increasing number of values of parameter $a$ with $1 < a \leq 4$. From $a = 3$ on, the behaviour of the solutions $x_n$ can become very complicated; see for details [28] and in particular [109]. The increasing values of $a$ where a higher-order periodic point emerges (for instance doubling periods 1, 2, 4, 8, etc) are called bifurcation points. At such a point the preceding periodic point becomes unstable and a new periodic point emerges.

This scenario in the form of period doubling with an infinite sequence leads to chaos in the limit. As a universal scenario it can also be applied to the flow induced by differential equations.

## 12.9 Nested Tori for Sprott A (or NE1)

Consider the Sprott A system

$$\dot{x} = y,\ \dot{y} = -x - yz,\ \dot{z} = y^2 - a, \tag{12.20}$$

with positive parameter $a$. Note that the system has no critical points (equilibria), but it has an invariant manifold given by $x = y = 0$, the $z$-axis with solutions, $z(t) = z(0) - at$. Localising to an $O(\varepsilon)$ neighbourhood of the $z$-axis by putting $x, y \to \varepsilon x, \varepsilon y$ and $a \to \varepsilon^2 a$ produces the system

$$\dot{x} = y,\ \dot{y} = -x - \varepsilon yz,\ \dot{z} = \varepsilon y^2 - \varepsilon a, \tag{12.21}$$

Transforming to slowly varying cylindrical variables $r, \phi, z$ we find

$$\begin{cases} \dot{r} &= -\varepsilon rz \sin^2(t+\phi), \\ \dot{\phi} &= -\frac{1}{2}\varepsilon z \sin(2t + 2\phi), \\ \dot{z} &= \varepsilon r^2 \sin^2(t+\phi) - \varepsilon a. \end{cases} \tag{12.22}$$

Averaging while keeping the same notation for variables we have for the components

$$\dot{r} = -\frac{1}{2}\varepsilon rz,\ \dot{\phi} = 0,\ \dot{z} = \varepsilon(\frac{1}{2}r^2 - a). \tag{12.23}$$

An equilibrium is $(r, \phi, z) = (\sqrt{2a}, \phi(0), 0)$. As expected for autonomous systems, see Remark 3.3, one of the eigenvalues of the Jacobian $Df^0$ is zero; the rank of the Jacobian at equilibrium is 2. We conclude with theorem 3.1 that a periodic solution exists in an $O(\varepsilon)$-neighbourhood of the equilibrium. However, the eigenvalues of the Jacobian at equilibrium are $\pm ai, 0$. In the case of purely imaginary eigenvalues the non-linear terms may stabilise or destabilise the equilibrium. However, it turns out that a second-order approximation does not add real terms $O(\varepsilon^2)$ to the eigenvalues, so the stability of this periodic solution remains unsolved by averaging.

System (12.20) was studied in [104, 105] where the authors noticed that the periodic solution at $O(\varepsilon)$ distance from the $z$-axis in the plane $z = 0$ is surrounded by invariant tori and is an example of dissipative KAM theory; the evidence is convincing but numerical. The relation with KAM theory was confirmed in [10] using reversibility and symmetries in the Sprott A system with conclusion that the periodic solution is surrounded by an infinite family of tori and is (neutrally) Lyapunov-stable; see Fig. 12.5. We give the conditions from [10]. An interesting discrete symmetry feature of (12.20) is:

**Lemma 12.1** *If* $(x(t), y(t), z(t))$ *is a solution of system* (12.20)*, then* $(-x(t), -y(t), z(t))$ *also is a solution.*

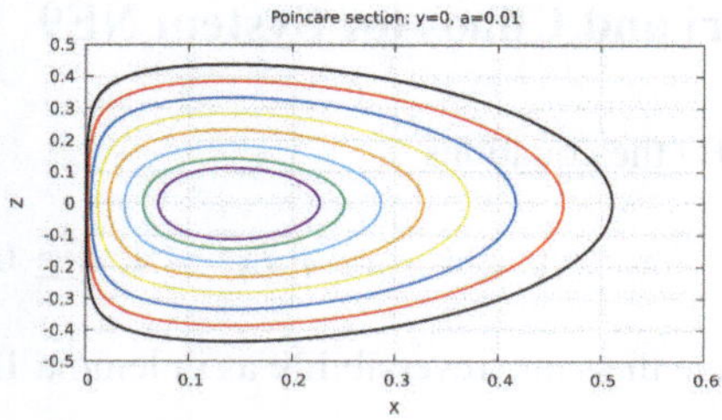


**Fig. 12.5** Poincaré section in the plane y = 0 of the Sprott system (12.20). For $a = 0.01$ an infinite number of invariant tori are found according to dissipative KAM theory near the $z$-axis. A small periodic solution serves as organising centre. Courtesy CHAOS [10]

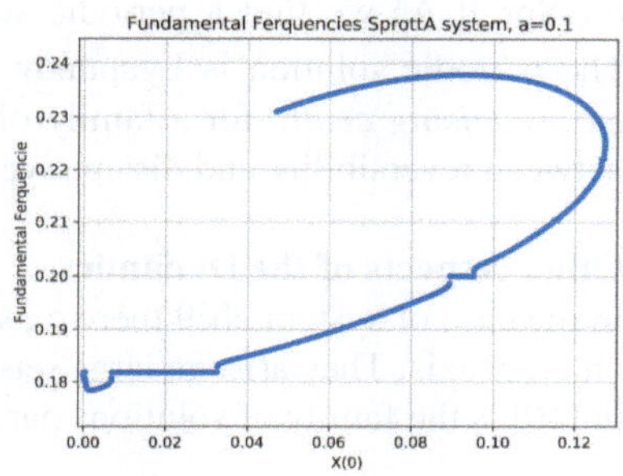


**Fig. 12.6** Behaviour of the prominent frequency moving transversally over the tori of the Sprott A system depicted in Fig. 12.5. We find gaps showing irregular behaviour; zooming in makes this more explicit. See [10]. Courtesy CHAOS [10]

This is verified by substitution. An important feature involving time reversal of (12.20) is:

**Lemma 12.2** *If* $(x(t), y(t), z(t))$ *is a solution of system* (12.20)*, then by putting* $\bar{x} = x, \bar{y} = -y, \bar{z} = -z$ *and reversing time* $\tau = -t$*,* $(\bar{x}(\tau), \bar{y}(\tau), \bar{z}(\tau))$ *is also a solution.*

Such time reversal symmetry is called $R$-symmetry in [96].

It is interesting to consider the possibility of gaps between the tori to look for an analogy of chaos between the tori arising from Cantor sets as in Hamiltonian systems. In Sect. 9.12.3 we considered the Laskar technique to determine the change of frequencies when moving along the tori as in Fig. 12.5. The result for the Sprott A system is given in Fig. 12.6. The jumps in the frequency pattern show the local break-up of the tori.

**Remark 12.1** Apart from the solutions on the $z$-axis the solutions of system (12.20) are bounded. A detailed analysis of periodic solutions on the nested tori and orbits further removed is given in [10]. Also canard behaviour can be observed.

In [81] chaos is observed for $a = 1$ at $x(0), y(0), z(0) = 0, 5, 0$.

## 12.10 Nested Tori and Chaos for System NE9

We have for system NE9 the equations

$$\dot{x} = y,\ \dot{y} = -x - yz,\ \dot{z} = -xz + 7x^2 - a,\ a \geq 0 \tag{12.24}$$

For system NE9 we have the same reversibility as in lemma 12.2 for Sprott A:

**Lemma 12.3** *If* $(x(t), y(t), z(t))$ *is a solution of system* (12.24), *then by putting* $\bar{x} = -x, \bar{y} = y, \bar{z} = -z$ *and reversing time* $\tau = -t$, $(\bar{x}(\tau), \bar{y}(\tau), \bar{z}(\tau))$ *is also a solution.*

We rescale $x, y, \to \varepsilon(x, y, z)$ and $a = \varepsilon^2 a_0$. Transforming and averaging (as for Sprott A) we find a periodic solution in the plane $z = 0$ with $r = \sqrt{\frac{2}{7}a_0}$. The periodic solution is Lyapunov stable as in the case of Sprott A and acts as an *organising centre* for a family of tori. This is again an example of the relation between reversibility and dissipative KAM theory, see for details [10].

**Other Aspects of the Dynamics**

In the case of system NE9 there exist unbounded solutions apart from the solutions on the $z$-axis. They arise in large sets in phase-space. An explicit example discussed in [10] is the family of solutions parameterised by $x(0)$:

$$x(t) = x(0) - \frac{t}{7},\ y(t) = -\frac{1}{7},\ z(t) = 7x(0) - t,\ a = 1. \tag{12.25}$$

An implication is that $z = 7x$, $y = -1/7$ is an invariant manifold in the case $a = 1$. According to [10] the manifold is asymptotically stable for $x > 0$, unstable for $x < 0$. As $x(t)$ decreases linearly with time on the manifold the stability change takes place at $t_c = 7x(0)$. Choosing the parameter $a$ smaller than 1 and starting on the invariant manifold periodic solutions emerge that partly follow the straight line manifold in $x, z$ projection; see Fig. 12.7. Between $a = 0.56$ and $a = 0.55$ the periodic solutions vanish and at $a = 0.55$ a strange attractor emerges with Kaplan-Yorke dimension $D_{KY} = 2.1544$.

In [81] chaos is observed for $a = 0.55$ at $x(0), y(0), z(0) = 0.5, 0, , 0$. Our analysis shows that the strange attractor appears in a neighbourhood of $x = 0$ after subsequent period doublings.

More chaotic sets can be identified in system NE9. In [10] a cascade of period doubling periodic solutions starting at $a = 0.2$ produces a chaotic attractor for $a = 0.20601$. A double torus loosing smoothness plays a part.

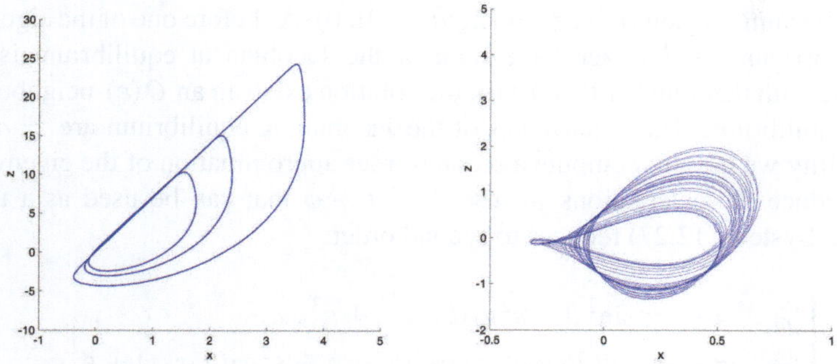

**Fig. 12.7** Left: three periodic solutions of the family near the invariant manifold $z = 7x, y = -1/7, a = 1$ of system NE9. Successively $a = 0.99, , 0.90, 0.80$ were chosen. Right: the strange attractor emerging after decreasing $a$, producing period doublings till $a = 0.55$. The attractor emerges and corresponds with starting at $x, y, z = 0.5, 0, 0$ in [81]

## 12.11 System NE8

Consider a related case of the list in [81] with a different outcome, system NE8:

$$\dot{x} = y, \; \dot{y} = -x - yz, \; \dot{z} = xy + \frac{1}{2}x^2 - a, \tag{12.26}$$

with positive parameter $a$. Note that again this system has no critical points (equilibria), it has an invariant manifold given by $x = y = 0$, the $z$-axis with solutions, $z(t) = z(0) - at$. Systems NE1, NE8 and NE9 have in common that we can formulate them as an oscillator equation with variable damping:

$$\ddot{x} + z\dot{x} + x = 0.$$

The damping "coefficient" z is different in the three cases; in particular NE8 does not satisfy the KAM theory requirements. We can see this already from the behaviour in an $O(\varepsilon)$ neighbourhood of the $z$-axis. Rescaling as for system (12.24) and transforming to slowly varying cylindrical variables $r, \phi, z$ as before we find

$$\begin{cases} \dot{r} &= -\varepsilon r z \sin^2(t + \phi), \\ \dot{\phi} &= -\frac{1}{2}\varepsilon z \sin(2t + 2\phi), \\ \dot{z} &= -\frac{\varepsilon}{2} r^2 \sin(2t + 2\phi) + \frac{\varepsilon}{2} r^2 \cos^2(t + \phi) - \varepsilon a. \end{cases} \tag{12.27}$$

Averaging while keeping the same notation for variables we have for the components of $f^0$

$$\dot{r} = -\frac{1}{2}\varepsilon r z, \; \dot{\phi} = 0, \; \dot{z} = \varepsilon(\frac{1}{4}r^2 - a). \tag{12.28}$$

An equilibrium is again $(r, \phi, z) = (2\sqrt{a}, \phi(0), 0)$. As before one of the eigenvalues of the Jacobian $Df^0$ is zero; the rank of the Jacobian at equilibrium is 2. We conclude with theorem 3.1 that a periodic solution exists in an $O(\varepsilon)$-neighbourhood of the equilibrium. The eigenvalues of the Jacobian at equilibrium are $\pm ai, 0$, so for stability we have to compute a second-order approximation of the eigenvalues.

To reduce the calculations we use $\theta = t + \phi$ that can be used as a timelike variable. System (12.27) reduces to second order

$$\begin{cases} \frac{dr}{d\theta} &= -\varepsilon rz \sin^2\theta + \varepsilon^2 g_1(\theta, r, z) + \varepsilon^3 \dots \\ \frac{dz}{d\theta} &= -\frac{\varepsilon}{2} r^2 \sin 2\theta + \frac{\varepsilon}{2} r^2 \cos^2\theta - \varepsilon a + \varepsilon^2 g_2(\theta, r, z) + \varepsilon^3 \dots \end{cases} \tag{12.29}$$

Averaging the $O(\varepsilon^2)$ terms we find $g_1^0 = 0,\ g_2^0 = -\frac{1}{8} r^2 z$.
In the notation of Theorem 4.1 we find

$$Df(\theta, r, z) = \begin{pmatrix} -z \sin^2\theta & -r \sin^2\theta \\ -r \sin 2\theta + r \cos^2\theta & 0 \end{pmatrix}, u^1(\theta, r, z)$$

$$= \begin{pmatrix} \frac{1}{4} rz \sin 2\theta \\ \frac{1}{4} r^2 (\cos 2\theta + \frac{1}{2} \sin 2\theta) \end{pmatrix}.$$

$$f_1^0(r, z) = \left(\frac{1}{16} r^3, -\frac{1}{4} r^2 z\right).$$

For the approximating solutions to second order we have

$$\frac{dr}{d\theta} = -\frac{1}{2}\varepsilon rz + \varepsilon^2 \frac{1}{16} r^3,\ \frac{dz}{d\theta} = \varepsilon(\frac{1}{4} r^2 - a) - \frac{\varepsilon^2}{4} r^2 z. \tag{12.30}$$

The equilibrium to second order becomes $(r, z) = (2\sqrt{a}, \frac{\varepsilon}{2} a)$. From Eq. (12.30) we compute the eigenvalues of this equilibrium to find that $-\frac{1}{8}\varepsilon a$ is added to the purely imaginary parts. At second order we have a small qualitative change; the periodic solution is stabilised. Numerical calculations confirm the result; see Fig. 12.8.

In [81] chaos is observed for $a = 1.3$ at $x(0), y(0), z(0) = 0, 0.1, 0$ with $D_{KY} =$ 2.0031 (nearly planar).

## 12.12 The Sprott B System, a Surprise

In [13] a variation of the thermostatic control systems, Sprott B was published. The system is

$$\dot{x} = y,\ \dot{y} = -x - yz,\ \dot{z} = x^2 + y^2 - a, \tag{12.31}$$

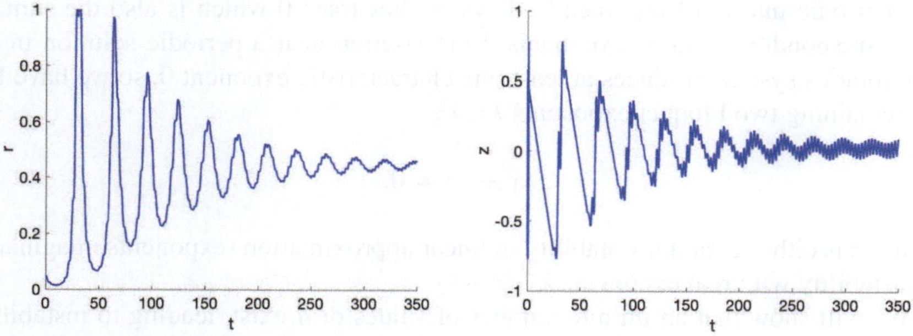


**Fig. 12.8** The $r(t), z(t)$ time-series of system NE8 (numerically) near the $z$-axis with $a = 0.05$, $x(0) = 0.05$, $y(0) = 0$, $z(0) = 0.5$, $r = \sqrt{x^2 + y^2}$. The flow contracts to a small periodic solution

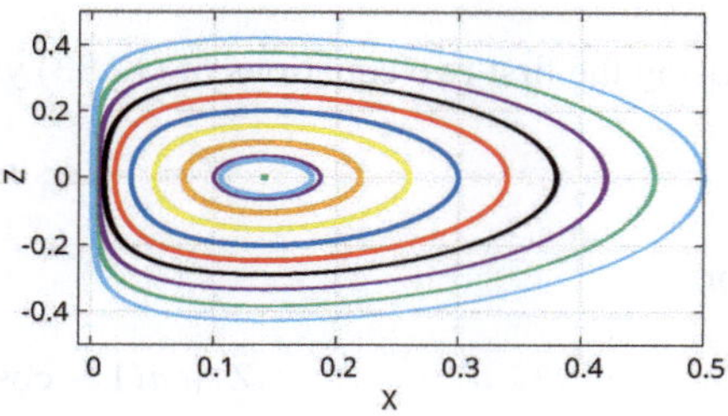


**Fig. 12.9** The Poincaré map of the $x, z$ transversal of system Sprott B (numerically) near the $z$-axis with $a = 0.01$. The small periodic solution (12.32) shows up as a point. Courtesy CHAOS 2024

with parameter $a > 0$. A surprising and useful aspect of the system is that it contains a family of easy-to-handle periodic solutions:

$$x(t) = \sqrt{a}\cos t,\ y(t) = -\sqrt{a}\sin t,\ z(t) = 0. \tag{12.32}$$

This family enables us to explore details of the dynamics of the Sprott B system that are usually hidden in other three-dimensional systems. Sprott B has $R$-symmetry, Lemmas 12.1 and 12.2 apply to the system. So, dissipative KAM theory applies too; we find again an infinite set of tori in the system for fixed $a$ and $z(t) = 0$ with the periodic solution (12.32) as organising centre; see Fig. 12.9.

Consider (as in [13]) a neighbourhood of the periodic solution by putting

$$x = \sqrt{a}\cos t + \varepsilon X,\ y = -\sqrt{a}\sin t + \varepsilon Y,\ z = \varepsilon Z.$$

System (12.31) transforms to

$$\begin{cases} \dot{X} = & Y, \\ \dot{Y} = & -X + \sqrt{a} Z \sin t - \varepsilon Y Z, \\ \dot{Z} = & 2\sqrt{a}(X\cos t - Y\sin t) + \varepsilon(X^2 + Y^2). \end{cases} \tag{12.33}$$

The periodic matrix of the linearised system has trace 0 which is also the sum of the corresponding Floquet exponents. Linearisation near a periodic solution in an autonomous system produces at least one characteristic exponent 0, so we have for the remaining two Floquet exponents $\lambda_1, \lambda_2$

$$\lambda_1 + \lambda_2 = 0.$$

So there is either Lyapunov stability in linear approximation (exponents imaginary) or instability with real exponents $\lambda_1 = -\lambda_2$.

We will show that an infinite number of values of $a$ exist, leading to instability of the periodic solution. Consider system (12.33) and differentiate the equation for Z and linearise. We find

$$\ddot{Z} = 2\sqrt{a}(\dot{X}\cos t - X\sin t - \dot{Y}\sin t - Y\cos t) + O(\varepsilon).$$

Using the first two equations of (12.33) yields

$$\ddot{Z} = -2a\sin^2 tZ + O(\varepsilon),$$

or

$$\ddot{Z} + a(1 - \cos 2t)Z = O(\varepsilon), \tag{12.34}$$

and after dividing by $a$ and introducing a new timescale $\tau = \sqrt{a}\,t$

$$\frac{d^2Z}{d\tau^2} + \left(1 - \cos(\frac{2}{\sqrt{a}}\tau)\right)Z = O(\varepsilon). \tag{12.35}$$

This linear equation is the Mathieu-equation. It is well known, see Chap. 11, that this equation contains an infinite number of instabilities of the trivial solution at an infinite, discrete number of $a$ values. The instabilities are found in so-called instability Mathieu tongues and are located in an infinite number of discrete segments of $a$ values. Experimentally we found, for instance, instability in the segment $0.658 \leq a \leq 1.800$ with the initial values $x(0) = \sqrt{a}, y(0) = -\sqrt{a}, z(0) = 0$ of Fig. 12.10. As the eigenvalues at the instabilities are real, the non-linear terms do not change the stability character but will produce only small quantitative changes.

The paper [13] contains many interesting technical details. As an illustration we discuss aspects of the dynamics in two instability intervals. They are computed by noting that the boundaries of the instability tongues contain $a$ values that correspond with periodic solutions so we can continue; for instance near $a = 1$, for smaller and for larger $a$ values to find periodic solutions. These $a$ values are the boundaries of the instability interval.

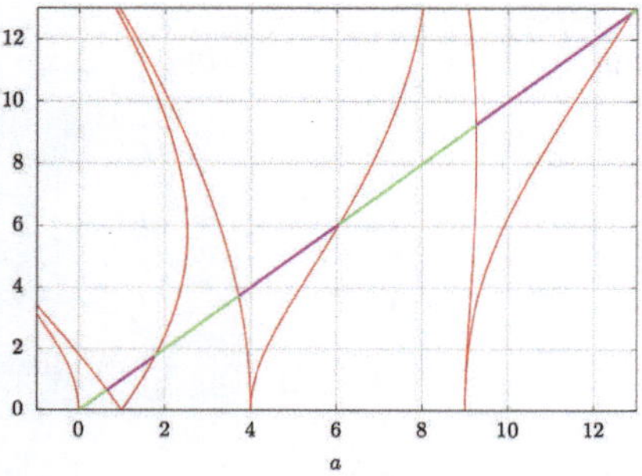


**Fig. 12.10** Instability tongues of the Mathieu-equation emanating from the horizontal $a$-axis at $a = 1, 4, 9$, etc.; the vertical axis indicates the amplitude of $Z(t)$ when starting near the periodic solution (12.32). The coloured straight line represents values when moving along the solutions of Eq. (12.34). The *magenta* segments within the unstable tongues indicate the unstable intervals for the parameter $a$. This ensures instability behaviour; outside the tongues neutral stability is maintained. Courtesy CHAOS (2024)

**Fig. 12.11** Almost-periodic solution with spikes of Sprott B in the case $a = 0.01$, $x(0 = z(0) = 0.01$, $y(0) = 0$. The solution is contained in a torus flattened near the $z$-axis; see Figs. 12.9 and 12.12

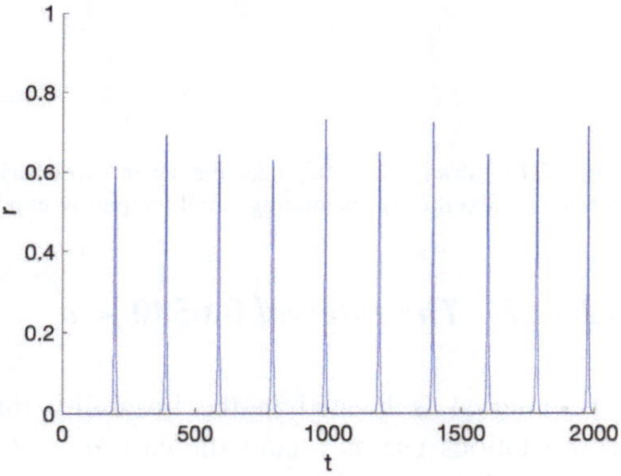


## 12.12.1 Canards Near the z-Axis of Sprott B, a ≪ 1

As for Sprott A (NE1) we have for $a \ll 1$ an infinite set of tori and canard behaviour near the $z$-axis. The asymptotic analysis runs as in Sect. 12.9. We show the spiking behaviour caused by the canards in Fig. 12.11. In Fig. 12.12 we show a three-dimensional picture of the spiked orbit of Fig. 12.11.

The period $2\pi$ contrasts with the recurrence time in the tori. For instance putting $y(0) = 0, z(0) = 0.5, a = 0.01$ and using the seven starting values $x(0) = 10^{-7}, 10^{-6}, \ldots, 10^{-1}$ successively we find recurrence times between 100 and 200. We expect that long-periodic solutions can be identified on the tori.

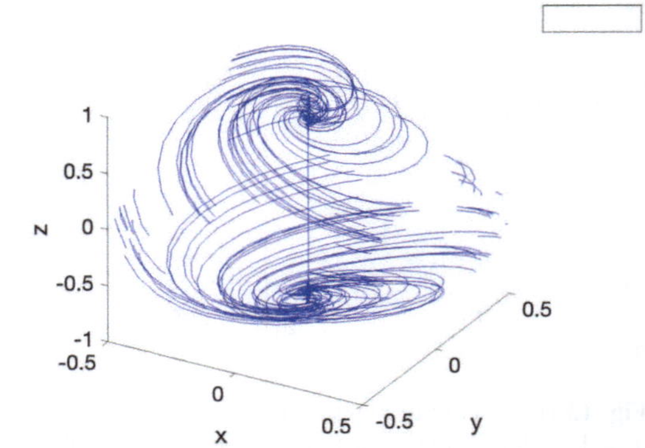


**Fig. 12.12** The same almost-periodic solution with spikes of Sprott B as in fig 12.11 but now in three dimensions $(x, y, z)$. The vertical line is a close approximation of the $z$-axis that represents the slow manifold of the Sprott B system. Near the poles we have an accumulation of varying input and exit values of the slow manifold

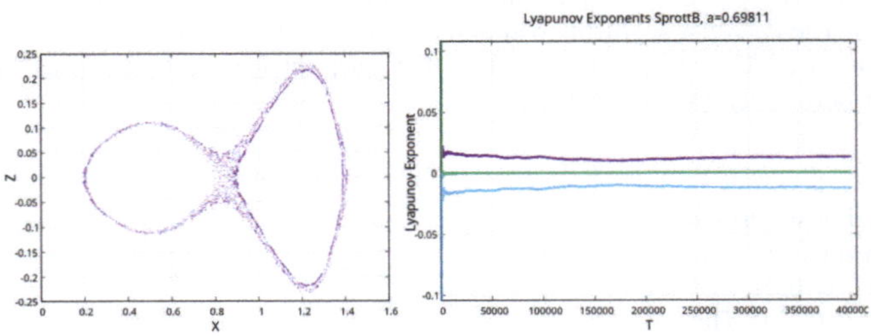


**Fig. 12.13** Strange attractor near the seven short-periodic solutions of Sprott B at the case $a = 0.69811$. Next the corresponding small Lyapunov-exponents. Courtesy CHAOS (2024)

## *12.12.2 The Interval* $0.6580 < a < 1.7796$*, Case* $a = 0.69811$

The interval is located in the instability tongue emanating from $a = 1$; the computations use as a start the case $a = 0.69811$. Using the Poincaré map of the $x, z$-plane into itself one can look for fixed points of the map; they correspond with period 1 solutions. It turns out that in [12] six more fixed points are found corresponding (together with solution (12.32)) to seven short-periodic solutions. One can look for long-period solutions by considering fixed points of maps $P^2$, $P^3$, $P^n$, etc; it turns out many of them exist. In these calculations numerical bifurcation techniques using [102] play a part.

Near these seven periodic solutions a strange attractor is found; see Fig. 12.13. The Lyapunov-exponents are 0.012719, $3 \times 10^{-6}$, $-0.012719$. The scenario to obtain a strange attractor in this case is the emergence of homoclinic tangles. In the centre of the strange attractor a saddle point of the Poincaré map can be found, corresponding with an unstable periodic solution. When the stable and unstable manifolds of this solution start to cross in the Poincaré map the emergence of the strange attractor takes place. This scenario was invented by Henri Poincaré; it is described in [110] vol. 3. The dynamics is nowadays also characterised by the emergence of a horseshoe map; see for introductions [66] and [28].

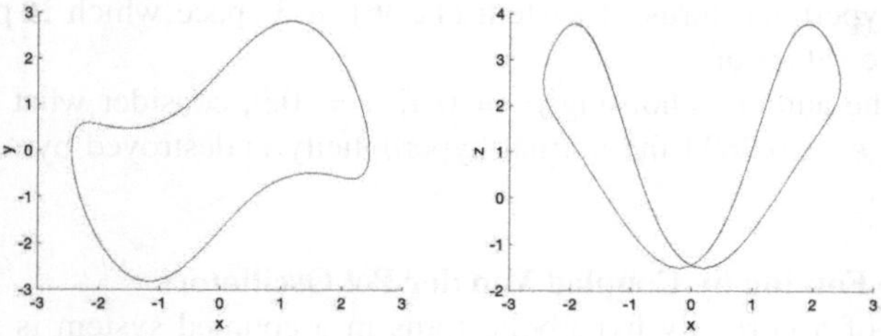

**Fig. 12.14** Stable periodic solution in the case $a = 4$ of the Sprott B system obtained when starting near periodic solution (12.32). Both the projection on the $x, y$ plane and on the $x, z$ plane are shown

### *12.12.3 The Next Instability Interval of Sprott B, Case a = 4*

Starting near the periodic solution (12.32) with amplitude two the solutions tend to another periodic solution depicted in Fig. 12.14. The variable $z(t)$ changes sign periodically, and the solution is self-excited.

Numerical experiments show the presence of more periodic solutions.

## 12.13 One or Two Forced Van der Pol Oscillators

There are natural extensions to non-autonomous systems for the emergence of tori by introducing the so-called stroboscopic map. We demonstrate this by an example derived from Broer, Osinga and Vegter [26]. See also for more background the monograph by Broer, Huitema and Sevryuk [25]. The following example can be extended to other cases.

Consider the forced Van der Pol oscillator

$$\ddot{x} + x = \mu(1 - x^2)\dot{x} + \varepsilon \cos \omega t$$

which we write as the system

$$\begin{cases} \dot{x} &= y, \\ \dot{y} &= -y + \mu(1 - x^2)y + \varepsilon \cos \tau, \\ \dot{\tau} &= \omega. \end{cases} \tag{12.36}$$

The $2\pi$-periodic forcing term $\varepsilon \cos \tau$ produces a stroboscopic map of the $x, y$-plane into itself. For $\varepsilon = 0$ this shows the map of the periodic solution of the Van der Pol equation as an invariant cycle; the closed orbit is normally hyperbolic. In the extended phase space $\mathbb{R}^2 \times \mathbb{R}/2\pi\mathbb{Z}$ this invariant circle for $\varepsilon = 0$ corresponds with a normally hyperbolic torus of system (12.36) in 3-space which is persistent for small, positive values of $\varepsilon$.

Actually the authors, choosing $\mu = 0.4, \omega = 0.9$, consider what happens if $\varepsilon$ increases. At $\varepsilon = 0.3634$ the normal hyperbolicity is destroyed by a saddle-node bifurcation.

**Extension to Forcing by Coupled Van der Pol Oscillators**
An example of a normally hyperbolic torus in a coupled system is found in the system

$$\begin{cases} \ddot{x} + x &= \mu(1 - x^2)\dot{x} + \varepsilon f(x, y), \\ \ddot{y} + \omega^2 y &= \mu(1 - y^2)\dot{y} + \varepsilon g(x, y), \end{cases} \tag{12.37}$$

with $\varepsilon$-independent positive constant $\omega$ and $\mu$ (fixed positive numbers, $O(1)$ with respect to $\varepsilon$) and smooth perturbations $f, g$. Omitting the perturbations $f, g$ we have two uncoupled normally hyperbolic oscillations. In general, if $\omega$ is irrational, the combined oscillations attract to a torus in 4-space, the product of the two periodic attractors, filled with quasi-periodic motion. Adding the perturbations $f, g$ cannot destroy this two-dimensional torus but only deforms it.

If $\mu$ is large enough, the tori are for $\varepsilon = 0$ built up from relaxation oscillations. The timescales of rotation are also different in this case; see [65].

## 12.14 A Flywheel on an Elastic Foundation

We have seen the phenomenon of locking into resonance. We consider now an interesting application involving passage through resonance and the possibility of (undesirable) capture into resonance. Consider a spring, modelling an elastic foundation, that can move in the vertical $x$ direction on which a rotating wheel is mounted; the rotation angle is $\phi$. The wheel has a small mass fixed on the edge that makes it slightly eccentric, a flywheel; see Fig. 12.15. The vertical displacement $x$ and the rotation $\phi$ are determined by the equations

$$\ddot{x} + x = \varepsilon(-x^3 - \dot{x} + \dot{\phi}^2 \cos\phi) + O(\varepsilon^2),$$
$$\ddot{\phi} = \varepsilon\left(\frac{1}{4}(2 - \dot{\phi}) + (1 - x)\sin\phi\right) + O(\varepsilon^2).$$

See [52] for the equations; we have added an appropriate scaling, assuming that the friction, the non-linear restoring force, the eccentric mass and several other forces are small. To obtain a standard form suitable for averaging we transform

$$x = r\sin\phi_2,\ \dot{x} = r\cos\phi_2,\ \phi = \phi_1,\ \dot{\phi}_1 = \Omega,$$

**Fig. 12.15** A rotating flywheel with a small eccentric mass is mounted on an elastic foundation. Courtesy A. Tondl

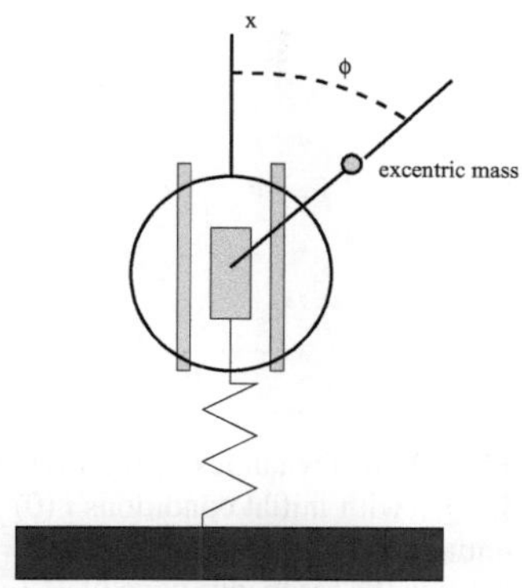


with $r > 0$, $\Omega > 0$. This introduces two angles and two slowly varying quantities:

$$\dot{r} = \varepsilon \cos\phi_2(-r^3 \sin^3\phi_2 - r\cos\phi_2 + \Omega^2 \cos\phi_1),$$

$$\dot{\Omega} = \varepsilon\left(\frac{1}{4}(2 - \Omega) + \sin\phi_1 - r \sin\phi_1 \sin\phi_2\right),$$

$$\dot{\phi}_1 = \Omega,$$

$$\dot{\phi}_2 = 1 + \varepsilon\left(r^2 \sin^4\phi_2 + \frac{1}{2}\sin 2\phi_2 - \frac{\Omega^2}{r}\cos\phi_1 \sin\phi_2\right).$$

The $O(\varepsilon^2)$ terms have been omitted. In the equation for $r$ and $\Omega$ to $O(\varepsilon)$ the angles are $\phi_1, \phi_2, \phi_1 + \phi_2, \phi_1 - \phi_2$. As $\phi_1$ and $\phi_2$ are monotonically increasing and so timelike the only resonance zone that can arise is at this order $\phi_1 - \phi_2 = 0$, which determines the resonance manifold $\Omega = 1$. Outside the resonance zone, a neighbourhood of $\Omega = 1$, we average over the angles to find the approximations given by

$$\dot{r} = -\frac{1}{2}\varepsilon r,$$

$$\dot{\Omega} = \frac{1}{4}\varepsilon(2 - \Omega).$$

This is already an interesting result. Outside the resonance zone $r(t)$ will decrease exponentially with time; on the other hand $\Omega(t)$ will tend to the value 2. If we start with $\Omega(0) < 1$, $\Omega(t)$ will after some time enter the resonance zone around $\Omega = 1$. How does this affect the dynamics? Will the system pass in some way through resonance or will it stay in the resonance zone, resulting in vertical oscillations that are undesirable for a mounted flywheel. Such phenomena are illustrated in Fig. 12.16 where capture in the resonance zone $\Omega = 1$ and transition is shown.

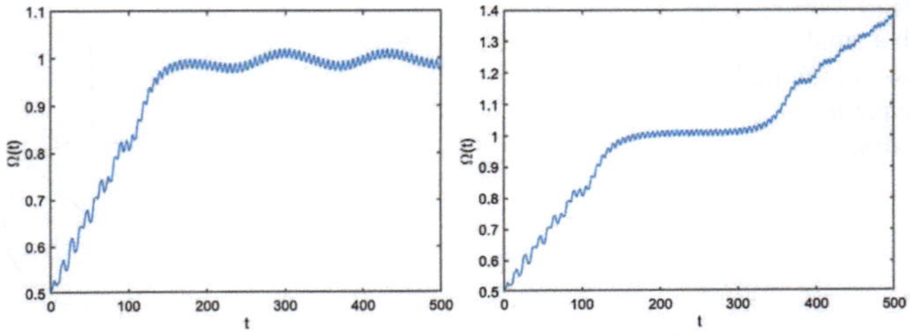


**Fig. 12.16** Dynamics of the Flywheel on an Elastic Foundation. Left: capture in resonance near $\Omega = 1$ with initial conditions $r(0) = 1, \Omega(0) = 0.5, \phi_1(0) = 1, \phi_2(0) = 0$ and $\varepsilon = 0.01$. With initial conditions $r(0) = 1, \Omega(0) = 0.5, \phi_1(0) = 0.9600, \phi_2(0) = 0$ we have a slow transition through the resonance zone $\Omega = 1$ (right picture)

The way to answer these questions systematically is to analyse what is going on in the resonance zone and find out whether there are attractors present. Following the analysis of localising into the resonance zone as before, see Chap. 7, we introduce the resonant combination angle $\psi = \phi_1 - \phi_2$ and the local variable

$$\omega = \frac{\Omega - 1}{\sqrt{\varepsilon}}.$$

Transforming the equations for $r$, $\Omega$ and the angles the leading terms are $O(\sqrt{\varepsilon})$; we find

$$\begin{aligned}
\dot{r} &= \varepsilon \cdots , \\
\dot{\omega} &= \sqrt{\varepsilon}\left(\frac{1}{4} + \sin\phi_1 - \frac{1}{2}r\cos\psi + \frac{1}{2}r\cos(2\phi_1 - \psi)\right) + \varepsilon \cdots , \\
\dot{\psi} &= \sqrt{\varepsilon}\omega + \varepsilon \cdots , \\
\dot{\phi}_1 &= 1 + \sqrt{\varepsilon}\omega.
\end{aligned}$$

We can average over the remaining angle $\phi_1$; as the equation for $r$ starts with $O(\varepsilon)$ terms, we have in the resonance zone that $r(t) = r(0) + O(\sqrt{\varepsilon})$. The equations for the first-order approximations of $\omega$ and $\psi$ are

$$\begin{aligned}
\dot{\omega} &= \sqrt{\varepsilon}\left(\frac{1}{4} - \frac{1}{2}r\cos\psi\right), \\
\dot{\psi} &= \sqrt{\varepsilon}\omega.
\end{aligned}$$

By differentiation of the equation for $\psi$ we can write this as the pendulum equation

$$\ddot{\psi} + \frac{1}{2}\varepsilon r(0)\cos\psi = \frac{1}{4}\varepsilon.$$

The timescale of the dynamics is clearly $\sqrt{\varepsilon}t$; there are two equilibria, one a centre point and the other a saddle. They correspond with periodic solutions in the resonance zone of the original system. The saddle is definitely unstable, and for the centre point we have to perform higher-order averaging, to $O(\varepsilon)$, to determine the stability. This analysis was carried out by Van den Broek [134]; see also [133]. The result is that by a second-order calculation in the small parameter the centre point in the resonance zone becomes an attracting focus so that the corresponding periodic solution is stable.

The next (difficult) step is to determine the set of initial values that lead to capture. In [134] an approach, formulated by Haberman in [68], was used as follows. Compute a second-order approximation of the solutions outside the resonance zone and determine the values assumed by the solutions on entering the resonance zone. Using the second order approximation in the resonance zone we find that the invariant manifolds of the saddle will have different 'energy' values; these will be used to localise initial values from entering the resonance zone that lead to transition of the zone and capture into resonance.

The implication is that for certain initial values the oscillator-flywheel might pass into resonance and stays there. Van den Broek [134] identified three sets of initial values leading to capture into resonance. The asymptotic estimates are close to the numerical results; in the case $\varepsilon = 0.01$ and with fixed $r(0) = 1, \Omega(0) = 0.5, \phi_2(0) = 0$ we find numerically the three sets for $\phi_1$ values leading to capture

$$[0.962, 1.227]\ [2.845, 3.148]\ [4.764, 4.923].$$

It is surprising that there are three sets; it raises the question whether higher-order approximations might produce even more sets, presumably of smaller size.

## 12.15 Torus Bifurcations in Two Coupled Oscillators

The analysis becomes more complicated if we do not start with a torus structure for $\varepsilon = 0$ to consider for perturbations. We will describe a few results for autonomous coupled oscillators of the paper [8]:

$$\begin{cases} \ddot{x} + x + \varepsilon(axy + \delta x^2\dot{x} + \gamma x^3) = 0, \\ \ddot{y} + 4y + \varepsilon(\kappa\dot{y} + bx^2) = 0. \end{cases} \tag{12.38}$$

The system turns out to have many interesting phenomena. We can rescale $x$ to obtain $a = 1$. The $x$-oscillator has a non-linear damping and the $y$-oscillator a linear one ($\kappa, \delta \geq 0$); the coupling is non-linear. The strategy is to find a periodic solution by averaging; to determine its stability is possible analytically but easier using numerical continuation, for instance, by MATCONT [102] while determining the Floquet (Lyapunov) exponents at each step. When the exponents correspond with a Neimark-Sacker bifurcation we expect the emergence of a 2-torus. A bonus in this case is that continuation also reveals destruction of tori and the emergence of a strange attractor. It will turn out that the parameter $\gamma$ is suitable for continuation, probably because the term $\gamma x^3$ changes the oscillation period of the $x$-oscillator in a non-linear way.

Transforming to amplitude-phase coordinates (1.6) $x = r_1 \cos(t + \phi_1)$, etc. and averaging we find with combination angle $\chi = 2\phi_1 - \phi_2$

$$\begin{cases} \dot{r}_1 &= \varepsilon r_1 (\frac{1}{4} r_2 \sin\chi - \frac{\delta}{8} r_1^2), \\ \dot{r}_2 &= \frac{\varepsilon}{2} r_2 (-b \frac{r_1^2}{4r_2} \sin\chi - \kappa), \\ \dot{\chi} &= \varepsilon (\frac{1}{2} r_2 \cos\chi + \frac{3}{4}\gamma r_1^2 - b \frac{r_1^2}{8r_2} \cos\chi). \end{cases} \tag{12.39}$$

Putting the right-hand sides of system (12.39) equal to zero we look for solutions outside the coordinate planes. From the first two equations we find non-trivial solutions if

$$\sin\chi = \pm\sqrt{\frac{-2\kappa\delta}{b}},$$

leading to the condition $b < -2\kappa\delta < 0$; $b$ has to be negative. The third equation in (12.39) implies $\cos\chi < 0$. We find for $\kappa > 0$ necessary conditions for equilibria:

$$r_2 = -\frac{b}{4\kappa} r_1^2 \sin\chi,\ r_1 = \sqrt{\frac{-8\kappa^2 \cot\chi}{12\kappa\gamma - b\sin 2\chi}}. \tag{12.40}$$

It is clear that we have the condition $\sin\chi \geq 0$. If $\kappa = 0$, we find a continuous family of equilibria and at this order of approximation not an isolated periodic solution. System (12.38) has a phase-locked periodic solution if $\kappa, \delta > 0, b < -2\kappa\delta$ and $\pi/2 < \chi < \pi$.

A surprising result in [8] obtained by using the Routh-Hurwitz criterion is that this isolated periodic solution is asymptotically stable. This result is valid for small $\varepsilon$ and would *exclude* a torus bifurcation, so it might produce new results when considering larger values of $\varepsilon$. At this point we combine again asymptotic analysis and numerical bifurcation methods, in our case MATCONT [102]. As an example we choose in system (12.38) $\varepsilon = 1, a = 0.5, b = -0.5, \delta = 0.4, \kappa = 0.1$. We use $\gamma$ as bifurcation parameter, starting with $\gamma = 0.2$ and slowly decreasing its value. Near the value $\gamma = 0.096$ we detect a Neimark-Sacker bifurcation, producing a stable torus that is not very close to the origin of phase-space. Decreasing $\gamma$ we find at smaller values, $\gamma = 0.0892$, torus destruction while a strange attractor emerges; see Fig. 12.17. In [8] there are more pictures, in particular Poincaré sections, see for instance Fig. 12.18. Also, more technical details like period doublings and Arnold tongues are discussed.

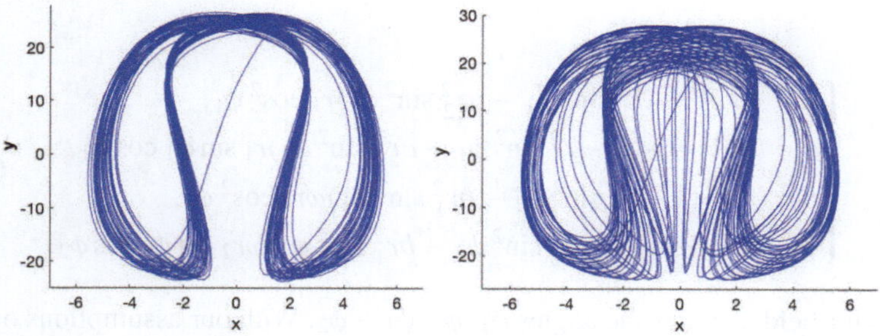


**Fig. 12.17** Two projections from 4-space on the $x, y$-plane of system (12.38). We took $a = 0.5, b = -0.5, \delta = 0.4, \kappa = 0.1$ and in both figures $x(0) = 1, \dot{x}(0) = 0, y(0) = 22, \dot{y}(0) = 0$. If $\gamma > 0.096$, we find a stable periodic solution, but at $\gamma = 0.096$ a Neimark-Sacker bifurcation produces a torus (left). If $\gamma = 0.0892$, we find a strange attractor (right) with Kaplan-Yorke dimension $D_{KY} = 2.32$. Note that the initial conditions produce some transient orbits

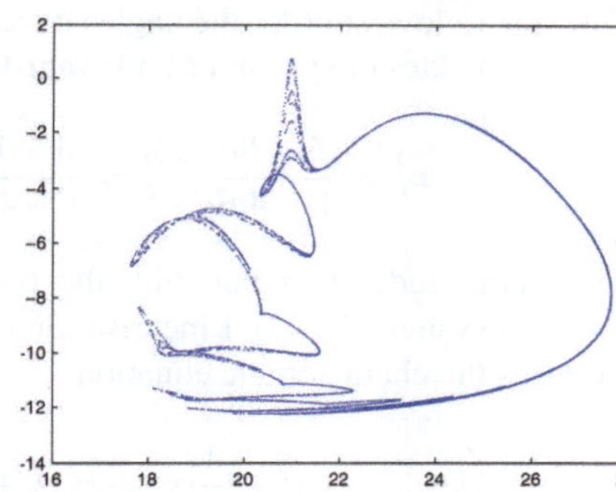

**Fig. 12.18** Projection $(y, \dot{y})$ of the Poincaré section with $x = 0$ as cross section. The torus no longer exists if $\gamma = 0.0892$; a strange attractor emerges instead with Kaplan-Yorke dimension $d_{KY}$2.32 Courtesy Z. Angew. Math. Phys. [8]

## 12.16 Interaction of Self-Excited Oscillations

In [71] Hale considered interaction of self-excited oscillators. Consider the system

$$\begin{cases} \ddot{x} + x &= \varepsilon(1 - x^2 - ay^2)\dot{x}, \\ \ddot{y} + \omega^2 y &= \varepsilon(1 - y^2 - bx^2)\dot{y}. \end{cases} \tag{12.41}$$

The normal mode planes are invariant manifolds containing a self-excited Van der Pol periodic solution. We consider the solutions in general position.

*Suppose that* $\omega \neq 1$ *and not* $\varepsilon$*- close to* 1; $0 < a, b \leq 1$.

We can characterise two angles in system (12.41) following the analysis of averaging over angles from Chap. 7. Transformation of (1.21)

$$x = r_1 \sin\phi_1, \dot{x} = r_1 \cos\phi_1, y = r_2 \sin\phi_2, \dot{y} = \omega r_2 \cos\phi_2$$

yields

$$\begin{cases} \dot{r}_1 &= \varepsilon(1 - r_1^2 \sin^2 \phi_1 - a r_2^2 \sin^2 \phi_2) r_1 \cos^2 \phi_1, \\ \dot{\phi}_1 &= 1 - \frac{\varepsilon}{r_1}(1 - r_1^2 \sin^2 \phi_1 - a r_2^2 \sin^2 \phi_2) r_1 \sin \phi_1 \cos \phi_1, \\ \dot{r}_2 &= \frac{\varepsilon}{\omega}(1 - r_2^2 \sin^2 \phi_2 - b r_1^2 \sin^2 \phi_1) \omega r_2 \cos^2 \phi_2, \\ \dot{\phi}_2 &= \omega - \frac{\varepsilon}{\omega}(1 - r_2^2 \sin^2 \phi_2 - b r_1^2 \sin^2 \phi_1) \omega r_2 \sin \phi_2 \cos \phi_2. \end{cases} \tag{12.42}$$

The vector field contains the angles $\phi_1, \phi_2, \phi_1 - \phi_2$. With our assumptions on $\omega$ we can average over the angles; there is no resonance manifold. We find after averaging

$$\begin{cases} \dot{r}_1 &= \frac{\varepsilon}{2} r_1 (1 - \frac{1}{4} r_1^2 - \frac{1}{2} a r_2^2), \dot{\phi}_1 = 1 + \varepsilon \ldots \\ \dot{r}_2 &= \frac{\varepsilon}{2} r_2 (1 - \frac{1}{4} r_2^2 - \frac{1}{2} b r_1^2), \dot{\phi}_2 = \omega + \varepsilon \ldots \end{cases} \tag{12.43}$$

Note that to lowest order the angles depend only on the basic frequencies $1, \omega$. The right-hand sides of system (12.43) vanish to $O(\varepsilon)$ if

$$r_1^2 = \frac{4 - 8a}{1 - 4ab}, \; r_2^2 = \frac{4 - 8b}{1 - 4ab}, \; 0 < a, b < \frac{1}{2}. \tag{12.44}$$

These amplitudes together with the two moving angles describe a torus of the averaged system (12.43). Linearisation of system (12.43) at the amplitudes (12.44) produces the characteristic equation

$$\lambda^2 + \frac{1}{2}(r_1^2 + r_2^2)\lambda + \frac{1}{4}(1 - ab) r_1^2 r_2^2 = 0.$$

We conclude that the torus is normally hyperbolic, persists for the original system (12.41) and is stable.

**Remark 12.2** Consider the special case $\omega$ not $\varepsilon$-close to 1 but $a = b = \frac{1}{2}$. In this case the torus of system (12.43) collapses to a sphere in phase-space described by $r_1^2 + r_2^2 = 4$. This is a degenerate case as $r_1, r_2$ restricted to the sphere are still free to choose. Higher-order approximations may change the dynamics qualitatively.

As an illustration of the analysis in the case $\omega = 1$ we reproduce a result from [148] ex. 12.13. Using transformation (1.6) to $r_1, r_2, \phi_1, \phi_2$ we find after first-order averaging

$$\begin{cases} \dot{r}_1 &= \frac{\varepsilon}{2} r_1 (1 - \frac{1}{4} r_1^2 - \frac{a}{2} r_2^2 + \frac{a}{4} r_2^2 \cos 2\chi), \\ \dot{r}_2 &= \frac{\varepsilon}{2} r_2 (1 - \frac{1}{4} r_2^2 - \frac{b}{2} r_1^2 + \frac{b}{4} r_1^2 \cos 2\chi), \\ \dot{\chi} &= -\frac{\varepsilon}{4}(b r_1^2 + a r_2^2) \sin 2\chi, \end{cases} \tag{12.45}$$

with $\chi = \phi_1 - \phi_2$. Putting $\sin 2\chi = 0$ we find periodic solutions.

We can also look for exact solutions by putting $y(t) = px(t)$; we find a continuous family of $p$-values producing exact Van der Pol periodic solutions in system (12.41) with $\omega = 1$. If, for instance, $a = b = \frac{1}{2}$, we have the family of exact solutions $y(t) = \pm x(t)$ in phase-space that contains attracting periodic solutions. For more details see [148].

## 12.17 Iteration by Integral Equations

We consider again Eq. (1.1) for $n$-dimensional ODEs

$$\dot{x} = f_0(t, x) + \varepsilon f_1(t, x) + O(\varepsilon^2).$$

As we have seen in the Introduction, simple expansion with respect to $\varepsilon$ to obtain approximations of the solutions runs into difficulties. The basic idea of the Poincaré-Lindstedt method, Sect. 3.3, is to use the Poincaré expansion theorem in combination with iteration via a contraction argument (see the Introduction, Chap. 1) on a suitable integral equation. The expansions produce approximation results on the timescale 1, so it is essential to focus on solutions that can be characterised on such a short timescale; in Sect. 3.3 we considered periodic solutions. The periodicity conditions are directly related to averaging.

We can use this idea also to approximate quasi-periodic solutions and bounded invariant manifolds of Eq. (1.1), for instance tori. In this case we have to start with a suitable solution of the equation if $\varepsilon = 0$ with $n$ free parameters $\lambda$; suppose this solution is $F(t, \lambda)$. We can formulate the equivalent Volterra integral equations or use variation of constants to take profit of the behaviour of the unperturbed solution $F(t, \lambda)$. We have some freedom in formulating the integral equations. In both cases we use iteration and Poincaré expansion while applying *quasi-periodic secularity conditions* on the parameters $\lambda$.

We will make the procedure more explicit for $N$ degree-of-freedom (dof) systems of the form

$$\ddot{x}_i + \omega_i^2 x_i = \varepsilon f(t, x, \dot{x}), \tag{12.46}$$

with $i = 1, 2, \ldots, N$, $x = (x_1, x_2, \ldots, x_N)$. Cases of particular interest are when some pairs of constant, positive frequencies $\omega_i, i = 1, 2, \ldots, N$ have irrational ratio's. For such a $2N$-dimensional system we use the variational equations from the Introduction. To be more explicit suppose amplitude-phase transformations (1.6) are suitable. Expanding the parameters $r_i, \phi_i$ in system (12.46) we have at zero order $x_i^0(t) = r_i^0 \cos(\omega_i t + \phi_i^0)$ and to next order

$$\tilde{x}(t) = r_i^0 \cos(\omega_i t + \phi_i^0) + \varepsilon \int_0^t \sin(\omega_i t - \omega_i s) f(s, x_i^0(s), \dot{x}_i^0)(s)) ds, \; i = 1, 2, \ldots, N. \tag{12.47}$$

If system (12.47) contains external forcing we can adapt the integral equation using appropriate variational transformations; see Example 12.1 below. For the solutions $x_i$ we will apply *quasi-periodic secularity conditions* that will produce at first-order $2N$ equations with $2N$ parameter values $r_i^0, \phi_i^0$. Solving the $2N$ (in general) transcendental equations and computing the Jacobian of the system we may find that its rank is $2N$. In this case the implicit function theorem applies, and we have pinpointed a quasi-periodic solution of system (12.46). The approximation of this solution is $\tilde{x}_i(t)$ with explicit parameter values substituted.

If the rank of the Jacobian matrix is $2N - 1$, we have to this order of approximation a branching of the solutions and in some cases a family of quasi-periodic solutions on a bounded manifold. This will often be a torus.

**Remark 12.3** If some of the equations of system (12.46) are autonomous, we expect a time-dependent shift of the phase. However, this is a complication that arises at $O(\varepsilon)$ and so for $O(\varepsilon^2)$ in the corresponding integral equation. We met this behaviour already in Sect. 3.3 for the Van der Pol equation where we have the approximation for the phase $\phi(t) = \phi_0 + O(\varepsilon)$ with for the periodic solution perturbation terms dependent on $t$.

As an example consider again a case of the forced Van der Pol equation of Sect. 5.3.4.

**Example 12.1**

$$\ddot{x} + x = \varepsilon(1 - x^2)\dot{x} + a\cos\sqrt{2}t. \tag{12.48}$$

If $\varepsilon = 0$, we have the solution

$$x^0(t) = r_0\cos(t + \phi_0) - a\cos\sqrt{2}t,$$

with constants $r_0, \phi_0$. Consider this 'unperturbed' solution as the lowest order approximation of the iteration scheme for the integral equation

$$x(t) = r\cos(t + \phi) - a\cos\sqrt{2}t + \varepsilon\int_0^t \sin(t - s)(1 - x^2(s))\dot{x}(s)ds, \tag{12.49}$$

with Poincaré expansions $r = r_0 + \varepsilon r_1 + O(\varepsilon^2)$, $\phi = \phi_0 + \varepsilon\phi_1 + O(\varepsilon^2)$. The second-order approximation is obtained by replacing $x$ by the lowest order approximation

$$\tilde{x}(t) = r_0\cos(t + \phi_0) - a\cos\sqrt{2}t + \varepsilon I(t), \tag{12.50}$$

with

$$I(t) = \int_0^t \sin(t - s)(1 - (r_0\cos(s + \phi_0) - a\cos\sqrt{2}s)^2)$$
$$(-r_0\sin(s + \phi_0) + a\sqrt{2}\sin\sqrt{2}s)ds.$$

Expanding $\sin(t-s) = \sin t \cos s - \cos t \sin s$ the integral splits into two parts. To find an approximate quasi-periodic solutions we have to choose $r_0, \phi_0$ so that the integrals produce no other type of solution. These are the *quasi-periodic secularity conditions*. In other words we have to choose the constants such that the generalised Fourier expansion of the integrands contain bounded oscillating terms only. Evaluating

$$\int_0^t \cos s(1-(r_0\cos(s+\phi_0)-a\cos\sqrt{2}s)^2)(-r_0\sin(s+\phi_0)+a\sqrt{2}\sin\sqrt{2}s)ds,$$

we find the condition

$$r_0 \sin\phi_0(1-\frac{1}{2}a^2-\frac{1}{4}r_0^2)=0.$$

Evaluating

$$\int_0^t \sin s(1-(r_0\cos(s+\phi_0)-a\cos\sqrt{2}s)^2)(-r_0\sin(s+\phi_0)+a\sqrt{2}\sin\sqrt{2}s)ds,$$

we find the condition

$$r_0 \cos\phi_0(1-\frac{1}{2}a^2-\frac{1}{4}r_0^2)=0.$$

We conclude that if $0 < a^2 < 2$, we have an $O(\varepsilon)$ approximate family of quasi-periodic solutions of the form

$$\tilde{x}(t) = 2\sqrt{1-\frac{1}{2}a^2}\cos(t+\phi_0) - a\sqrt{2}t. \tag{12.51}$$

If $a^2 > 2$, we have a periodic solution of the form $-a\cos\sqrt{2}t$. The 'standard' Van der Pol periodic solution has been destroyed by quasi-periodic forcing.

We will analyse some examples of the form (12.46) with $N = 2$.

## 12.18 Hale's Example by Iteration

Consider iteration for integral equations in the case of Hale's example 12.16 for system (12.41):

$$\ddot{x} + x = \varepsilon(1-x^2-ay^2)\dot{x},$$
$$\ddot{y} + \omega^2 y = \varepsilon(1-y^2-bx^2)\dot{y}.$$

Frequency $\omega$ is irrational and not $\varepsilon$-close to 1, $0 < a, b < 1$. The equivalent integral equations are

$$\begin{cases} x(t) = r\cos(t+\phi) + \varepsilon \int_0^t \sin(t-s)(1 - x^2(s) - ay^2(s))ds, \\ y(t) = R\cos(\omega t + \psi) + \varepsilon \int_0^t \sin(\omega t - \omega s)(1 - y^2(s) - bx^2(s))ds \end{cases} \tag{12.52}$$

We exclude the normal modes and expand $x(t)$, $y(t)$ and the parameters $r = r_0 + \varepsilon r_1 + O(\varepsilon^2)$, $\phi = \phi_0 + \varepsilon\phi_1 + O(\varepsilon^2)$, $R = R_0 + \varepsilon R_1 + O(\varepsilon^2)$, $\psi = \psi_0 + \varepsilon\psi_1 + O(\varepsilon^2)$. At lowest order we find for the $x$-approximation

$$\tilde{x}(t) = r_0\cos(t+\phi_0) - \varepsilon \int_0^t \sin(t-s)(1 - r_0^2\cos^2(s+\phi_0) \\ -aR_0^2\cos^2(\omega s + \psi_0))r_0\sin(s+\phi_0)ds.$$

The integral consists of two parts:

$$-\varepsilon r_0 \sin t \int_0^t (1 - r_0^2\cos^2(s+\phi_0) - aR_0^2\cos^2(\omega s + \psi_0))\frac{1}{2}\sin(2s + 2\phi_0)ds.$$

and

$$\varepsilon r_0 \cos t \int_0^t (1 - r_0^2\cos^2(s+\phi_0) - aR_0^2\cos^2(\omega s + \psi_0))\sin^2(s+\phi_0)ds.$$

The first integral is uniformly bounded and gives no secularity condition, and the second integral gives the quasi-periodic secularity condition

$$\frac{1}{2}r_0^2 + aR_0^2 = 4. \tag{12.53}$$

At lowest order we have the $y$-approximation

$$\tilde{y}(t) = R_0\cos(\omega t + \psi_0) - \varepsilon \int_0^t \sin(\omega t - \omega s)(1 - R_0^2\cos^2(\omega s + \psi_0) \\ -br_0^2\cos^2(s+\phi_0))R_0\sin(\omega s + \psi_0)ds.$$

The integral has two parts:

$$-\varepsilon R_0 \sin\omega t \int_0^t (1 - R_0^2\cos^2(\omega s + \psi_0) - br_0^2\cos^2(s+\phi_0))\frac{1}{2}\sin(2\omega s + 2\psi_0)ds.$$

and

$$\varepsilon R_0 \cos \omega t \int_0^t (1 - R_0^2 \cos^2(\omega s + \psi_0) - br_0^2 \cos^2(s + \phi_0)) \sin^2(\omega s + \psi_0) ds.$$

The first integral is uniformly bounded and gives no condition, and the other integral gives the second quasi-periodic secularity condition

$$\frac{1}{2} R_0^2 + br_0^2 = 4. \tag{12.54}$$

In $r_0$, $R_0$ parameter-space the two secularity conditions produce two ellipses. Only positive values of $r_0$, $R_0$ are useful; we find one solution if $0 < a < \frac{1}{2}, 0 < b < \frac{1}{2}$ and if $\frac{1}{2} < a < 1, \frac{1}{2} < b < 1$; these values correspond with two-dimensional tori in phase-space. If $a = b = \frac{1}{2}$, we find an invariant sphere.

The first-order approximations for $x(t)$, $y(t)$ can be obtained by performing the integrations of $\tilde{x}(t)$, $\tilde{y}(t)$ using the quasi-periodic secularity conditions. The results agree with averaging for this problem. The parameters $\phi_0$, $\psi_0$ are still free, so for the approximation we have a two-parameter family of solutions.

## 12.19 Iteration of Two Coupled Oscillators with Forcing

Consider the system of two coupled oscillators with damping and forcing and non-linear coupling governed by the terms with coefficient $a \neq 0$.

$$\begin{cases} \ddot{x} + x &= \varepsilon(-\mu \dot{x} + axy^2 + h_1 \sin t), \\ \ddot{y} + 2y &= \varepsilon(-\mu \dot{y} + ax^2 y + h_2 \sin \sqrt{2}t). \end{cases} \tag{12.55}$$

Assume $\mu \geq 0$, $h_1, h_2$ are constant amplitudes. Consider the equivalent integral equations

$$\begin{cases} x(t) &= r \cos(t + \phi) + \varepsilon \int_0^t \sin(t - s)(-\mu \dot{x}(s) + ax(s)y^2(s) + h_1 \sin s) ds, \\ y(t) &= R \cos(\sqrt{2}t + \psi) + \frac{\varepsilon}{\sqrt{2}} \int_0^t \sin(\sqrt{2}t - \sqrt{2}s)(-\mu \dot{y}(s) + ax^2(s)y(s) \\ & \qquad + h_2 \sin \sqrt{2}s) ds \end{cases} \tag{12.56}$$

As in the preceding example we exclude the normal modes and expand $x(t)$, $y(t)$ and the parameters $r = r_0 + \varepsilon r_1 + O(\varepsilon^2)$, $\phi = \phi_0 + \varepsilon \phi_1 + O(\varepsilon^2)$, $R = R_0 + \varepsilon R_1 + O(\varepsilon^2)$, $\psi = \psi_0 + \varepsilon \psi_1 + O(\varepsilon^2)$. At each order of approximation we have four parameter values. Using as lowest order approximation

$x_0(t) = r_0 \cos(t + \phi_0)$, $\dot{x}_0(t) = -r_0 \sin(t + \phi_0)$, $y_0(t) = R_0 \cos(\sqrt{2}t + \psi_0)$, $\dot{y}_0(t) = -\sqrt{2} R_0 \sin(\sqrt{2}t + \psi_0)$ we find for the next step

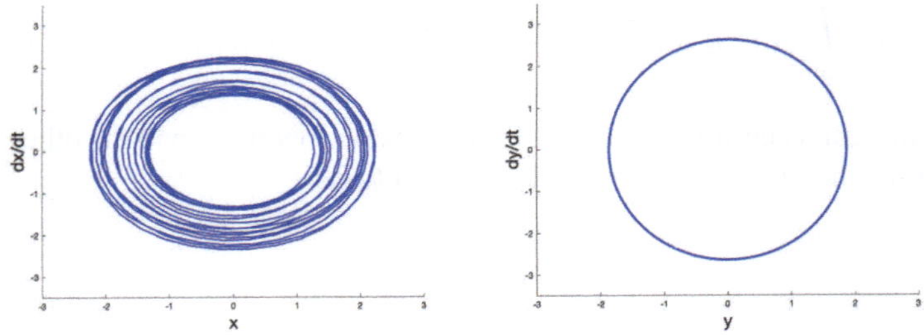


**Fig. 12.19** Two projections from 4-space on the $x, \dot{x}$-plane and the $y, \dot{y}$-plane of system (12.55). Corresponding with the quasi-periodic secularity conditions we took $\mu = 1, a = -0.5\sqrt{3}, h_1 = 1.37315, h_2 = -3.26606, \varepsilon = 0.05$ and $x(0) = 0.6866, \dot{x}(0) = -1.1892, y(0) = 1.7321, \dot{y}(0) = 1$

$$\tilde{x}(t) = r_0 \cos(t + \phi_0) + \varepsilon I_1(t), \tilde{y}(t) = R_0 \cos(\sqrt{2}t + \psi_0) + \frac{\varepsilon}{\sqrt{2}} I_2(t)$$

with integral expressions

$$I_1(t) = \int_0^t \sin(t-s)(\mu r_0 \sin(s+\phi_0) + a r_0 \cos(s+\phi_0) R_0^2 \cos^2(\sqrt{2}s+\psi_0) + h_1 \sin s) ds,$$

and for $I_2(t)$

$$\int_0^t \sin(\sqrt{2}t - \sqrt{2}s)(\mu \sqrt{2} R_0 \sin(\sqrt{2}s + \psi_0) + a r_0^2 \cos^2(s + \phi_0) R_0 \cos(\sqrt{2}s + \psi_0)$$
$$+ h_2 \sin \sqrt{2}s) ds.$$

Expanding $\sin(t - s) = \sin t \cos s - \cos t \sin s$ and using that $r_0$, $R_0$ are positive we find for $I_1(t)$ the quasi-periodic secularity conditions

$$\mu \sin \phi_0 + \frac{a}{2} R_0^2 \cos \phi_0 = 0, \mu r_0 \cos \phi_0 - \frac{a}{2} r_0 R_0^2 \sin \phi_0 + h_1 = 0. \quad (12.57)$$

Expanding $\sin(\sqrt{2}t - \sqrt{2}s)$ we find for $I_2(t)$ the quasi-periodic secularity conditions

$$\mu\sqrt{2} \sin \psi_0 + \frac{a}{2} r_0^2 \cos \psi_0 = 0, \mu\sqrt{2} R_0 \cos \psi_0 - \frac{a}{2} r_0^2 R_0 \sin \psi_0 + h_2 = 0. \quad (12.58)$$

We have to exclude $\sin \phi_0 = 0$ and $\sin \psi_0 = 0$ as these values lead to normal mode solutions. Also $\cos \phi_0$ and $\cos \psi_0$ should not vanish. In Fig. 12.19 we show two projections of a torus corresponding with parameter values of the quasi-periodic secularity conditions.

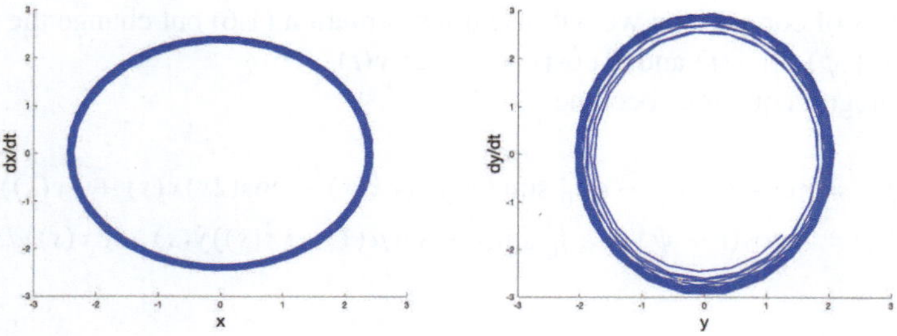


**Fig. 12.20** Two projections from 4-space on the $x, \dot{x}$-plane and the $y, \dot{y}$-plane of system (12.55). Corresponding with the quasi-periodic secularity conditions we took $\mu = 1, a = -0.5, h_1 = -2^{1.75}, h_2 = -4, \varepsilon = 0.05$ and $x(0) = 2^{3/4}, \dot{x}(0) = -2^{3/4}, y(0) = \sqrt{2}, \dot{y}(0) = -\sqrt{2}$

We compute the Jacobian $J$ of the four secularity conditions to test the uniqueness of the parameter values and the corresponding solutions. We find with abbreviations $\cos\phi_0 = C_f, \sin\phi_0 = S_f, \cos\psi_0 = C_p, \sin\psi_0 = S_p$

$$J = \begin{pmatrix} 0 & \mu C_f - \frac{a}{2}R_0^2 S_f & a R_0 C_f & 0 \\ \mu C_f - \frac{a}{2}R_0^2 S_f & -\mu r_0 S_f - \frac{a}{2} r_0 R_0^2 C_f & -a r_0 R_0 S_f & 0 \\ a r_0 C_p & 0 & 0 & \mu\sqrt{2}C_p - \frac{a}{2}r_0^2 S_p \\ -a r_0 R_0 S_p & 0 & \mu\sqrt{2}C_p - \frac{a}{2}r_0^2 S_p & -\mu\sqrt{2}R_0 S_p - \frac{a}{2}r_0^2 R_0 C_p \end{pmatrix}.$$

We use the parameter values of Fig. 12.19. We expect that we can use the implicit function theorem if the rank of the matrix is 4. A MATLAB calculation produces that the determinant of $J$ does not vanish, so the implicit function theorem applies producing for these parameter values a quasi-periodic solution of system (12.55).

An example with different parameter values is shown in Fig. 12.20. The quasi-periodic secularity conditions have been used to produce a torus of system (12.55). In this case a MATLAB calculation of the corresponding Jacobian $J$ shows rank 3; we have a branching producing a one-parameter family of quasi-periodic solutions at this order of approximation.

## 12.20 Parametrically Excited and Self-Excited Oscillations

Consider as a simple model of coupled parametric- and self-excitation

$$\begin{aligned} \ddot{x} + \varepsilon\kappa\dot{x} + (1 + \varepsilon\cos 2t)x + \varepsilon c y &= 0, \\ \ddot{y} - \varepsilon\mu(1 - y^2)\dot{y} + y + \varepsilon c x &= 0. \end{aligned}$$

As an alternative approach we will use iteration and the quasi-periodic secularity condition. Averaging produces in a natural way the combination angle $\chi = \phi_1 - \phi_2$.

For reasons of comparison we will use transformation (1.6) but change the notation to $r\cos(t+\phi)$ for $x(t)$ and $R\cos(t+\psi)$ for $y(t)$.

The integral equations become

$$\begin{cases} x(t) &= r\cos(t+\phi) - \varepsilon\int_0^t \sin(t-s)(\kappa\dot{x}(s) + \cos(2s)x(s) + cy(s))ds, \\ y(t) &= R\cos(t+\psi) + \varepsilon\int_0^t \sin(t-s)(\mu(1-y^2(s))\dot{y}(s) - cx(s))ds. \end{cases} \tag{12.59}$$

where we still have to replace $x$, $y$, etc. by transformation (1.6). As in earlier examples we expand the amplitudes and phases taking into account Remark 12.3. Substitution of the lowest order approximations $x_0(t) = r_0(\cos(t+\phi_o)$, $y_0(t) = R_0\cos(t+\psi_0)$ produces for $x(t)$ the secularity conditions:

$$\begin{cases} -\kappa r_0\sin\phi_0 + \frac{1}{2}r_0\cos\phi_0 + cR_0\cos\psi_0 &= 0, \\ -\kappa r_0\cos\phi_0 + \frac{1}{2}r_0\sin\phi_0 - cR_0\sin\psi_0 &= 0. \end{cases} \tag{12.60}$$

Eliminating $r_0/R_0$ we find the condition independent of $c \neq 0$:

$$\frac{\cos\psi_0}{\kappa\,\sin\phi_0 - 0.5\cos\phi_0} = \frac{-\sin\psi_0}{\kappa\,\cos\phi_0 - 0.5\sin\phi_0}. \tag{12.61}$$

The condition (12.61) for $r_0/R_0$ is satisfied if

$$(\phi_0, \psi_0) = (\frac{\pi}{4}, -\frac{\pi}{4}),\ (\phi_0, \psi_0) = (-\frac{\pi}{4}, \frac{\pi}{4}). \tag{12.62}$$

We can obtain these values also by requiring system (12.60) to have non-trivial solutions for $r_0$, $R_0$. The secularity conditions for $y(t)$ become

$$\begin{cases} \mu R_0\sin\psi_0(1-\frac{1}{4}R_0^2) + cr_0\cos\phi_0 &= 0, \\ \mu R_0\cos\psi_0(1-\frac{1}{4}R_0^2) - cr_0\sin\phi_0 &= 0. \end{cases} \tag{12.63}$$

Choosing the second case of the angles in (12.62) we have

$$\kappa r_0 + 0.5r_0 + cR_0 = 0,\ \mu R_0(1-\frac{1}{4}R_0^2) + cr_0 = 0,\ \frac{r_0}{R_0} = \frac{-c}{\kappa+0.5}. \tag{12.64}$$

Solving for positive solutions of $r_0$, $R_0$ produces a unique periodic solution if the discriminant $\Delta = det J$ of the Jacobian $J$ does not vanish in $(r_0, R_0)$. We find

$$r_0 = \frac{-2c}{\kappa+0.5}\sqrt{1-\frac{c^2}{\mu(\kappa+0.5)}},\ R_0 = 2\sqrt{1-\frac{c^2}{\mu(\kappa+0.5)}}. \tag{12.65}$$

For the determinant of Jacobian $J$ of the equations for $r_0$, $R_0$ in (12.64) we find in this case

$$\Delta = (\kappa + 0.5)\mu(1 - \frac{3}{4}R_0^2) - c^2. \tag{12.66}$$

For values of the parameters with vanishing $\Delta$ a branching of a solution may take place. The condition is not necessary; the local behaviour of the solutions near the parameter values depends on the type of branching phenomenon. For instance a branching from the trivial solutions takes place for the parameter values satisfying (12.66)

$$\kappa = 1.3, \mu = 4, c = 2.6833\ldots$$

For the first case of the angles in Eq. (12.62) we have

$$-\kappa r_0 + 0.5 r_0 + c R_0 = 0, \ -\mu R_0(1 - \frac{1}{4}R_0^2) + c r_0 = 0, \ \frac{r_0}{R_0} = \frac{-c}{-\kappa + 0.5}, \tag{12.67}$$

leading to

$$r_0 = -\frac{2c}{-\kappa + 0.5}\sqrt{1 + \frac{c^2}{\mu(-\kappa + 0.5)}}, \ R_0 = 2\sqrt{1 + \frac{c^2}{\mu(-\kappa + 0.5)}}. \tag{12.68}$$

The results agree with the expressions obtained by averaging. For the determinant $\Delta$ of the Jacobian $J$ we find

$$\Delta = (-\kappa + 0.5)\mu(1 - \frac{3}{4}R_0^2) + c^2. \tag{12.69}$$

Zeroes of $\Delta$ may correspond with branching phenomena. We have, for instance, branching from the trivial solutions satisfying Eq. (12.69) if‘

$$\kappa = 1, \mu = 4, c = \sqrt{2}.$$

In both cases of the angles the analysis of eigenvalues provides additional information.

# Chapter 13
# Instability Induced by Dissipation

**Abstract** When studying rotating fluid masses it was discovered in the nineteenth century that adding dissipation could destabilise the equilibria. This can be illustrated simpler by the rotating spherical bowl of H. Lamb and more generally by the rotating surface formulated by L.E.J. Brouwer; in the second case also the rotating saddle is included. An explication of the phenomena was presented by Bottema; this is also related to Arnold's treatment of structural stability of matrices. An application in mechanics is given for a rotor system with external excitation.

It is a widely accepted notion that Lyapunov stable equilibria become asymptotically stable when dissipation in the form of damping or friction is added to the system. The notion is not correct. In a simple form this idea derives from considering the harmonic oscillator where the origin is stable and on adding a damping term becomes asymptotically stable. This phenomenon of damping producing asymptotic stability carries over to quite general Hamiltonian systems with for one dof an expansion of the Hamiltonian near the origin:

$$H(p,q) = \frac{1}{2}(p^2 + \omega^2 q^2) + \ldots$$

We have $\omega > 0$ and the dots representing as usual a Taylor expansion starting with cubic terms. The equation of motion becomes

$$\ddot{q} + \omega^2 q = \cdots$$

with the dots standing for quadratic and higher-order terms. The origin of phase-plane is Lyapunov (neutrally) stable. Adding a linear dissipative term in the form $\mu\dot{x}$ one assumes $\mu > 0$ and Coulomb damping by passage through a fluid or gas and alternatively a form of friction between solid materials like a suspension point or a spring that reduces the velocity. There exists more complicated and also very different types of dissipation; however, the equation with linear dissipation becomes in this case



F. Verhulst, *Averaging for Nonlinear Dynamics with Applications and Numerical Bifurcations*, Applied Mathematical Sciences 223,
https://doi.org/10.1007/978-3-032-12745-7_13

$$\ddot{q} + \mu\dot{q} + \omega^2 q = \cdots$$

Introducing polar coordinates with radius $r = \sqrt{p^2 + \omega^2 q^2}$ we have by differentiation

$$\dot{r} = -\mu\frac{\dot{q}^2}{r} + \cdots$$

The higher-order terms do not stop the decrease of $r(t)$ near the origin so the equilibrium corresponding with the origin of the phase-plane is asymptotically stable. This idea of 'dissipation makes equilibria stable' is quite persistent in the literature.

This persistence is surprising as for a long time examples with two or more dof were known where dissipation *does* destabilise equilibria. Examples showing such phenomena in rotor systems were published in the twentieth century but already in 1879 Thomson and Tait [128] showed that an unstable conservative system that is stable when adding gyroscopic forces could be destabilised by small damping forces. A classic example is the theory of rotating fluid masses developed in the nineteenth century by famous scientists as Maclaurin, Jacobi, Riemann and later Poincaré. The term *bifurcation* arose from this field as rotation of a fluid mass can produce by instability a splitting into two fluid masses, a bifurcation. An example is given by the secular instability of the Maclaurin spheroids due to fluid viscosity that acts dissipative. In 1885 Poincaré [111] gave conditions for secular stability of rotating systems; he concluded to secular stability of the rotating Jacobi ellipsoids. There exists a huge amount of papers on this subject, an authoritative survey including the mathematical analysis is presented in [82].

In modern theories also magnetic diffusivity or losses due to radiation play a part.

In meteorology where thermal and viscous diffusivities are natural one meets these phenomena in so-called baroclinic waves; see also [88] for a recent survey.

Historical remarks and references can be found in [87] and [89]; as late as 1952 destabilisation phenomena by dissipation in mechanical systems were wrongly presented as a paradox.

## 13.1 A Particle in a Rotating Bowl

One of the simplest physical models showing instability by dissipation was proposed by H. Lamb in a treatise on stability in [95]; he considered the stability of a particle in a rotating, spherical bowl close to the bottom of the bowl. The topologist Brouwer took up this problem in 1918 [34]; he also admitted negative curvatures at the bottom, but he ignored most of the known stability theory of that time. Consider a spherical bowl described by a smooth surface S rotating around a vertical axis with constant angular velocity $\omega$. According to [111] an equilibrium corresponds with a stationary point of

$$T_k = T + K$$

where $T$ is the potential energy and $K$ the kinetic energy of the cyclic motions. Lamb [95] calls this expression the kinetic potential. In the case of the rotating spherical bowl with radius $a$, gravitational constant $g$, mass of the particle $M$ and angular distance from the lowest point $\theta$ we have for the kinetic potential

$$T_k = -Mga\cos\theta - \frac{1}{2}M\omega^2 a^2 \sin\theta. \tag{13.1}$$

Stationary points are $\theta = 0$ (bottom of the bowl) and $\theta$ such that $\cos\theta = g/(\omega^2 a)$. At $\theta = 0$ the second derivative of the kinetic potential is

$$Ma(g - \omega^2 a),$$

so the well at the bottom of the bowl is stable if $\omega^2 < g/a$, unstable if $\omega^2 > g/a$. In the unstable case of the well the position $\cos\theta = g/(\omega^2 a)$ is stable. Adding dissipation does not change the instability. To be explicit consider Euclidean coordinates in Brouwer's formulation [34]. We choose the equilibrium position at the bottom of the bowl at $(x, y) = (0, 0)$. Neglecting dissipation the equations of motion linearised near $(0, 0)$ are according to [34]:

$$\begin{cases} \ddot{x} - 2\omega\dot{y} + (gk_1 - \omega^2)x &= 0, \\ \ddot{y} + 2\omega\dot{x} + (gk_2 - \omega^2)y &= 0. \end{cases} \tag{13.2}$$

The constants $k_1, k_2$ are the curvatures of $S$ at $(0, 0)$ in the $x, y$ directions and $g$ is the gravitational constant. If $k_1 = k_2 = k > 0$ as considered by Lamb the bowl has a well at the bottom. Brouwer considers also curvatures that are unequal in the $(x, y)$ directions and negative values of the curvatures.

The model described by system (13.2) is typical for a number of rotating mechanical systems. Consider, for instance, the case of the circular, restricted three-body problem in celestial mechanics. The triangular libration points $L_4$ and $L_5$ are equilibria discovered by Lagrange. The stability problems of these Lagrangian equilibria are of the same form as for the rotating bowl with $k_1 k_2 < 0$.

Another related mechanical problem is the so-called Jeffcott rotor. See for these problems the discussion of the model in [89] and for the general relevance of the rotating bowl model the extensive discussion in ch. 1 of [88].

**Remark 1** We will consider the rotating bowl to study the combination of gyroscopic effects and positive damping. Simultaneously positive and negative damping is also relevant for applications, but will not be considered here; see for more information [88].

### 13.1.1 The Symmetric Well, $k_1 = k_2 = k$

The eigenvalues of the matrix of coefficients near the origin are:

If $gk \neq \omega^2$, four purely imaginary eigenvalues $\lambda$ from $\lambda^2 = -(gk + \omega^2 \pm \sqrt{(gk+\omega^2)^2 - (gk-\omega^2)^2}$.

If $gk = \omega^2$, eigenvalues $0, 0, \pm 2\omega i$.

Introducing small damping $O(\varepsilon)$ to the motion on the rotating bowl we have the system

$$\begin{cases} \ddot{x} - 2\omega\dot{y} + \varepsilon\dot{x} + (gk - \omega^2)x &= 0, \\ \ddot{y} + 2\omega\dot{x} + \varepsilon\dot{y} + (gk - \omega^2)y &= 0. \end{cases} \tag{13.3}$$

The damping coefficient $\varepsilon$ is positive and equal in the $x, y$-directions. The characteristic (eigenvalue) equation takes the form

$$\lambda^4 + a_1\lambda^3 + a_2\lambda^2 + a_3\lambda + a_4 = 0, \tag{13.4}$$

with $a_1, a_3 = O(\varepsilon)$ and $a_2, a_4 = O(1)$. If $\omega^2 > gk$, we find two positive real parts of the eigenvalues; if $\omega^2 < gk$, all real parts are negative. Brouwer [34] obtained these results analytically, a MATLAB calculation confirms them.

The resulting instability derives from both damping and the gyroscopic effects in the rotating bowl; a general treatment in [21]-[22] clarifies this; see next section.

We will focus on the critical case $\omega^2 = gk$ where we have two purely imaginary eigenvalues and two zeroes. We put

$$\omega^2 = gk + a\varepsilon.$$

to obtain an *unfolding* of this case; system (13.3) becomes

$$\begin{cases} \ddot{x} - 2\omega\dot{y} &= -\varepsilon\dot{x} + a\varepsilon x, \\ \ddot{y} + 2\omega\dot{x} &= -\varepsilon\dot{y} + a\varepsilon y. \end{cases} \tag{13.5}$$

If $a < 0$, we expect stability at the origin, if $a > 0$ instability. The matrix to be analysed for the stability of the origin is

$$\begin{pmatrix} 0 & 1 & 0 & 0 \\ a\varepsilon & -\varepsilon & 0 & 2\omega \\ 0 & 0 & 0 & 1 \\ 0 & -2\omega & a\varepsilon & -\varepsilon \end{pmatrix}, \tag{13.6}$$

with characteristic equation for the eigenvalues

$$P(\lambda) = \lambda^4 + 2\varepsilon\lambda^3 + (\varepsilon^2 - 2a\varepsilon + 4\omega^2)\lambda^2 + a^2\varepsilon^2. \tag{13.7}$$

In this specific case the eigenvalues of the polynomial $P(\lambda)$ can be explicitly expressed using radicals, resulting in a straightforward closed-form solution:

$$\begin{cases} \lambda_1 &= \frac{1}{2}\left(-\sqrt{4a\varepsilon + (\varepsilon - 2i\omega)^2} - \varepsilon + 2i\omega\right), \\ \lambda_2 &= \frac{1}{2}\left(\sqrt{4a\varepsilon + (\varepsilon - 2i\omega)^2} - \varepsilon + 2i\omega\right), \\ \lambda_3 &= \frac{1}{2}\left(-\sqrt{4a\varepsilon + (\varepsilon + 2i\omega)^2} - \varepsilon - 2i\omega\right), \\ \lambda_4 &= \frac{1}{2}\left(\sqrt{4a\varepsilon + (\varepsilon + 2i\omega)^2} - \varepsilon - 2i\omega\right). \end{cases} \tag{13.8}$$

With $\omega = O(1)$ a power series expansion of the eigenvalues in terms of $\varepsilon$ yields explicit expressions for each eigenvalue, clearly separating their real and imaginary components. We have

$$\lambda_2 = 2\omega i + O(\varepsilon), \lambda_3 = -2\omega i + O(\varepsilon).$$

Up to order $O(\varepsilon^2)$ we obtain the following expressions for the eigenvalues:

$$\begin{cases} \lambda_1 &= \frac{a\varepsilon^2}{4\omega^2} + \left(\frac{a\varepsilon}{2\omega} + \frac{a^2\varepsilon^2}{8\omega^2}\right) i + O(\varepsilon^3), \\ \lambda_2 &= -\varepsilon - \frac{a\varepsilon^2}{4\omega^2} + \left(2\omega - \frac{a\varepsilon}{2\omega} - \frac{a^2\varepsilon^2}{8\omega^3}\right) i + O(\varepsilon^3), \\ \lambda_3 &= -\varepsilon - \frac{a\varepsilon^2}{4\omega^2} - \left(2\omega - \frac{a\varepsilon}{2\omega} - \frac{a^2\varepsilon^2}{8\omega^3}\right) i + O(\varepsilon^3), \\ \lambda_4 &= \frac{a\varepsilon^2}{4\omega^2} - \left(\frac{a\varepsilon}{2\omega} + \frac{a^2\varepsilon^2}{8\omega^2}\right) i + O(\varepsilon^3). \end{cases} \tag{13.9}$$

The Puiseux expansion of the eigenvalues produces no fractional exponents of $\varepsilon$, so we have for the dynamics the timelike variables $t, \varepsilon t, \varepsilon^2 t, \varepsilon^3 t$, etc. Note also that the instability starts at $O(\varepsilon^2)$ and $a > 0$.

### 13.1.2 The Rotating Saddle, $ab < 0$

In the asymmetric case $ab < 0$ we have a rotating saddle at the bottom of the bowl with the following matrix:

$$\tilde{A} = \begin{pmatrix} 0 & 1 & 0 & 0 \\ a\varepsilon & -\varepsilon & 0 & 2\omega \\ 0 & 0 & 0 & 1 \\ 0 & -2\omega & b\varepsilon & -\varepsilon \end{pmatrix}, \tag{13.10}$$

with $\omega > 0$, $ab < 0$ and $0 < \varepsilon \ll 1$. The eigenvalues of the matrix $\tilde{A}$ are the roots of the characteristic polynomial:

$$\tilde{P}(\lambda) = \lambda^4 + 2\varepsilon\lambda^3 + \left(-a\varepsilon - b\varepsilon + \varepsilon^2 + 4\omega^2\right)\lambda^2 + \lambda\left(-a\varepsilon^2 - b\varepsilon^2\right) + ab\varepsilon^2. \tag{13.11}$$

To obtain an expansion of the eigenvalues in terms of the small parameter $\varepsilon$ one can as above put

$$\lambda = \sum_{i=0}^{n} \beta_i \varepsilon^i$$

and determine the coefficients $\beta_i$ by substituting this expression into the characteristic equation and matching powers of $\varepsilon$. Applying this expansion to the characteristic polynomial $\tilde{P}$ yields two possible asymptotic series for the eigenvalues $\lambda$, each corresponding to a distinct solution branch. We find

$$\lambda_1 = \frac{\sqrt{|ab|}}{2\omega}\varepsilon + \frac{(a+b)\left(\sqrt{|ab|}+2\omega\right)}{16\omega^3}\varepsilon^2 + O(\varepsilon^3),$$
$$\lambda_2 = -\frac{\sqrt{|ab|}}{2\omega}\varepsilon - \frac{(a+b)\left(\sqrt{|ab|}-2\omega\right)}{16\omega^3}\varepsilon^2 + O(\varepsilon^3).$$

We conclude to instability of the rotating saddle in this model.

### 13.1.3 *Averaging*

A more detailed view of the dynamics can be obtained by averaging the system (13.3) to second order. If $\omega^2 = gk$, this suggests the transformation $x, \dot{x}, y, \dot{y} \to A, B, C, D$ by

$$\begin{cases} x &= A + B\cos 2\omega t + C\sin 2\omega t,\ \dot{x} = -2\omega B\sin 2\omega t + 2\omega C\cos 2\omega t, \\ y &= D - B\sin 2\omega t + C\cos 2\omega t,\ \dot{y} = -2\omega B\cos 2\omega t - 2\omega C\sin 2\omega t. \end{cases} \tag{13.12}$$

By variation of constants we can derive variational equations for $A, B, C, D$ and average to $O(\varepsilon^2)$, but after our calculation of the eigenvalues this produces for stability questions not much new information.

## 13.2 Some Related Problems

We discuss briefly some related problems where gyroscopic effects and dissipation lead to similar interesting phenomena.

### 13.2.1 Lamb's Rotating Pendulum

In [95] one considers a symmetric pendulum that can swing in a vertical plane. The pendulum hangs from a vertical spindle that rotates with a constant angular velocity $\omega$. If $\omega = 0$ the vertical position of the pendulum at rest is a stable equilibrium. When $\omega$ increases the vertical equilibrium becomes unstable if $\omega$ exceeds a critical value. It starts moving in a widening cone until its rotation equals $\omega$.

The dynamics is described by a two dof system that is analysed in [95]. The position of the pendulum is described by two angles, $\theta, \phi$, where $\theta$ is the inclination to the vertical and $\phi$ gives the horizontal angle (azimuth). If $I_1, I_2$ are the principal moments of inertia, $h$ is the distance of the suspension point to the centre of gravity of the pendulum, and $m$ is the mass of the pendulum, we find for the kinetic potential

$$T_k = -mgh\cos\theta - \frac{1}{2}\omega^2(I_1\sin^2\theta + I_2\cos^2\theta). \tag{13.13}$$

The stationary points are

$$\theta = 0,\ \cos\theta = \frac{mgh}{(I_1 - I_2)\omega^2}.$$

For small values of $\omega$ the $\theta = 0$ position is stable; when the speed of rotation reaches $0 < mgh/(I_1 - I_2)\omega^2 \leq 1$ the position $\theta = 0$ becomes unstable and the rotation of the pendulum will be conical. A small amount of friction of the spindle does not keep the vertical position stable. For an illustration by Lamb see Fig. 13.1. Small modifications of the model system do not essentially change the phenomena.

A more complicated model in [95] describes a metal, cylindrical shaft that rotates horizontally between two fixed bearings. Exceeding a critical rotation velocity circular vibrations may appear depending on imperfections of the symmetry and influenced by dissipation.

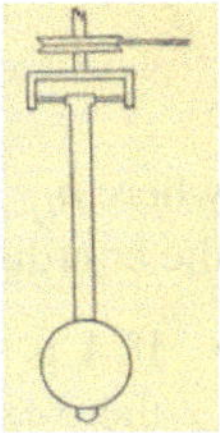

**Fig. 13.1** Lamb's rotating pendulum, picture from [95]

### 13.2.2 The Bowl with Fluctuating Rotation

A modification of the rotating bowl problem can be obtained by introducing small fluctuations of the rotation speed $\omega$; see [149]. Replace $\omega^2$ by $\omega^2 + 2a\varepsilon \cos \nu t$. The modulation frequency is $\nu$ and $a$ is a positive constant. The symmetric case given by system (13.3) changes to

$$\begin{cases} \ddot{x} - 2\omega\dot{y} + \varepsilon\dot{x} + (gk - \omega^2)x & = \varepsilon 2a(\frac{1}{\omega}\dot{y} \cos \nu t + x \cos \nu t), \\ \ddot{y} + 2\omega\dot{x} + \varepsilon\dot{y} + (gk - \omega^2)y & = \varepsilon 2a(-\frac{1}{\omega}\dot{x} \cos \nu t + y \cos \nu t). \end{cases} \tag{13.14}$$

Using quasi-periodic averaging a number of bifurcations can be found for different combination resonances. Stabilisation of the single well solutions from system (13.3) is possible.

## 13.3 General Explanation by Bottema

As mentioned before there was some confusion in the literature of mechanical engineering about the observed presence of instability caused by dissipation. One called the instabilities even paradoxes. In the papers [21]-[22] these problems were solved by Bottema in a general approach for two dof non-conservative systems. Bottema's approach looks less general than the discussion on the structural stability of matrices by Arnold in [4] ch. 6, but it is quite effective for mechanical engineering; see also section 13.5.

Consider a linear system with two dof and coordinates $x, \dot{x}, y, \dot{y}$; the kinetic energy is $T = (\dot{x}^2 + \dot{y}^2)/2$. The system is supposed to be autonomous with system (13.3) as a special case; the equations near the equilibrium $x = \dot{x} = y = \dot{y} = 0$ are

$$\begin{cases} \ddot{x} + a_{11}x + a_{12}y + b_{11}\dot{x} + b_{12}\dot{y} & = 0, \\ \ddot{y} + a_{21}x + a_{22}y + b_{21}\dot{x} + b_{22}\dot{y} & = 0, \end{cases} \tag{13.15}$$

where $a_{ij}$ and $b_{ij}$ are constants, $\mathbf{A} := (a_{ij})$ is the matrix of the forces depending on the coordinates $x, y$, $\mathbf{B} := (b_{ij})$ of those depending on the velocities. We note:

- If **A** is symmetrical and disregarding the damping associated with the matrix **B**, there exists a potential energy function $V = (a_{11}x^2 + 2a_{12}xy + a_{22}y^2)/2$.
- If **A** is anti-symmetrical, the forces are circulatory.
- If the matrix **B**is symmetrical, we have a non-gyroscopic damping force, which is positive when the dissipative function $(b_{11}x^2 + 2b_{12}xy + b_{22}y^2)/2$ is positive definite.
- If **B** is anti-symmetrical the forces depending on the velocities are purely gyroscopic.

The equation for the eigenvalues is again of the form (13.4); splitting the matrices **A**, **B** in sums of symmetric and anti-symmetrical parts, the coefficients $a_1, \ldots, a_4$ are determined completely by the traces of these matrices. A technical analysis enables the author of [21] and [22] to determine necessary and sufficient conditions for the eigenvalues to have negative and or positive parts and so the stability of the equilibrium at the origin of system (13.15); see also [89]. In particular we have for this equilibrium at the origin:

- If system (13.15) is conservative and stable, the stability does not change when gyroscopic terms are added.
- If system (13.15) is frictionless and both non-conservative and gyroscopic forces are added, the equilibrium will be unstable.
- If system (13.15) is non-conservative and there are both damping and gyroscopic forces, we have the general case; the equilibrium can be stable or unstable.
- Part of the results for two dof can be generalised to three dof.

The basic results obtained by Bottema in [21] and [22] are still not well known in mechanical engineering. An example of an interesting paper on the stability of rotating systems with damping is [100]. As an application, it deals with the stability problem of a rotating elastic shaft with small damping forces and asymmetric imperfections. The linearisation results agree with [21] and [22]; in a way it also continues the work of Lamb [95].

## 13.4 Tondl's Forced Rotor

In [119] Tondl developed a model for a vertical bar topped by a disk with mass $M$, rotating with constant velocity $\omega$, flexible horizontal support and vertical periodic excitation $\varepsilon \cos \omega_0 t$ at the foot $A$ of the bar. The model was inspired by excitation of rotating bars in lorries, buses and various machinery. The dynamics is in fact described by an autoparametric system as in Chap. 11, but the analysis by averaging shows in addition remarkable friction and instability effects (see Fig. 13.2).

### *13.4.1 The Rotor Without Friction*

We will consider first the stability of the rotor without friction (or damping) by assuming small oscillations around $\theta = 0$, so considering the stability of the upright position. If $x, y$ indicate the coordinates in a horizontal plane the derivation in [119] leads to the linearised parametric equations around $(x, \dot{x}, y, \dot{y}) = (0, 0, 0, 0)$:

$$\begin{cases} \ddot{x} + 2\alpha\dot{y} + (1 + 4\varepsilon\eta^2 \cos 2\eta t)x = 0, \\ \ddot{y} - 2\alpha\dot{x} + (1 + 4\varepsilon\eta^2 \cos 2\eta t)y = 0, \end{cases} \tag{13.16}$$

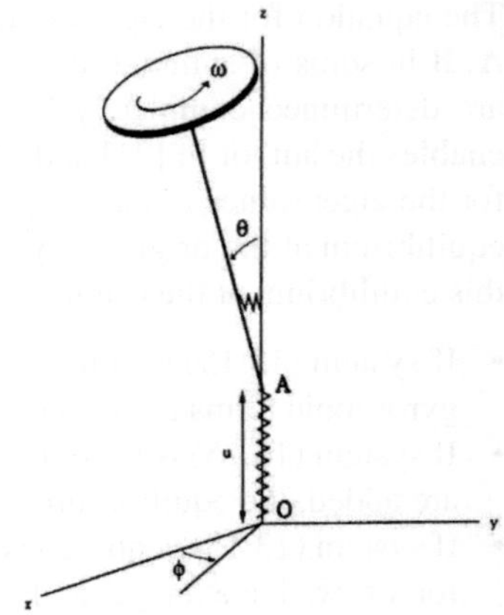


**Fig. 13.2** Rotating bar with horizontal flexible connections and disk at the top; the bar experiences small vertical excitation $u$

with $\alpha$ proportional to the angular velocity $\Omega$ and $\eta$ is the rescaled frequency $\omega_0$. We introduce the complex variable $z = x + yi$ and obtain from system (13.16):

$$\frac{d^2z}{dt^2} - 2i\alpha\frac{dz}{dt} + (1 + 4\varepsilon\eta^2 \cos 2\eta t)z = 0. \tag{13.17}$$

We return to real variables by introducing $v = exp.(-i\eta t)$, $\eta t = \tau$ to obtain

$$\frac{d^2v}{d\tau^2} - \left(\frac{1+\alpha^2}{\eta^2} + 4\varepsilon \cos 2\tau\right) v = 0. \tag{13.18}$$

Eq. (13.18) is a Mathieu-equation for real variable $v$. Putting $\varepsilon = 0$ the equation has the frequencies

$$\omega_1 = \sqrt{\alpha^2 + 1} + \alpha,\ \omega_2 = \sqrt{\alpha^2 + 1} - \alpha. \tag{13.19}$$

The trivial solution is unstable for Mathieu-equation ($\varepsilon > 0$) if we have the so-called primary resonance:

$$\sqrt{1 + \alpha^2} = \eta. \tag{13.20}$$

We have in fact instability if $\sqrt{1+\alpha^2} = n\eta, n = 1, 2, \ldots$ The instability tongues have boundaries that can be calculated as in Chap. 11.

The primary resonance ($n = 1$) arises if we have the *sum resonance*:

$$\omega_1 + \omega_2 = 2\eta. \tag{13.21}$$

### 13.4.2 The Rotor with Friction

With linear friction (or damping) added in system (13.16) Eq. (13.17) becomes

$$\frac{d^2z}{dt^2} - 2i\alpha\frac{dz}{dt} + (1 + 4\varepsilon\eta^2\cos 2\eta t)z + 2\varepsilon\kappa\dot{z} = 0. \tag{13.22}$$

The primary resonance takes place at $\eta = \eta_0 + \varepsilon\sigma, \omega_1 + \omega_2 = 2\eta_0$; we introduced $\sigma$ as we are interested in the instability interval around $\eta_0$; does it change because of damping? Handling the linear, complex equation (13.22) is slightly more complicated than Eq. (13.17). Using

$$z(t) = z_1 e^{i\omega_1 t} + z_2 e^{i\omega_1 t},$$

we obtain after substituting for the complex quantities expressions for $z_1, z_2$ and will use the expressions to substitute in addition (see for details [119]):

$$z_1 = v_1 e^{i\varepsilon\sigma\tau},\ z_2 = v_2 e^{-i\varepsilon\sigma\tau}.$$

The resulting equations are suitable for averaging

$$\begin{cases} \frac{dv_1}{dt} &= \frac{\varepsilon}{\eta_0}\left(-(\omega_1\kappa + i\sigma\eta_0)v_1 + \kappa\omega_2 v_2 e^{-2i\eta_0 t}\right. \\ &\quad \left. + i\eta_0^2(v_1(e^{2i\eta_0 t} + e^{-2i\eta_0 t}) + v_2(1 + e^{-4i\eta_0 t}))\right), \\ \frac{dv_2}{dt} &= \frac{\varepsilon}{\eta_0}\left(\kappa\omega_1 v_1 e^{2i\eta_0 t}\right. \\ &\quad \left. -(\omega_2\kappa - i\sigma\eta_0)v_2 - i\eta_0^2(v_1(1 + e^{4i\eta_0 t}) + v_2(e^{2i\eta_0 t} + e^{-2i\eta_0 t}))\right). \end{cases} \tag{13.23}$$

The averaged system becomes

$$\frac{dv_1}{dt} = \frac{\varepsilon}{\eta_0}(-(\omega_1\kappa + i\sigma\eta_0)v_1 + i\eta_0^2 v_2),\ \frac{dv_2}{dt} = \frac{\varepsilon}{\eta_0}(-i\eta_0^2 v_1 - (\omega_2\kappa - i\sigma\eta_0)v_2). \tag{13.24}$$

For this linear system the characteristic equation for the eigenvalues is

$$\lambda^2 + 2\eta_0\kappa\lambda + \kappa^2 - 2i\alpha\kappa\sigma\eta_0 + \sigma^2\eta_0^2 - \eta_0^4 = 0, \tag{13.25}$$

with (complex) roots

$$\lambda = -\eta_0\kappa \pm \sqrt{(\alpha\kappa + i\eta_0\sigma)^2 + \eta_o^4}.$$

The real parts of the eigenvalues determine the stability; we have instability if the real parts are positive. This is the case if

$$\sigma^2 < \eta_o^4 - \kappa^2. \tag{13.26}$$

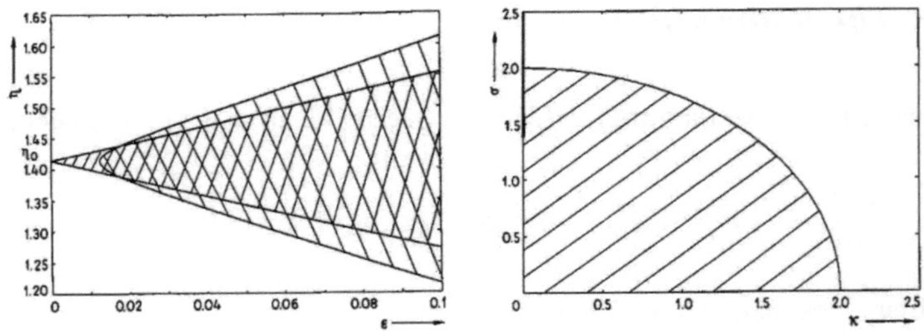


**Fig. 13.3** Stability and instability regions for system (13.22). Left: instability tongue that is enlarged by damping and right: corresponding $\kappa, \sigma$ diagram with thick line indicating stability. Courtesy John Wiley, [119]

We have stability for all values of $\sigma$ if $\eta_0^4 - \kappa^2 < 0$. Suppose $\eta_0^4 - \kappa^2 > 0$. In this case there are remarkable consequences. The instability tongue is not only lifted of the $\eta$-axis but also wider than in the case $\kappa = 0$; the instability region has increased if damping is introduced; see Fig. 13.3, left. In the same figure on the right we show the instability in a $(\kappa, \sigma)$ diagram; the hashed part indicates instability if $\kappa > 0$ and the thick line stability if $\kappa = 0$.

## 13.5 Putting the Results in an Algebraic Framework

Inspired by the results for Tondl's rotor of system (13.16) the results were linked to the theory of structural stability of matrices and bifurcation theory in [79]. Consider the system

$$\dot{x} = A_0 x + \varepsilon f(x, \mu, t), x \in \mathbb{R}^4, \mu \in \mathbb{R}^p. \tag{13.27}$$

The dimension for the parameter $\mu$ has still to be specified. The idea is that based on Arnold [4] ch. 6 the bifurcation phenomena observed for Tondl's rotor are examples of generic singularities in real matrices. Such a result links the rotor phenomena to stability by linearisation of much more general vector fields near equilibrium.

Assuming that the vector field $f$ is $2\pi$-periodic in $t$ and that matrix $A_0$ has purely imaginary eigenvalues only, a first step will be to apply averaging to obtain a suitable autonomous system. Two frequencies $\omega_1, \omega_2$ play a part. The four parameters used to test the structural stability are the detunings of the frequencies $\omega_1, \omega_2$ and the two parameters representing the damping of the oscillators. It is shown that the matrices with damping are structurally stable and are a versal deformation of the case with damping zero. In this way the mechanical phenomena are understood in terms of modern bifurcation theory.

Such an approach with a slightly different emphasis was systematically carried out in [88]. The linear algebra of gyroscopic systems with damping, Hamiltonian and reversible systems, adjoint and non-self-adjoint boundary eigenvalue problems and many other problem fields ask for an algebraically adjusted analysis with many interesting questions.

The rotating well of Lamb and rotating saddle of Brouwer show similar phenomena.

For coupled oscillators this general approach was anticipated by Bottema in the papers [21, 22]. We conclude here with an example:

$$\begin{cases} \ddot{x} + x - ay + \varepsilon\dot{x} + \varepsilon\dot{y} &= 0, \\ \ddot{y} + ax + y + \varepsilon\dot{x} + \varepsilon\dot{y} &= 0. \end{cases} \tag{13.28}$$

If $a = 0, \varepsilon \geq 0$ the equilibrium $(0, 0, 0, 0)$ is stable; if $a = 1, \varepsilon > 0$ it is unstable.

# Chapter 14
# Diffusion and Waves

**Abstract** A problem field where averaging for PDEs can be applied is advection-diffusion of material in a tidal basin. The model can be extended with the presence of reactions and sources in the tidal basin. Another application is a wave equation on a bounded domain with a cubic non-linearity; this leads to interesting resonances. Other applications are to the cubic Klein-Gordon equation on a square, a self-excited wave equation and wave equations with parametric excitation.

As noted earlier the theory of partial differential equation (PDEs) is complicated and research is in full progress; to illustrate the ideas our examples will be simplified as much as possible. This is also useful as much of the available research on PDE problems is presented in rather abstract form with quantitative aspects referred to numerical approaches. Quantitatively oriented exceptions are the theories of stability and bifurcations where localisation enables us to apply advanced linear algebra theory. For wave equations with one space dimension an approach is often to look for travelling waves with similarity solutions of the form $\phi(x - ct)$. This leads to equations for $\phi$ and stability questions. See for instance [83].

We will use asymptotic methods that may reveal basic patterns and adds to the numerics by showing dependence on parameters. We will demonstrate this by discussing diffusion and non-linear wave equations, both conservative and dissipative. The validity of approximations is based on the theorems of Chap. 8.

A large part of the illustrations is for PDEs on bounded spatial domains. This is relevant for problems in acoustics and the mechanics of fluids in pipes and containers to name a few applications. In a number of physical, chemical and biological problems, formulations for unbounded domain are more natural. In section 14.10 we add some material on unbounded domains. In ch. 14 of [148] one can find additional examples and references to the literature.



F. Verhulst, *Averaging for Nonlinear Dynamics with Applications and Numerical Bifurcations*, Applied Mathematical Sciences 223,
https://doi.org/10.1007/978-3-032-12745-7_14

## 14.1 Advection-Diffusion in a Tidal Basin

As an application one considers in [91] the transport of material (chemicals or sediment) by advection and isotropic diffusion in a tidal basin. In this case the advective flow is because of the tides nearly periodic, but diffusive effects are small and take a long time. The inspiration for the model was the transport of chemicals and natural sediment in the North Sea, a large tidal basin surrounded by six countries. Political controversies about the effect of chemical deposits played a part. The formulation applies of course to similar tidal basins elsewhere. On a two-dimensional domain and after rescaling length and time parameters the problem can be formulated as

$$\frac{\partial C}{\partial t} + \nabla.(uC) - \varepsilon \Delta C = 0, \quad C(x, y, 0) = \gamma(x, y), \tag{14.1}$$

where $C(x, y, t)$ is the two-dimensional concentration of the transported material on a domain $D$, the flow $u = u_0(x, y, t) + \varepsilon u_1(x, y)$ is given, $u_0$ is $T$-periodic in time and represents the tidal flow and $\varepsilon u_1$ is a small reststream caused by the input of rivers that is not much changing with time. As the diffusion process takes a period of time close to a year it is clearly slower than the tidal period of roughly 12 hours. We are interested in a long timescale approximation. This was also a motivation for the analysis as numerical approximation schemes for this problem turned out to be rather unstable on long timescales.

### *14.1.1 The Reduced Problem*

As usual in these models we assume the flow of water to be divergence-free; the unperturbed ($\varepsilon = 0$) flow problem is then given by

$$\frac{\partial C_0}{\partial t} + u_0 \nabla C_0 = 0, \quad C_0(x, y, 0) = \gamma(x, y), \tag{14.2}$$

a first-order equation which can be integrated along the characteristics (often called streamlines) with solution of the form $C_0 = \gamma(Q(t)(x, y))$. This analysis presupposes a flow operator that can be inverted. In the spirit of variation of constants we introduce the change of variables

$$C(x, y, t) = F(Q(t)(x, y), t) \tag{14.3}$$

In actual tidal basins with irregular geometry this is analytically difficult and the reduced problem still has to be solved numerically; one may wonder what then the use is of this technique. The answer is that one needs solutions on a long timescale and that numerical integration of a PDE on a domain $D$ for Eq. (14.1) where the fast periodic oscillations have been eliminated is a much safer procedure.

### 14.1.2 The Perturbed Problem

Inspired by our discussion of ordinary differential equation (ODE) problems we expect $F$ to be slowly time-dependent when introducing (14.3) into the original equation (14.1). Using again the technical assumption that the flow $u_0 + \varepsilon u_1$ is divergence-free we find as in Chap. 8.3 an explicit slowly varying equation of the form (8.5)

$$\frac{\partial F}{\partial t} = \varepsilon G(x, y, t).$$

This equation can be averaged and the approximating equation is time-independent. For the solutions we have an $O(\varepsilon)$ approximation on the timescale $1/\varepsilon$. Assuming almost-periodicity of the time dependence in Eq. (14.1) the estimate is weaker. Note that the assumption of divergence-free flow is not essential, but it simplifies the calculations.

**Remark 14.1** In the analysis presented thus far we have considered unbounded domains. To study the equation on spatially bounded domains, adding boundary conditions does not present serious obstacles to the techniques and the proofs.

In [91] applications are presented with explicit examples where the slowly varying equation is averaged to obtain a time-independent parabolic problem.

## 14.2 Tidal Diffusion with Reactions and Sources

An extension of the advection-diffusion problem formulation has been obtained in [76]. It is natural to add to the model of Eq. (14.1) reactions of chemicals or sediment using a reaction term f(C). The chemicals tend to get dissolved or the sediments react with the flow. Secondly we include localised sources indicated by $B(x, y, t)$ which in the case of tidal basins can be interpreted as illegal, periodic dumping of chemicals or sediment in the basin.

Following [76] the equation becomes

$$\frac{\partial C}{\partial t} + u_0.\nabla C + \varepsilon u_1.\nabla C = \varepsilon \mu \Delta C + \varepsilon f(C) + \varepsilon B(x, y, t), \tag{14.4}$$

for $t \geq 0$ with initial value $C(x, y, 0) = \gamma(x, y)$; we replaced $u$ in (14.1) by $u_0(x, y, t) + \varepsilon u_1(x, y)$ where $u_0$ represents the dominant periodic part of the tides,

$u_1$ stands for the small reststream and the diffusion $\varepsilon\mu\Delta C$ is small because of the second parameter $\mu$. The reaction term $f(C)$ will in general be non-linear, for instance, $f(C) = aC^2$ or $f(C) = aC^5$ depending on the type of chemical reaction. The source term $B(x, y, t)$ is periodic in t. Using again variation of constants we obtain from Eq. (14.4) a perturbation equation in the same way as shown above, but with a more complicated operator $L(t)$. As the tidal period of $v_0(x, y, t)$ is roughly 12 hours, it is natural to assume a common period $T$ with the dumping of chemicals indicated by $B(x, y, t)$. Averaging produces an approximation $\tilde{C}$ of the solution $C$ of the initial value problem (14.4).
Interestingly, the result is stronger than in the case without source term $B$. One can prove that $\tilde{C}$ converges in time to the solution $\tilde{C}_0$ of a time-independent boundary value problem. We have that $C$ converges to a $T$-periodic solution that is $\varepsilon$-close to $\tilde{C}_0$ for all time. This behaviour is analogous to the case of forced ODEs as we have seen in Chap. 6. The proof is based on maximum principles and the use of suitable sub- and supersolutions; more details are given in [76].

**Application to a Circular Domain**
To simplify the analysis we consider a circular domain and omit the reststream $u_1(x, y)$. We put for the tidal flow a tangential field of the form

$$u_0(x, y, t) = (-y \sin kt, x \sin kt), \tag{14.5}$$

with $k$ determining the tidal period. The flow reverses direction periodically. With $f(C) = aC^2$ Eq. (14.4) becomes on domain $x^2 + y^2 < R^2$

$$\frac{\partial C}{\partial t} + u_0.\nabla C = \varepsilon\mu\Delta C + \varepsilon a C^2 + \varepsilon B(x, y, t), \tag{14.6}$$

with Dirichlet condition $C(x, y, t) = 0$ for $t \geq 0$ at $x^2 + y^2 = R^2$. Using the two-dimensional Dirac-delta function $\delta$ we choose for the dumping of chemicals $B$ point sources located at $(-\frac{13}{20}R, \frac{13}{20}R)$ and $(\frac{13}{20}R), \frac{13}{20}R)$. The averaging is technically complicated, involving averaging over Dirac-delta functions; it is carried out in [76]. The averaged diffusion equation can now be studied numerically without difficulties; the final distribution of the concentration $C$ is shown in Fig. 14.1.

## 14.3 The Wave Equation with a Cubic Non-linearity

The terminology to describe waves is inspired by the terms used in the theory of linear waves, in the case of the non-linear wave given by Eq. (14.7) in the case $\varepsilon = 0$. A solution of the form $\phi = a\cos(\omega t - kx)$ has an *amplitude a* and *phase velocity* $W = \omega/k$. A superposition of such wave solutions

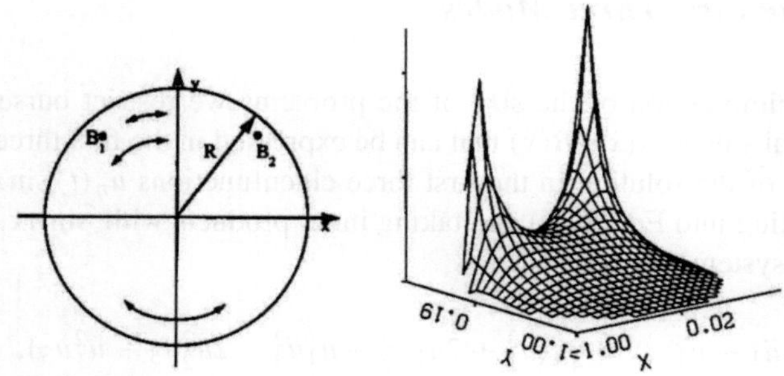


**Fig. 14.1** Tidal diffusion with two periodic sources of pollution $B_1$, $B_2$ in a circular domain (left). The flow converges to a periodic solution with approximation $\tilde{C}_0$ by averaging (right). For the parameters we have $k = 12$, $R = 1$, $\mu = 0.1$, $a = 1$. Courtesy Wiley [76]

$$\phi = a\cos(\omega_1 t - k_1 x) + a\cos(\omega_2 t - k_2 x)$$

is called a *wave packet*. If a medium is homogeneous the phase velocity will be constant for the waves ($\omega_1/k_1 = \omega_2/k_2$); the medium and the waves will be called non-dispersive. An example is the propagation of electromagnetic waves in vacuum. On the other hand, sound waves have phase velocities that depend strongly on the density and temperature of the medium; such waves are called dispersive. The varying wave length $\lambda = 2\pi/k$ for sound waves in an inhomogeneous dispersive medium will cause changes in a wave packet. As we shall see non-linearities in the wave equations will also change the picture.

A rather artificial (but not simple) non-linear wave equation is inspired by [127] and [90]. Consider the equation

$$u_{tt} - u_{xx} = \varepsilon u^3, \quad t \geq 0, 0 < x < \pi, \tag{14.7}$$

with boundary conditions $u(0, t) = u(\pi, t) = 0$ and initial values $u(x, 0) = A(x)$, $u_t(x, 0) = B(x)$ which are supposed to be sufficiently smooth. The eigenfunctions of the linearised problem are $u_n(t)\sin nx$, $n = 1, 2, 3, \ldots$; the spectrum is fully resonant. We expand the solution

$$u(t, x) = \sum_{n=1}^{\infty} u_n(t)\sin nx.$$

According to Remark 8.1 we have with a cubic non-linearity and analytic initial values expressed by a finite superposition of eigenfunctions to take a number $N$ of time-dependent eigenfunctions that increases if $\varepsilon \to 0$; in this case explicitly $N = O(|log\,\varepsilon|)$. A finite superposition of $N$ eigenfunctions will be called again a wave packet.

### 14.3.1 The First Three Modes

To obtain an impression of the size of the problems we restrict ourselves first to consider initial values $A(x)$, $B(x)$ that can be expressed in the first three modes and an expansion of the solution in the first three eigenfunctions $u_n(t)\sin nx$. We find after substituting into Eq. (14.7) and taking inner products with $\sin nx$, $n = 1, 2, 3$ the three dof system

$$\begin{cases} \ddot{u}_1 + u_1 &= \frac{3}{4}\varepsilon(u_1^3 + 2u_1u_2^2 - u_1u_3^2 - 2u_2^2u_3 - u_1^2u_3), \\ \ddot{u}_2 + 4u_2 &= \frac{3}{4}\varepsilon u_2(2u_1^2 + u_2^2 + 2u_3^2 + 2u_1u_3), \\ \ddot{u}_3 + 9u_3 &= \frac{3}{4}\varepsilon(2u_1^2u_3 + 2u_2^2u_3 + u_3^3 + u_1u_2^2 - \frac{1}{3}u_1^3). \end{cases} \tag{14.8}$$

The frequencies $\omega$ of the linearised system are 1, 2, 3, so near the origin of phase-space the system is in $1 : 2 : 3$ resonance. However, as we shall see, this is not the generic $1 : 2 : 3$ resonance. Moreover, as the whole spectrum $\{n^2\}, n = 1, 2, \dots$ produces resonant frequencies, increasing the number of modes adds to the number of resonances.

System (14.8) has two exact normal modes: the $u_2$ normal mode ($u_1 = u_3 = 0$) and the $u_3$ normal mode ($u_1 = u_2 = 0$). In addition the system has a two-dof invariant manifold consisting of the modes $u_1, u_3$ ($u_2 = \dot{u}_2 = 0$). We discuss approximations of the three-mode system by introducing slowly varying variables $r, \phi$ by transformation (1.6) $u_n = r_n \cos(nt + \phi_n)$, $\dot{u}_n = -nr_n \sin(nt + \phi_n)$, $n = 1, 2, 3$ and averaging. To discuss the dynamics of system (14.8) we present first the averaged system for the amplitudes:

$$\begin{cases} \dot{r}_1 &= -\frac{3}{32}\varepsilon r_3\left(r_2^2\sin(\phi_1 + \phi_3 - 2\phi_2) - r_1^2\sin(3\phi_1 - \phi_3)\right), \\ \dot{r}_2 &= 0, \\ \dot{r}_3 &= -\frac{1}{96}\varepsilon r_1\left(3r_2^2\sin(\phi_1 + \phi_3 - 2\phi_2) - r_1^2\sin(\phi_3 - 3\phi_1)\right). \end{cases} \tag{14.9}$$

Analysis of the equations for the combination angles may produce periodic solutions as in Chap. 9. The averaged (normalised) system for three modes (14.9) is integrable and as a Hamiltonian system unexpected. The first-order approximation keeps $r_2(t)$ to $O(\varepsilon)$ constant, a higher-order precision calculation or including more than three modes may produce less trivial results for $r_2(t)$. The exchange of energy in system (14.9) takes place between the first and third mode, and the combination angles are $(\phi_1 + \phi_3 - 2\phi_2)$ and $(\phi_3 - 3\phi_1)$.

Interesting questions concern the stability of the normal modes of system (14.9) and secondly, when adding a fourth mode $u_4$ with frequency $\omega_4 = 4$ to the expansion, will we find interaction with mode 2?

In the notation of [90] we have $M$ from the initial conditions and $N$ for the number of modes. The implication is that with $M = 2, N = 3$ (as above) we have an $O(\varepsilon^{1/4})$ approximation on the timescale $1/\varepsilon^{1/2}$; with $M = 2, N = 4$ we have an $O(\varepsilon^{1/3})$ approximation on the timescale $1/\varepsilon^{5/9}$, thus far a rather weak result.

We will look for the stability of normal modes and dynamics in 1 : 3 and 1 : 2 resonances.

### 14.3.2 The Two-Dof 1 : 3 Resonances

Consider the dynamics of the system spanned by the two modes. $u_n(t)\sin nx$ and $u_{3n}(t)\sin 3nx$, $n = 1, 2, \ldots$. In system (14.8) $u_3$ was found as an exact normal mode, in the averaged system (14.9) both $u_1$ and $u_3$ are normal modes. We find after substituting into Eq. (14.7) and taking inner products with $\sin nx$ and $\sin 3nx$ the two dof system with equations of motion

$$\begin{cases} \ddot{u}_n + n^2 u_n &= \frac{3}{4}\varepsilon(u_n^3 - u_n^2 u_{3n} + 2u_n u_{3n}^2), \\ \ddot{u}_{3n} + (3n)^2 u_{3n} &= \frac{3}{4}\varepsilon(-\frac{1}{3}u_n^3 + 2u_n^2 u_{3n} + u_{3n}^3). \end{cases} \tag{14.10}$$

Constructing variational equations in amplitude-phase variables and averaging over time we find

$$\dot{r}_n = \frac{3}{32n}\varepsilon r_{3n} r_n^2 \sin\chi, \; \dot{r}_{3n} = -\frac{1}{96n}\varepsilon r_n^3 \sin\chi, \tag{14.11}$$

and with combination angle $\chi = 3\phi_n - \phi_{3n}$

$$\dot{\chi} = -\frac{3}{32n}\varepsilon\left(\frac{23}{3}r_n^2 + 11r_{3n}^2 - \frac{r_n}{r_{3n}}(3r_{3n}^2 - \frac{1}{3}r_n^2)\cos\chi\right). \tag{14.12}$$

Stationary solutions for the amplitude $r_n$ and corresponding normal mode $u_n$ exist if $\chi = 0, \pi$ and $d\chi/dt = 0$; as $d\chi/dt$ has no zeroes for general position periodic solutions they do not exist in the averaged system.

System (14.10) is Hamiltonian. As we know from Chap. 9 the averaged (normal) form of a time-independent two-dof Hamiltonian system is integrable. The normal form integrals are

$$r_n^2 + 9r_{3n}^2 = 2E_0, \; \frac{3}{4}(r_n^4 + r_{3n}^4) + 2r_n^2 r_{3n}^2 - \frac{1}{3}r_n^3 r_{3n}\cos\chi = I_0, \tag{14.13}$$

with $E_0, I_0$ constants.

To consider the stability of orbits we can use the integrals or consider a perturbation of the normal mode in the form $u_1 = \psi(t) + \varepsilon w_1, u_3 = \varepsilon w_3$. For this relatively simple system a third approach is to use numerical experiments; we conclude to stability of the normal modes in the $u_n, u_{3n}$ systems.

The averaged equations show that the contributions for high frequencies ($n$ large) are small in size.

### 14.3.3 The Two-Dof 1 : 2 Resonances

As a second test case we consider the projection

$$u_{2n}(t)\sin 2nx + u_{4n}(t)\sin 4nx,\, n = 1, 2, \ldots.$$

Substituting into the wave equation (14.7) and taking inner products with $\sin 2nx$, $\sin 4nx$ produces the system

$$\begin{cases} \ddot{u}_{2n} + (2n)^2 u_{2n} &= \frac{3}{4}\varepsilon u_{2n}(u_{2n}^2 + 2u_{4n}^2), \\ \ddot{u}_{4n} + (4n)^2 u_{4n} &= \frac{3}{4}\varepsilon u_{4n}(2u_{2n}^2 + u_{4n}^2). \end{cases} \tag{14.14}$$

The equations are derived from the (symmetric) Hamiltonian

$$H = \frac{1}{2}(\dot{u}_{2n}^2 + (2n)^2 u_{2n}^2 + \dot{u}_{4n}^2 + (4n)^2 u_4^2) - \frac{3}{4}\varepsilon(\frac{1}{4}u_{2n}^4 + \frac{1}{4}u_{4n}^4 + u_{2n}^2 u_{4n}^2). \tag{14.15}$$

Averaging over time produces $\dot{r}_{2n} = \dot{r}_{4n} = 0$ and no combination angle in the normal form. So by first-order averaging we have no general position periodic orbits. The $u_{2n}$- and $u_{4n}$-normal modes are stable.

## 14.4 The Dispersive, Cubic Klein-Gordon Equation

As an often studied prototype of a non-linear wave equation with dispersion consider the non-linear (cubic) Klein-Gordon equation

$$u_{tt} - u_{xx} + u = \varepsilon u^3, \quad t \geq 0, 0 < x < \pi \tag{14.16}$$

with boundary conditions $u(0, t) = u(\pi, t) = 0$ and initial values $u(x, 0) = \phi(x)$, $u_t(x, 0) = \psi(x)$ which are supposed to be sufficiently smooth. The problem has been studied by many authors, often by ingenious formal approximation procedures as in [86].

What do we know qualitatively? It follows from the analysis in [90] that we have existence and uniqueness of solutions on the timescale $1/\varepsilon$ and for all time if we put a minus sign on the right-hand side. This information has practical use as in some problems, for instance, in non-linear diffusion equations, one may find unbounded solutions.

We start with the eigenfunction expansion (8.14) with

$$v_n(x) = \sin(nx), \lambda_n^2 = n^2 + 1, n = 1, 2, \cdots$$

for the eigenfunctions and eigenvalues. The spectrum is not resonant for eigenvalues that are independent of $\varepsilon$. Substituting the expansion $\sum u_n(t) v_n x$ into the equation (14.16) and taking the $L_2$ inner products with $v_n(x)$ for $n = 1, 2, \cdots$ produces an infinite number of coupled ordinary differential equations of the form

$$\ddot{u}_n + (n^2 + 1) u_n = \varepsilon f_n(u), \ \ n = 1, 2, \cdots, \infty$$

with

$$f_n(u) = \sum \sum \sum_{n_1, n_2, n_3 = 1}^{\infty} c_{n_1 n_2 n_3} u_{n_1} u_{n_2} u_{n_3}.$$

As the spectrum is non-resonant (see [127]) we can easily average the complete system or, alternatively, to any truncation number $N$. The rather trivial result is that the actions are constant to the order of approximation

$$\dot{r}_i = 0, \ \dot{\phi}_i = \sigma_i(r_1(0), r_2(0), \ldots, r_N(0)), \ i = 1, 2, \ldots, N,$$

with $N$ integrals of motion; the angles are varying slowly as a function of the energy level of the modes. Considering the theory summarised before in Chap. 8 we can make the following observations with regard to the asymptotic character of the estimates:

- In [127] it was proved that, depending on the smoothness of the initial values $(\phi, \psi)$, we need $N = O(\varepsilon^{-\beta})$ modes ($\beta$ a positive constant) to obtain an $O(\varepsilon^{\alpha})$ approximation ($0 < \alpha \leq 1$) on the timescale $1/\varepsilon$.
- If the initial values can be expressed in a finite number of eigenfunctions $v_n(x)$ (as in the preceding example), it follows from section 8.4 that the error is $O(\varepsilon)$ on the timescale $1/\varepsilon$.
- In [142] a method is developed to prove an $O(\varepsilon)$ approximation on the timescale $1/\varepsilon$ which is applied to the non-linear Klein-Gordon equation with a quadratic non-linearity ($\varepsilon u^2$) and some generalisations. A few problems arising in the numerics of these cases are discussed. The analysis in the case of such non-linearities has many different aspects; we refer for this to the general literature on non-linear wave theory.
- As the first-order results are not very interesting in [127] a second-order approximation was constructed. It turns out that there exists a small interaction between modes with number $n$ and number $3n$ which probably involves much longer timescales than $1/\varepsilon$. This is still an open problem.
- In [24] one considers the non-linear Klein-Gordon equation (14.16) in the rather general form

$$u_{tt} - u_{xx} + V(x)u = \varepsilon f(u), \quad t \geq 0, 0 < x < \pi \tag{14.17}$$

with $V$ a periodic, even function and $f(u)$ an odd polynomial in $u$. Assuming rapid decrease of the amplitudes in the eigenfunction expansion (8.14) and diophantine (non-resonant) conditions on the spectrum, it is proved that *infinite*-dimensional invariant tori persist in the non-linear wave equation (14.17) corresponding with almost-periodic solutions. The proof involves a perturbation expansion which is valid on a long timescale.

- In [15] one considers the non-linear Klein-Gordon equation (14.16) in the more general form

$$u_{tt} - u_{xx} + mu = \varepsilon \phi(x, u), \quad t \geq 0, 0 < x < \pi, \tag{14.18}$$

and the same boundary conditions. The function $\phi(x, u)$ is polynomial in $u$, analytic and periodic in $x$ and odd in the sense that $\phi(x, u) = -\phi(-x, -u)$. Under a certain non-resonance condition on the spectrum it is shown in [15] that the solutions remain close to finite-dimensional invariant tori, corresponding with quasi-periodic motion on timescales longer than $1/\varepsilon$.

## 14.5 The Cubic Klein-Gordon Equation on a Square

More than one spatial dimension is natural in real-life problems and this can produce a much richer dynamics. The embedded double resonance, discussed in Chap. 7, arises quite naturally in systems with two or more dof; we will discuss a non-linear wave example analysed for two spatial dimensions mainly based on [108]. For the solutions one may have mirror (discrete), point and/or rotational symmetries, reflected by the eigenfunctions that emerge.

Consider the cubic Klein-Gordon equation on a square:

$$u_{tt} - u_{xx} - \alpha u_{yy} + \beta u = \varepsilon u^3, \quad (x, y) \in [0, \pi] \times [0, \pi], \tag{14.19}$$

with positive constants $\alpha, \beta$ and smooth initial conditions. Consider the solution in the Hilbert space:

$$H_0^k = \{u \in L^2(0, \pi)\}$$

and the expansion in $H_0^k$

$$u(x, y, t) = \sum_{k,l=1}^{\infty} u_{kl}(t) \sin kx \sin ly, \quad \omega_{kl}^2 = k^2 + \alpha l^2 + \beta. \tag{14.20}$$

The eigenvalues can produce resonant frequencies; for illustration we take $\alpha = \beta = 1$. One obtains ODE systems by projection on a finite number of modes; the ODEs are of the form

$$\ddot{u}_{kl} + \omega_{kl}^2 u_{kl} = \varepsilon f_{kl}(u),\ k, l = 1, 2, \ldots$$

A key role will be played by $1:1$ resonances in the cases $\omega_{kl} = \omega_{lk}$.

### 14.5.1 Stability of a Normal Mode

Suppose the initial energy is in a mode with frequency $\omega_{kl}$ only (a normal mode) and we have only one different mode with the same frequency $\omega_{kl} = \omega$. We will consider the pure $1:1$ interaction of the two modes with equal frequency. The two-mode expansion (14.20) produces after projection two coupled second-order ODEs with equal frequency. Transforming $u_i = r_i \cos(\omega t + \phi_i)$, $\dot{u}_i = -\omega r_i \sin(\omega t + \phi_i)$, $i = 1, 2$ and keeping $r_1$ for the mode with non-zero energy, we can apply averaging. With $\chi = 2(\phi_1 - \phi_2)$ we have following [108] the averaged equations:

$$\begin{cases} \dot{r}_1 &= -\varepsilon \frac{3}{32\omega} r_1 r_2^2 \sin\chi, \\ \dot{r}_2 &= \varepsilon \frac{3}{32\omega} r_1^2 r_2 \sin\chi, \\ \dot{\phi}_1 &= -\frac{\varepsilon}{8\omega}\left(\frac{3}{4} r_2^2 \cos\chi + \frac{27}{16} r_1^2 + \frac{3}{2} r_2^2\right), \\ \dot{\phi}_2 &= -\frac{\varepsilon}{8\omega}\left(\frac{3}{4} r_1^2 \cos\chi + \frac{27}{16} r_2^2 + \frac{3}{2} r_1^2\right) \end{cases} \tag{14.21}$$

For the combination angle $\chi$ we have

$$\dot{\chi} = -\frac{\varepsilon}{4\omega}(r_2^2 - r_1^2)\left(\frac{3}{4}\cos\chi - \frac{3}{16}\right). \tag{14.22}$$

The averaged system has two first integrals, the quadratic $r_1^2 + r_2^2 = 2E_0$ and an integral involving $\chi$. Periodic solutions with constant amplitude are found if $\sin 2\chi = 0$ and $\chi$ is stationary; this is the case if $r_1 = r_2 = \sqrt{E_0}$. According to [108] these periodic solutions are Lyapunov stable if $\chi = 0, \pi$ and unstable if $\chi = \pi/2, 3\pi/2$.

The $r_1$ normal mode is present in the averaged system and on eliminating $r_2$ with the quadratic integral we find instability of the normal mode. By symmetry this holds also for the $r_2$ normal mode; isolated normal modes with $k \neq l$ are unstable because of the $1:1$ resonance. See Fig. 14.2 with various orbits in a $r_1$, $\chi$ diagram.

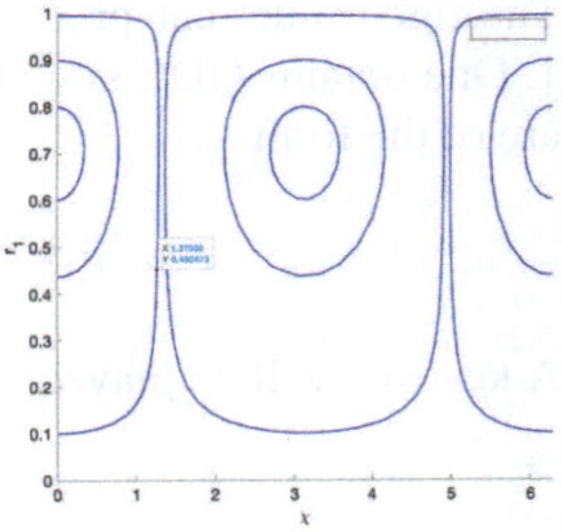

**Fig. 14.2** Reduction to a $r_1, \chi$-diagram of the $1:1$ interaction of two modes approximated by system (14.21). The reduction is possible because of two independent integrals. The $r_1$ and $r_2$ normal modes are unstable

## 14.5.2 *A Detuned* 1 : 2 : 2 *Resonance*

An example to illustrate the increase of phenomena when going from one to two space dimensions is the three-mode case with $(k, l)$ respectively $(3, 3)$, $(5, 7)$, $(7, 5)$ and so the frequencies $\omega_{33}^2 = 19$, $\omega_{57}^2 = \omega_{75}^2 = 75$. This can be handled as a detuned $1:2:2$ or $\frac{1}{2}, 1, 1$ resonance. Other resonance cases are presented in [155]. We find the three dof system

$$\begin{cases} \ddot{u}_{33} + \frac{1}{4}u_{33} = & -\frac{1}{300}u_{33} + \varepsilon\frac{1}{100}(\frac{3}{4}u_{33}^3 + u_{33}u_{57}^2 + u_{33}u_{75}^2), \\ \ddot{u}_{57} + u_{57} = & \varepsilon\frac{1}{100}(\frac{3}{4}u_{57}^3 + u_{33}^2u_{57} + u_{57}u_{75}^2), \\ \ddot{u}_{75} + u_{75} = & \varepsilon\frac{1}{100}(\frac{3}{4}u_{75}^3 + u_{33}^2u_{75} + u_{57}^2u_{75}), \end{cases} \tag{14.23}$$

with $-\frac{1}{300}u_{33}$ as the detuning term. Averaging produces to first order for the amplitudes

$$\begin{aligned} \dot{r}_{33} &= 0, \\ \dot{r}_{57} &= -\varepsilon c r_{57} r_{75}^2 \sin 2(\phi_{57} - \phi_{75}), \\ \dot{r}_{75} &= +\varepsilon c r_{57}^2 r_{75} \sin 2(\phi_{57} - \phi_{75}). \end{aligned}$$

The $1:1$ resonance of $u_{57}, u_{75}$ is effective; the mode $u_{33}$ is not involved in the first-order approximate dynamics, so the dynamics of modes $u_{57}$ and $u_{75}$ dominates to $O(\varepsilon)$ on the timescale $1/\varepsilon$. However, the resonance zones where

$$\sin 2(\phi_{57} - \phi_{75}) = 0$$

present a different picture. In such a resonance zone the modes $u_{57}$ and $u_{75}$ vary very little; we need a second-order calculation if we want to take into account quantities that have the same magnitude of variation. Guided by the location of one of the resonance zones we consider a numerical two-dimensional projection of the amplitudes in a resonance zone; this produces quadratic curves corresponding with the $2:4:4$ resonance.

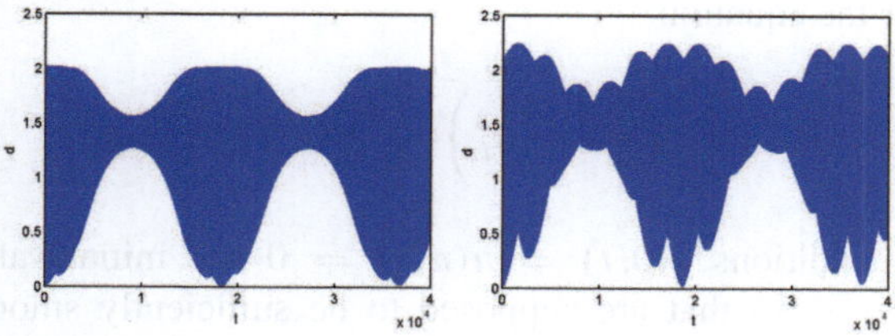

**Fig. 14.3** Recurrence from solutions of system (14.23) in 18,000 timesteps for passage through a resonance zone (Euclidean distance $d$ to initial value as a function of time) with $u_{57}(0) = 0.1, u_{75}(0) = 1$ and $u_{33}(0) = 0.1, 0.5$. Left: a dynamics where the $u_{33}$ mode has little energy. Right: a case with considerable energy for the $u_{33}$ mode; in this case we have well-developed tori around the stable periodic solutions causing more obstruction in the passage through the resonance zone and so more complicated recurrence. Courtesy Acta Appl. Math. [155]

In Fig. 14.3 we show the recurrence when passing a resonance zone by computing the Euclidean distance $d(t)$ to two given sets of initial conditions. The $1 : 1$ resonance zones of the $u_{57}$ and $u_{75}$ modes are characterised by the first-order averaging normal form as

- Equal amplitudes of modes $u_{57}, u_{75}$
- Periodic solutions with phase differences $0, \pi/2, \pi, 3\pi/2$
- Two stable and two unstable periodic solutions on the energy manifold

The $2 : 4 : 4$ resonance is obtained from a higher-order calculation (not presented) and is an *embedded double resonance* in these zones. We observe that a higher-order asymptotic calculation produces here new qualitative phenomena.

As we have seen in Chap. 7 the complexity of the resonance zones will generally increase with the dimension. We will find stable and unstable periodic solutions producing homo- and heteroclinic tangles. Solutions that pass through a resonance zone will be delayed by the complexity of the dynamics, but they will always return to the zones, winding for some time around the tori associated with the stable periodic solutions and being perturbed by the tangles. This delay of orbits in resonance zones is called *quasi-trapping* in [163].

Apart from a measure zero number of orbits, passage *through* the resonance zone is the rule and both quasi-trapping and recurrence will take place. By studying the Euclidean deviation (or distance) $d(t)$ from the initial conditions in phase-space as a function of time one may get an indication of the dynamics of the resonance zone inviting further study. Relatively long recurrence times suggest complexity.

## 14.6 Self-Excitation, the Keller-Kogelman Problem

An interesting example of a non-linear equation with dispersion and dissipation, generated by a self-excitation (Rayleigh) term, was presented in [85] using multiple-timing. Consider the equation

$$u_{tt} - u_{xx} + u = \varepsilon \left( u_t - \frac{1}{3} u_t^3 \right), \; t \geq 0, 0 < x < \pi, \tag{14.24}$$

with boundary conditions $u(0,t) = u(\pi,t) = 0$ and initial values $u(x,0) = \phi(x), u_t(x,0) = \psi(x)$ that are supposed to be sufficiently smooth. As before, putting $\varepsilon = 0$, we have for the eigenfunctions and eigenvalues

$$v_n(x) = \sin(nx), \; \lambda_n = \omega_n^2 = n^2 + 1, \; n = 1, 2, \cdots,$$

and again we propose to expand the solution of the initial boundary value problem for the equation (14.24) in a Fourier series with respect to these eigenfunctions of the form (8.14). We will use projection and averaging. Substituting the expansion into the differential equation we have

$$\sum_{n=1}^{\infty} \ddot{u}_n \sin nx + \sum_{n=1}^{\infty} (n^2 + 1) u_n \sin nx = \varepsilon \sum_{n=1}^{\infty} \dot{u}_n \sin nx - \frac{\varepsilon}{3} (\sum_{n=1}^{\infty} \dot{u}_n \sin nx)^3.$$

When taking inner products we have to Fourier analyse the cubic term. This produces many terms, and it is clear that we will not have exact normal mode solutions, as for instance mode $m$ will excite mode $3m$.

At this point we can start averaging, and it becomes important that the spectrum is not resonant. In particular, we have in the averaged equation for $u_n$ only terms arising from $\dot{u}_n^3$ and $\sum_{i \neq n}^{\infty} \dot{u}_i^2 \dot{u}_n$. The other cubic terms do not survive the first-order averaging process; the part of the equation for $n = 1, 2, \cdots$ that produces non-trivial terms is

$$\ddot{u}_n + \omega_n^2 u_n = \varepsilon \left( \dot{u}_n - \frac{1}{4} \dot{u}_n^3 - \frac{1}{2} \sum_{i \neq n}^{\infty} \dot{u}_i^2 \dot{u}_n \right) + \cdots,$$

where the dots stand for non-resonant terms. With the non-resonant terms included this is an infinite system of ordinary differential equations that is still fully equivalent to the original problem.

We can now perform the actual averaging in a notation that contains only minor differences with that of [85]. Transforming in the usual way $u_n(t) = a_n(t) \cos \omega_n t + b_n(t) \sin \omega_n t$, $\dot{u}_n(t) = -\omega_n a_n(t) \sin \omega_n t + \omega_n b_n(t) \cos \omega_n t$; to obtain the standard form we find after averaging equations for the approximate $(a_n, b_n)$ given by

$$2\dot{a}_n = \varepsilon a_n \left( 1 + \frac{n^2 + 1}{16} (a_n^2 + b_n^2) - \frac{1}{4} \sum_{k=1}^{\infty} (k^2 + 1)(a_k^2 + b_k^2) \right),$$

$$2\dot{b}_n = \varepsilon b_n \left(1 + \frac{n^2+1}{16}(a_n^2 + b_n^2) - \frac{1}{4}\sum_{k=1}^{\infty}(k^2+1)(a_k^2 + b_k^2)\right).$$

This system shows fairly strong (although not complete) decoupling because of the non-resonant character of the spectrum. Because of the self-excitation, we have no conservation of energy. Putting $a_n^2 + b_n^2 = E_n,\ n = 1, 2, \cdots$, multiplying the first equation with $a_n$ and the second equation with $b_n$ and adding the equations, we have

$$\dot{E}_n = \varepsilon E_n \left(1 + \frac{n^2+1}{16}E_n - \frac{1}{4}\sum_{k=1}^{\infty}(k^2+1)E_k\right).$$

We have immediately a non-trivial result: starting in a mode with zero energy this mode will not be excited on a timescale $1/\varepsilon$. The implication is that if the initial conditions are represented by only $N$ eigenfunctions $u_n(t)v_n(x), n = 1, 2, \ldots$, no other eigenfunctions will be excited on the timescale $1/\varepsilon$.

Another observation is that if we have initially only one non-zero mode, say for $n = m$, the equation for $E_m$ becomes

$$\dot{E}_m = \varepsilon E_m \left(1 - \frac{3}{16}(m^2+1)E_m\right).$$

We conclude that with one initial mode we have stable equilibrium at the value

$$E_m = \frac{16}{3(m^2+1)}.$$

If we start with initial conditions in a finite number of modes the error is $O(\varepsilon)$; see Sect. 8.4. For related qualitative results see [93].

## 14.7 A Parametrically Excited Linear Wave Equation

Consider the one-dimensional linear wave equation

$$u_{tt} - c^2 u_{xx} + \varepsilon\beta u_t + (\omega_0^2 + \varepsilon\phi(t))u = 0,\ t \geq 0, 0 < x < \pi, \tag{14.25}$$

with boundary conditions $u_x(0,t) = u_x(\pi,t) = 0$ and small periodic or almost-periodic parametric excitation $\varepsilon\phi(t)$ with $\omega_0 > 0$ and small damping coefficient $\varepsilon\beta$. In [113] the experimental motivation is discussed; it models for instance a one-dimensional chain of oscillators with vertical parametric excitation or the linearised behaviour of water waves in a channel with vertical forcing.

Using the eigenfunctions $v_n(x) = \cos nx$ for the Neumann problem (if $\varepsilon = 0$ with eigenvalues $\omega_n^2 = \omega_0^2 + n^2c^2, n = 0, 1, 2, \ldots$) we expand the solution as

$$u(x,t) = \sum_{0}^{\infty} u_n(t) \cos nx.$$

Substituting into Eq. (14.25) and taking $L_2$-inner products with $v_n(x)$ produces the system:

$$\ddot{u}_n + \omega_n^2 u_n = -\varepsilon\beta\dot{u}_n - \varepsilon\phi(t)u_n,\ n = 0, 1, 2, \ldots \tag{14.26}$$

with suitable initial conditions. System (14.26) is fully equivalent with Eq. (14.25). The normal mode solutions satisfy the system and correspond with two-dimensional invariant manifolds.

We conclude that if the term $\phi(t)u_n$ is not resonant, all the modes tend to zero. We consider now a prominent resonance case.

**The Mathieu Case $\phi(t) = \cos 2t$**
A resonance case occurs if one of the eigenvalues equals 1 or is $\varepsilon$-close to 1. Assume that for certain $m \neq 0$ we have $\omega_m^2 = 1 + \varepsilon\delta$ with $\delta$ an $O(1)$ detuning. We find after first-order averaging using transformation (1.6) for $n = 0, 1, 2, \ldots$ but $n \neq m$

$$\dot{r}_n = -\varepsilon\frac{\beta}{2}r_n,\ \dot{\psi}_n = 0. \tag{14.27}$$

If $n \neq m$ the modes are decaying. If $n = m$ we find

$$\dot{r}_m = -\varepsilon\frac{r_m}{2}(-\beta + \frac{1}{2}\sin 2\psi_m),\ \dot{\psi}_m = \frac{1}{2}\varepsilon(\delta + \frac{1}{2}\cos 2\psi_m). \tag{14.28}$$

If $\beta > 1/2$ the damping exceeds the excitation and the mode $u_m$ also decays. If $0 < \beta < 1/2$ we find two solutions with

$$\sin 2\psi_m = 2\beta.$$

They correspond with periodic solutions if simultaneously

$$\delta + \frac{1}{2}\cos 2\psi_m = 0.$$

Elimination of the trigonometric functions produces the relation between $\beta$ and $\delta$:

$$\beta^2 + \delta^2 = \frac{1}{4},$$

which describe the boundaries of a well-known prominent Floquet instability tongue in parameter-space. In the linear equation (14.25) we find that nearly all modes decay, but in the case of resonant parametric excitation one mode can be unstable as we know from Floquet theory. This evokes the need for non-linear modelling and analysis in the applications that follow.

## 14.8 Parametrical Excitation of Non-linear Waves

In [113] and additional analysis in [7] a non-linear version of a parametrically excited wave equation is studied:

$$u_{tt} - c^2 u_{xx} + \varepsilon \beta u_t + (\omega_0^2 + \varepsilon \phi(t))u = \varepsilon b u^3, \; t \geq 0, 0 < x < \pi, \tag{14.29}$$

with boundary conditions $u_x(0, t) = u_x(\pi, t) = 0$ and small periodic excitation $\varepsilon \phi(t)$, with $\omega_0 > 0$ and small damping coefficient $\varepsilon \beta$.

Equation (14.29) models a one-dimensional non-linear chain of oscillators with vertical parametric excitation of water waves in a channel with vertical forcing.

Using again the eigenfunctions $v_n(x) = \cos nx$ with eigenvalues $\omega_n^2 = \omega_0^2 + n^2 c^2, n = 0, 1, 2, \ldots$ we expand the solution as

$$u(x, t) = \sum_0^\infty u_n(t) \cos nx.$$

Substituting into Eq. (14.29) and taking $L_2$-inner products with $v_n(x)$ produces the system

$$\ddot{u}_n + \omega_n^2 u_n = -\varepsilon \beta \dot{u}_n - \varepsilon \phi(t) u_n + \varepsilon f_n(\mathbf{u}), \; n = 0, 1, 2, \ldots \tag{14.30}$$

with $(\mathbf{u}) = (u_0, u_1, u_2, \ldots)$. The components $f_n$ are cubic expressions; the normal mode solutions do not satisfy Eq. (14.30). As an interesting resonance case most attention is given to the choice $\phi(t) = \gamma \cos 2t$. Expanding to three modes we have for the cubic terms

$$\begin{cases} g_0 &= u_0^3 + \frac{3}{2} u_0 u_1^2 + \frac{3}{2} u_0 u_2^2 + \frac{3}{4} u_1^2 u_2, \\ g_1 &= \frac{3}{4} u_1^3 + 3 u_0^2 u_1 + \frac{3}{2} u_1 u_2^2 + 3 u_0 u_1 u_2, \\ g_2 &= \frac{3}{4} u_2^3 + 3 u_0^2 u_2 + \frac{3}{2} u_1^2 u_2 + \frac{3}{4} u_0 u_1^2. \end{cases} \tag{14.31}$$

The analysis in [113] and [7] deals with many different cases using averaging and numerics. In the numerical analysis 3, 10 and 20 modes are considered. For this complex problem averaging to first and second order produces the skeleton part of the dynamics. Details of the bifurcations and extension for larger values of $\varepsilon$ have been studied by numerical exploration guided by the analytic results. We summarise a few aspects.

- If the wave speed $c$ is $O(\varepsilon)$ and $\omega_0$ is an $O(1)$ quantity we have by truncation to a finite number of modes the $1 : 1 : 1 : \ldots : 1$ resonance.
- If the dispersion is small ($\omega_0 = O(\varepsilon)$) system (14.30) is fully resonant. This problem is unsolved.
- The choice $\phi(t) = \gamma \cos 2t$ is considered. If one of the eigenvalues is close to $1/2$ we have the first Floquet resonance. If $\beta$ is small enough , in fact $\beta < |\gamma|$, one mode will give non-trivial solutions as in the linear case of the preceding subsection, and the other modes are not resonant and will decay.
- In [7] the interaction of modes is discussed in more detail. In the case of the combined $1 : 2$ and $1 : 1 : 1$ resonances first- and second-order averaging is used followed by numerical bifurcations analysis using MATCONT. Averaging to second order is necessary as the terms act for the $1 : 1 : 1$ resonance primarily on the cubic terms; the long expressions resulting from second-order averaging are presented in an appendix of [7]. One finds from this analysis Hopf bifurcation of limit cycles corresponding with Neimark-Sacker bifurcation to 2-tori and 3-tori; in addition there are many other bifurcations revealing complex dynamics.
- Simulations with more than three modes show the presence of more complexity and chaos. In this region second-order averaging does not suffice to show all details.

## 14.9 Parametrical Excitation of Two-Dimensional Non-linear Waves

We will consider the non-linear wave problem of the preceding section on a square domain with special interest in new phenomena. We will find an infinite number of $1 : 1$ resonances, interesting accidental resonances and complex resonance zones.

Consider the parametrically excited non-linear wave equation formulated in [113] in the one-dimensional case; we will now consider from [132] the equation on a square with boundary $S$:

$$u_{tt} - c^2(u_{xx} + u_{yy}) + \mu u_t + (\omega_0^2 + \beta \cos(\Omega t))u = \alpha u^3, \tag{14.32}$$

where $t \geq 0$ and $0 < x < \pi, 0 < y < \pi$. The Neumann boundary values are $\partial u/\partial n|_S = 0$.

The parameters $\mu, \beta$ are positive and small in a way to be specified.

The system of equations and conditions model the surface deflections $u(x, y, t)$ of a fluid in a square basin with parametric excitation and damping, and $c$ is the wave speed.

Resonant non-linear waves in two spatial dimensions were also considered in [108] and [155]. We associate with the system the eigenfunctions

$$v_{mn}(x, y) = \cos mx \cos ny, m, n = 0, 1, 2 \ldots \tag{14.33}$$

with the space-dependent operator producing the eigenvalues

$$\omega_{mn}^2 = \omega_0^2 + (m^2 + n^2)c^2, \omega_{mn} = \omega_{nm} = \omega. \tag{14.34}$$

The solutions of Eq. (14.32) with boundary conditions can be approximated by projection to a finite sum of eigenfunctions followed by averaging approximation.

The choice of eigenfunctions is determined by the initial values of Eq. (14.32) while keeping an eye on the resonances of the eigenvalues. It turns out that for the geometry considered here, there are an infinite number of 1 : 1 resonances. This will require our main attention. In addition we will briefly look at prominent accidental resonances.

### 14.9.1 *The Two-Mode* 1 : 1 *Resonance*

We propose a two-mode expansion with

$$u_p(x, y, t) = u_1(t) \cos mx \cos ny + u_2(t) \cos nx \cos my, m, n = 0, 1, 2 \ldots, m \neq n. \tag{14.35}$$

Put $\omega_0 = 1$ and rescale $u = \sqrt{\varepsilon}\bar{u}$ (and its derivatives likewise) in Eq. (14.32) with $\varepsilon$ a small positive parameter; we omit the bars. Substituting expansion (14.35) into Eq. (14.32) and taking inner products with the eigenfunctions we find with $\omega_{mn} = \omega, m \neq n$:

$$\begin{cases} \ddot{u}_1 + \omega^2 u_1 &= -\mu\dot{u}_1 - \beta u_1 \cos(\Omega t) + \varepsilon\alpha(\frac{9}{16}u_1^3 + \frac{3}{4}u_1 u_2^2), \\ \ddot{u}_2 + \omega^2 u_2 &= -\mu\dot{u}_2 - \beta u_2 \cos(\Omega t) + \varepsilon\alpha(\frac{9}{16}u_2^3 + \frac{3}{4}u_1^2 u_2). \end{cases} \tag{14.36}$$

We choose $\Omega = 2\omega$ to study prominent Floquet resonances; rescale $\mu = \varepsilon\bar{\mu}, \beta = \varepsilon\bar{\beta}$ after which we omit the bars. System (14.36) contains the 1 : 2 Floquet resonance and in addition the 1 : 1 resonance of the Hamiltonian interaction force.

Note that because of the symmetry of system (14.36) $u_1(t) = \pm u_2(t)$ satisfies the system.

The coordinate planes $u_1, \dot{u}_1$ and $u_2, \dot{u}_2$ are invariant under the phase-flow; we start with the analysis of these *normal mode planes*.

The analysis for both coordinate planes runs exactly along the same lines with symmetric results, so we consider only the $u_1, \dot{u}_1$ plane. We put $u_2 = \dot{u}_2 = 0$ and introduce amplitude-phase coordinates as usual by

$$u_1 = r_1 \cos(\omega t + \psi_1), \dot{u}_1 = -r_1\omega \sin(\omega t + \psi_1).$$

Deriving the equations for $r_1, \psi_1$ and averaging over time we find the first-order averaged system:

$$\dot{r}_1 = \frac{\varepsilon}{2} r_1 (-\mu + \frac{\beta}{2\omega} \sin 2\psi_1),\ \dot{\psi}_1 = \frac{\varepsilon}{4\omega} (\beta \cos 2\psi_1 - \alpha \frac{27}{32} r_1^2). \tag{14.37}$$

A critical point corresponding with an equilibrium of system (14.37) is given by

$$\beta \sin 2\psi_1 = 2\mu\omega,\ \beta \cos 2\psi_1 = \alpha \frac{27}{32} r_1^2,\ 0 < \frac{2\mu\omega}{\beta} < 1. \tag{14.38}$$

We can eliminate the phase angle to find for the critical point:

$$r_1^2 = r_0^2 = \frac{32}{27\alpha} \sqrt{\beta^2 - 4\mu^2\omega^2}. \tag{14.39}$$

Computing eigenvalues at the critical point shows that the periodic solution is stable within the invariant coordinate plane. For the eigenvalues we have

$$\lambda_{1,2} = -\mu \pm \sqrt{5\mu^2 - \frac{\beta^2}{\omega^2}}. \tag{14.40}$$

If $\beta > \sqrt{5}\mu\omega$, the periodic solution is complex and stable in the coordinate plane, and if $2\mu\omega < \beta < \sqrt{5}\mu\omega$, the periodic solution is stable with real eigenvalues. If $\beta = 2\mu\omega$, the periodic solution vanishes.

An important question is whether the periodic solution is stable or unstable in the full four-dimensional system. For $u_2, \dot{u}_2$ near zero we should not use polar coordinates. Instead we introduce with (1.16) in system (14.36) the slowly varying variables $a, b$ by

$$u_2 = a \cos \omega t + \frac{b}{\omega} \sin \omega t,\ \dot{u}_2 = -a\omega \sin \omega t + b \cos \omega t.$$

Introducing amplitude-phase variables by (1.6) for $u_1$ and $a, b$ variables for $u_2$ in system (14.36) we have to average the system. To determine the stability of the normal mode periodic solution we compute the matrix of coefficients of the averaged system for $r_1, \psi_1, a, b$ and find the eigenvalues of its gradient, the Jacobian at the periodic $u_1(t)$ for $a = b = 0$. This means that we can leave out the quadratic and cubic expressions in $a, b$. For the averaged system in the variables $r_1, \psi_1, a, b$ we find

$$\begin{cases} \dot{r}_1 &= \frac{\varepsilon}{2} r_1 (-\mu + \frac{\beta}{2\omega} \sin 2\psi_1) + \dots, \\ \dot{\psi}_1 &= \frac{\varepsilon}{2} (\frac{\beta}{2\omega} \cos 2\psi_1 - \alpha \frac{27}{64\omega} r_1^2) + \dots, \\ \dot{a} &= \frac{\varepsilon}{2} (-\mu a + \frac{\beta}{2\omega^2} b + \alpha \frac{3}{16\omega} r_1^2 (a \sin 2\psi_1 + 2\frac{b}{\omega} - \frac{b}{\omega} \cos 2\psi_1)) + \dots, \\ \dot{b} &= \frac{\varepsilon}{2} (-\mu b - \frac{\beta}{2} a + \alpha \frac{3}{16} r_1^2 (2a + a \cos 2\psi_1 - \frac{b}{\omega} \sin 2\psi_1)) + \dots. \end{cases} \tag{14.41}$$

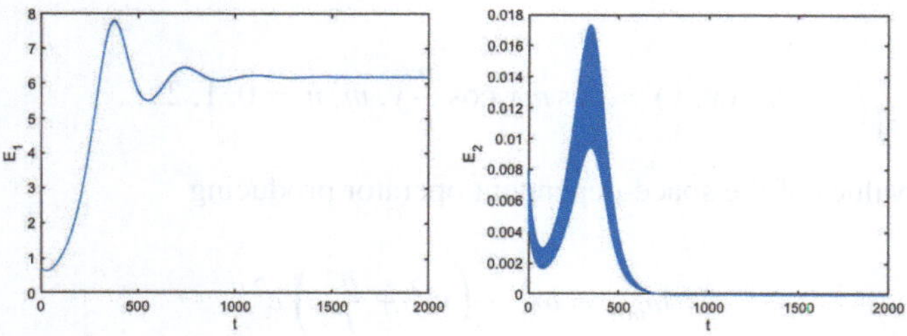


**Fig. 14.4** The behaviour of the solutions of system (14.36) near the invariant $u_1, \dot{u}_1$ coordinate plane is shown in [132] by plotting $E_1(t) = 0.5(\dot{u}_1^2(t) + 6u_1^2(t))$ and $E_2(t) = 0.5(\dot{u}_2^2(t) + 6u_2^2(t))$ for the parametrically excited oscillators. The initial conditions are $u_1(0) = 0.5, \dot{u}_1(0) = 0, u_2(0) = \dot{u}_2(0) = 0.05$; $\omega^2 = 6, \mu = 0.01, \beta = 0.1, \alpha = 0.05$

where the dots stand for the omitted higher-order terms in $a, b$. The Jacobian at the periodic solution in the coordinate plane becomes when omitting the factor $\varepsilon/2$

$$\begin{pmatrix} 0 & \frac{r_0}{\omega}\beta\cos 2\psi_1 & 0 & 0 \\ -\frac{27\alpha}{32\omega}r_0 & -2\mu & 0 & 0 \\ 0 & 0 & -\mu + \frac{3\alpha}{16\omega}r_0^2\sin 2\psi_1 & \frac{\beta}{2\omega^2} + \frac{3\alpha}{16\omega^2}r_0^2(2-\cos 2\psi_1) \\ 0 & 0 & -\frac{\beta}{2} + \frac{3\alpha}{16}r_0^2(2+\cos 2\psi_1) & -\mu - \frac{3\alpha}{16\omega}r_0^2\sin 2\psi_1 \end{pmatrix}. \tag{14.42}$$

The four eigenvalues are splitting up in two groups; the first two correspond with the eigenvalues of Eq. (14.40), and the second group produces the eigenvalues $\lambda_{3,4}$ with $\lambda_3 + \lambda_4 = -2\mu$. We find

$$\lambda_{3,4} = -\mu \pm \sqrt{\frac{13}{108}\frac{\beta^2}{\omega^2} - \frac{88}{81}\mu^2 - \frac{128}{81}\frac{\omega^2\mu^4}{\beta^2}}. \tag{14.43}$$

$\lambda_{3,4}$ depends on the parameters $\mu, \beta, \omega, \alpha$. We conclude that the two periodic normal mode solutions of the 1 : 1 resonances are asymptotically stable if $(39/2)^{1/4}\beta \leq \omega\mu$. See Fig. 14.4.

In [132] a detailed analysis is given for general position orbits of the 1 : 1 resonance with criteria for the existence and stability of periodic solutions. Both in the dissipative case $\mu > 0$ and the volume-preserving case $\mu = 0$ the general position periodic solutions are unstable, suggesting a selection in the wave equation to normal mode solutions.

Additional results are given in [132] for coupled 1 : 1 resonant systems and cases like the 1 : 1 : 3 and 1 : 1 : 1 resonances.

A natural question is how the results obtained for the non-linear wave equation with parametric excitation (14.32) in [132] is affected by the geometry of the domain. The square domain produces resonances like 1 : 1 because of its symmetry. One can choose for instance a rectangle with unequal sides like $0 < x < \pi$, $0 < y < L\pi$, , $L \neq 1$. The eigenfunctions become in the case of Neumann conditions

$$v_{mn}(x, y) = \cos mx \cos \frac{n}{L} y, m, n = 0, 1, 2 \ldots$$

with eigenvalues of the space-dependent operator producing

$$\omega_{mn}^2 = \omega_0^2 + \left(m^2 + \frac{n^2}{L^2}\right) c^2.$$

A two-mode expansion with

$$u_p(x, y, t) = u_1(t) \cos m_1 x \cos \frac{n_1}{L} y + u_2(t) \cos m_2 x \cos \frac{n_2}{L} y,$$

gives two perturbed equations like system (14.36) but with three different frequencies $\omega_1, \omega_2, \Omega$. In general there will be no resonance or a resonance will take place between two frequencies. There are many interesting cases to be analysed for this problem.

How to perform the analysis for a geometrically more general, bounded two-dimensional domain also poses interesting questions.

## 14.10 Wave Phenomena on Unbounded Domains

In the introduction to this chapter we mentioned already the monograph [83]. An interesting approach was given by Whitham [161] who analysed wave equations by averaging corresponding Lagrangians. It is a natural approach but mathematically formal; also to obtain a higher-order calculation is not trivial in this formulation.

We discuss briefly equations that were considered in this chapter for bounded domains.

An often used idea is to look for particular solutions as a function of the characteristics obtained for linear wave equations. As a well-known example consider first the equation

$$u_{tt} - u_{xx} + u = 0, \ -\infty < x < \infty. \tag{14.44}$$

We assume that a wave moving to the right as time evolves will be a function $\Phi$ of $\theta = (kx - \omega t)$. Substituting $\Phi(\theta)$ into Eq. (14.44) we find

$$(\omega^2 - k^2) \frac{d^2 \Phi}{d\theta^2} + \Phi = 0.$$

Choosing for simplicity $\omega^2 - k^2 = 1$, the solutions are

$$u(t,x) = \Phi(t,x) = a_0 \cos(kx - \omega t + \phi_0),$$

with parameters $a_0, \phi_0$. The initial conditions are clearly $u(0,x) = a_0 \cos(kx + \phi_0)$, $u_t(0,x) = a_0\omega \sin(kx + \phi_0)$, for instance, with support $0 \leq x \leq 2\pi$.

This solution describes a cosine wave form that moves with phase velocity $k/\omega$; it makes physical sense if the solution is stable. Note that the wave form extends for $t > 0$ to all values of $x$ in the sense that the wave form shows cosine oscillation when passing a fixed value of $x$.

**The Cubic Klein-Gordon Equation**

$$u_{tt} - u_{xx} + u = \varepsilon u^3, \ -\infty < x < \infty. \tag{14.45}$$

As for the linear equation (14.44) we look for solutions of the form $u = \Phi(kx - \omega t)$ that will describe waves moving to the right. Initial conditions and the parameters $k, \omega$ have to be adapted later. Substituting $\Phi$ into Eq. (14.45) we have with $\theta = kx - \omega t$

$$(\omega^2 - k^2)\frac{d^2\Phi}{d\theta^2} + \Phi = \varepsilon \Phi^3.$$

Assuming that $\omega^2 - k^2 > 0$ this is the Duffing equation that we have seen in Chaps. 2 and 3. To keep the expressions simple we choose again $\omega^2 - k^2 = 1$. Assuming that $\theta$ is timelike we can construct approximations by averaging for $\Phi(\theta)$. We find from Chaps. 2 and 3 a three-parameter family of first-order $\varepsilon$-approximations with a modulation of the phase speed valid on the timescale $1/\varepsilon$:

$$\Phi = a_0 \cos(kx - \omega t + \frac{3}{8\omega} a_0^2 \varepsilon t), \ \omega^2 - k^2 > 0. \tag{14.46}$$

At this stage the stability of the approximations is an open problem. Because of its periodicity the wave form extends at $t = 0$ for all values of $x$; for $t > 0$ the wave form oscillates for fixed values of $x$.

**Averaging over the Characteristics**

We mention some interesting ideas developed in [39]. The authors consider a single wave equation with non-linearities of the form

$$u_{tt} - u_{xx} = \varepsilon H(u_t, u_x), \ -\infty < x < \infty$$

with initial conditions. For various choices of the non-linearities $H$ the solutions are approximated in [39] by a multiple timescales approximations which to first order is related to averaging.

The paper was an inspiration for the approach in [51] to transform to characteristics of the linearised equation and apply averaging. Consider

$$u_{tt} - u_{xx} = \varepsilon f(u, u_t), \ -\infty < x < \infty. \tag{14.47}$$

Following [39] and references there we note that if $\varepsilon = 0$, the solutions of the wave equation can be written as functions of the characteristics $\xi = x - t$ and $\eta = x + t$. With given initial conditions we can write the solution as an explicit function of $\xi, \eta$. Introducing the characteristics as new independent variables we find

$$\frac{\partial^2 u}{\partial \xi \partial \eta} = f(\ldots), \tag{14.48}$$

with dots representing the transformed arguments of function $f$. Using this equation as starting point and *assuming* that the characteristics are timelike we can apply general averaging (Chap. 5). Examples in [148] and [51] are cases where $f = -u_t$ (damped waves), the cubic Klein-Gordon equation and the Rayleigh wave equation with $f = -u_t + \frac{1}{3}u_t^3$.

# References


1. Abadi, On self-excited auto-parametric systems. Nonlinear Dynam. **24**, 147–166 (2001)
2. C. Abdulwahed, F. Verhulst, Recurrent canards producing relaxation oscillations. CHAOS (2021). https://doi.org/10.1063/5.0040726
3. G.M. Andersen, J.F. Geer, Power series expansions for the frequency and period of the limit cycle of the van der Pol equation. SIAM J. Appl. Math. **42**, 678–693 (1982)
4. V.I. Arnold, *Geometric methods in the theory of ordinary differential equations* (Springer, New York, 1983) (orig. Russian ed. 1978)
5. V.I. Arnold, V.V. Kozlov, A.I. Neishtadt, Mathematical aspects of classical and celestial mechanics, in *Dynamical Systems III*, ed. by V.I. Arnold (Springer, New York, 1988)
6. T. Bakri, R. Nabergoj, A. Tondl, F. Verhulst, Parametric excitation in nonlinear dynamics. Int. J. Nonlinear Mech. **39**, 311–329 (2004)
7. T. Bakri, H.G.E. Meijer, F. Verhulst, Emergence and bifurcations of Lyapunov manifolds in nonlinear wave equations. J. Nonlinear Sci. **19**, 571–596 (2009). https://doi.org/10.1007/s00332-009--9045-2
8. T. Bakri, F. Verhulst, Bifurcations and quasi-periodic dynamics: torus breakdown. Z. Angew. Math. Phys. **65**, 1053–1076 (2014)
9. T. Bakri, Y.A. Kuznetsov, F. Verhulst, Torus bifurcations in a mechanical system. J. Dyn. Diff. Equ. **27**, 371–403 (2015)
10. T. Bakri, F. Verhulst, Time-reversal, tori families and canards in the Sprott A and NE9 systems. CHAOS (2022). https://doi.org/10.1063/5.0097508
11. T. Bakri, F. Verhulst, From A. Tondl's Dutch contacts to Neimark-Sacker-bifurcation. Appl. Comp. Mech. **16**, (2022). https://doi.org/10.24132/acm.2022.770
12. T. Bakri, F. Verhulst, *Emergence and approximation of tori*, in *Advances in Nonlinear Dynamics and Control of Mechanical and Physical Systems*, vol. 301, ed. by Belhaq (Springer Proc. Physics, New York, 2024), pp. 1–22
13. T. Bakri, F. Verhulst, The Sprott B system. Chaos **34**, 103116 (2024). https://doi.org/10.1063/5.0212565
14. T. Bakri, F. Verhulst, A note on the Kaplan-Yorke dimension. Int. J. Bif. Chaos (2025). https://doi.org/10.1142/S0218127425501330
15. D. Bambusi, On long time stability in Hamiltonian perturbations of nonresonant linear pde's. Nonlinearity **12**, 823–850 (1999)
16. P.W. Bates, K. Lu, C. Zeng, Existence and persistence of invariant manifolds for semiflows in Banach space. Memoirs AMS **135**, 645 (1998)





F. Verhulst, *Averaging for Nonlinear Dynamics with Applications and Numerical Bifurcations*, Applied Mathematical Sciences 223,
https://doi.org/10.1007/978-3-032-12745-7

17. P.W. Bates, K. Lu, C. Zeng, Persistence of overflowing manifolds for semiflow. Comm. Pure Appl. Math. **52**, 983–1046 (1999)
18. A. Ben Lemlih, J.A. Ellison, Method of averaging and the quantum anharmonic oscillator. Phys. Rev. Lett. **55**, 1950–1953 (1985)
19. A.I. Bobenko, S. Kuksin, The nonlinear Klein-Gordon equation on an interva l as a perturbed Sine-Gordon equation. Comment. Meth. Helvetici **70**, 63–112 (1995)
20. N.N. Bogoliubov, Yu.A. Mitropolsky, *Asymptotic Methods in the Theory of Nonlinear Oscillations* (Gordon and Breach, New York, 1961)
21. O. Bottema, On the stability of the equilibrium of a linear mechanical system. Z. Angew. Math. Phys. **6**, 97–104 (1955)
22. O. Bottema, The Routh-Hurwitz condition for the biquadratic equation. Indagationes Mathematicae **18**, 403–406 (1956)
23. H. Bohr, *Almost-periodic functions*, reprint (Chelsea, New York, 1947)
24. J. Bourgain, Construction of approximative and almost periodic solutions of perturbed linear Schrödinger and wave equations. GAFA **6**, 201–230 (1996)
25. H.W. Broer, G.H. Huitema, M.B. Seyvruk, *Quasi-periodic motions in families of dynamical systems*. Lect. Notes in Math., vol. 1645 (Springer, New York, 1996)
26. H.W. Broer, H.M. Osinga, G. Vegter, Algorithms for computing normally hyperbolic invariant manifolds. ZAMP **48**, 480–524 (1997)
27. H.W. Broer, M.B. Sevryuk, *KAM theory: quasi-periodicity in dynamical systems*, in Handbook of Dynamical Systems, vol. 3, Ch. 6, ed. by H.W, Broer, B. Hasselblatt, F. Takens (Elsevier, New York, 2010), pp. 251–344
28. H. Broer, F. Takens, *Dynamical Systems and Chaos*. Appl. Math. Sciences, vol. 172 (Springer, New York, 2011)
29. Robert F. Brown, *A Topological Introduction to Nonlinear Analysis*, 3rd edn. (Springer, New York, 2014)
30. R. Bruggeman, F. Verhulst, The inhomogenous Fermi-Pasta-Ulam chain. Acta Appl. Math. **152**, 111–145 (2017)
31. R. Bruggeman, F. Verhulst, *Dynamics of a chain with 4 particles, alternating masses and nearest-neighbour interaction*, in ed. by M. Belhaq. Recent Trends in Applied Nonlinear Mechanics and Physics, vol. 199 (Springer Proc. in Physics, New York, 2018)
32. R. Bruggeman, F. Verhulst, Near-integrability and recurrence in FPU chains with alternating masses. J. Nonlinear Sci. **29**, 183–206 (2019)
33. R. Bruggeman, F. Verhulst, *Optical-acoustical interaction in FPU Chains with alternating masses* (2020). https://doi.org/10.48550/arXiv.2004.03876
34. L.E.J. Brouwer, Beweging van een materieel punt op den bodem eener draaiende vaas onder den invloed der zwaartekracht (in Dutch). Nieuw Archief v. Wisk. 2e reeks **12**, 407–419 (2018). (English transl. The motion of a particle on the bottom of a rotating vessel under the influence of a rotating force, Brouwer Collected Works, North-Holland Publ. 1976)
35. P.F. Byrd, M.B. Friedman, *Handbook of Elliptic Integrals for Engineers and Scientists* (Springer, New York, 1971)
36. G.M. Chechin, V.P. Sakhnenko, Interaction between normal modes in nonlinear dynamical systems with discrete symmetry. Exact results. Physica D **117**, 43–76 (1998)
37. G.M. Chechin, N.V. Novikova, A.A. Abramenko, Bushes of vibrational normal modes for Fermi-Pasta-Ulam chains. Physica D **166**, 208–238 (2002)
38. G.M. Chechin, D.S. Ryabov, K.G. Zhukov, Stability of low-dimensional bushes of vibrational modes in the Fermi-Pasta-Ulam chains. Physica D **203**, 121–166 (2005)
39. S.C. Chikwendu, J. Kevorkian, A perturbation method for hyperbolic equations with small nonlinearities. SIAM J. Appl. Math. **22**, 235–258 (1972)
40. S.-N. Chow, J.A. Sanders, On the number of critical points of the period. J. Diff. Equ. **64**, 51–66 (1986)
41. S.-N. Chow, J.K. Hale, *Methods of Bifurcation Theory* (Springer, New York, 1982)
42. O. Christov, Non-integrability of first order resonances in Hamiltonian systems with three degrees of freedom. Celes. Mech. Dyn. Astr. **112**, 149–167 (2012)

43. O. Christov, Non-integrability of first order resonances in Hamiltonian systems in three degrees of freedom. Celes. Mech. Dyn. Astron. **112**, 149–167 (2012)
44. C. Ciocci, A. Litvak-Hinenzon, H.W. Broer. Survey on dissipative KAM theory including quasi-periodic bifurcation theory. Geometric Mech. Symmetry Peyresq Lect. **306**, 303–355 (2005)
45. E.A. Coddington, N. Levinson, *Theory of Ordinary Differential Equations* (McGraw-Hill Book, New York, 1955)
46. E.A. Coddington, N. Levinson, *Theory of Ordinary Differential Equations* (McGraw-Hill Book, New York, 1955)
47. R.L. Devaney, Homoclinic orbits in hamiltonian systems. J. Differ. Equ. **21**, 431–438 (1976)
48. R.E. Lee DeVille, C. Eugene Wayne, Reduced equations for models of laminated materials in thin domains. I. Asymptotic Anal. **42**, 263–309 (2005)
49. R.E. Lee DeVille, Reduced equations for models of laminated materials in thin domains. II. Asymptotic Anal. **42**, 311–346 (2005)
50. J.J. Duistermaat, Non-integrability of the 1 : 1 : 2-resonance. Ergodic Theory Dynam. Syst. **4**, 553–568 (1984)
51. W. Eckhaus, New approach to the asymptotic theory of nonlinear oscillations and wave propagation. J. Math. Anal. Appl. **49**, 575–611 (1975)
52. R.M. Evan Iwanowski, *Resonance Oscillations in Mechanical Systems* (Elsevier, New York, 1976)
53. J.D. Farmer, E. Ott, J.A. Yorke, The dimension of chaotic attractors. Physica D **7**, 153–180 (1983). Included in [80]
54. S. Fatimah, M. Ruijgrok, Bifurcation in an autoparametric system in 1:1 internal resonance with parametric excitation. Int. J. Non-linear Mech. **37**, 297–308 (2002)
55. P. Fatou, Sur le mouvement d'un système soumis á des forces á courte période. Bull. Soc. Math. **56**, 98–139 (1928)
56. M. Fečkan, A Galerkin-averaging method for weakly nonlinear equations. Nonlinear Anal. **41**, 345–369 (2000)
57. M. Fečkan, *Galerkin-averaging method in infinite-dimensional spaces for weakly nonlinear problems*, in ed. by H.R. Grosinho, M. Ramos, C. Rebelo, L. Sanches, Progress in Nonlinear Differential Equations and Their Applications, vol. 43 (Birkhäuser Verlag, Basel, 2001)
58. J.P. Fink, W.S. Hall, A.R. Hausrath, A convergent two-time method for periodic differential equations. J. Diff. Equ. **15**, 459–498 (1974)
59. J. Ford, The Fermi-Pasta-Ulam problem: paradox turns discovery. Phys. Rep. **213**, 271–310 (1992)
60. G. Fournier, J. Mawhin, On periodic solutions of forced pendulum-like equations. J. Diff. Equ. **60**, 381–395 (1985)
61. L. Galgani, A. Giorgilli, A. Martinoli, S. Vanzini, On the problem of energy partition for large systems of the Fermi-Pasta-Ulam type: analytical and numerical estimates. Physica D **59**, 334–348 (1992)
62. M. Golubitsky, I. Stewart, *The Symmetry Perspective* (Birkhäuser Verlag, Basel, 2000)
63. G. Gorelik, A. Vitt, Oscillations of an elastic pendulum as an example for two parametrically excited vibratory sysstems (Russ.). J. Techn. Phys. USSR col. **3**, (1933)
64. P. Grassberger, I. Procaccia, Measuring the strangeness of strange attractors. Physica D **9**, 189–208 (1983). Included in [80]
65. J. Grasman, *Asymptotic Methods of Relaxation Oscillations and Applications*. Appl. Math. Sci., vol. 63 (Springer, New York, 1987)
66. J. Guckenheimer, Ph. Holmes, *Nonlinear Oscillations, Dynamical Systems and Bifurcations of Vvector Fields*. Appl. Math. Sciences, vol. 42, 5th printing (Springer, New York, 1996)
67. F.G.. Gustavson, On constructing formal integrals of a Hamiltonian system near an equilibrium point. Astron. J. **71**, 670–686 (1969)
68. R. Haberman, Energy bounds for the slow capture by a center in sustained resonance. SIAM J. Appl. Math. **43**, 244–256 (1983)

69. J.D. Hadjidemetriou, Two-body problem with variable mass: a new approach. Icarus **2**, 440–451 (1963)
70. J.K. Hale, *Oscillations in Nonlinear Systems* (MacGraw-Hill, New York, 1963). Reprint Dover Publ., New York (1992)
71. J.K. Hale, *Ordinary Differential Equations* (Wiley-Interscience, New York, 1969)
72. G. Haller, *Chaos Near Resonance*. Appl. Math. Sci., vol. 138 (Springer, New York, 1999)
73. H. Hanßmann, R. Mazrooei-Sebdani, F. Verhulst, The1 : 2 : 4 resonance in a particle chain. Indag. Math. **32**, 101–120 (2021)
74. H. Hanßmann, R. Mazrooei-Sebdani, Relative equilibria of a 4- particle ring close to the 1:2:4 resonance. J. Nonlin. Sci. **35**(4):76, 1–46 (2025). https://doi.org/10.1007/s00332-025-10173-4
75. M. Henon, C. Heiles, The applicability of the third integral oi motion: some numerical experiments. Astron. J. **69**, 73–79 (1964)
76. J.J. Heijnekamp, M.S. Krol, F. Verhulst, Averaging in non-linear transport problems. Math. Methods Appl. Sci. **18**, 437–448 (1995)
77. M. Hirsch, C. Pugh, M. Shub, *Invariant Manifolds*. Lecture Notes in Math., vol. 583 (Springer, New York, 1977)
78. I. Hoveijn, F. Verhulst, Chaos in the 1 : 2; 3 Hamiltonian normal form. Physica D **44**, 397–406 (1990)
79. I. Hoveijn, M. Ruijgrok, The stability of parametrically forced coupled oscillators in sum resonance. Z. Angew. Math Phys **46**, 384–392 (1995)
80. B. Hunt, J.A. Kennedy, T.-Y. Li, H.E. Nusse (eds.), *The Theory of Chaotic Attractors* (Springer, New York, 2004)
81. S. Jafari, J.C. Sprott, S. Golpayegani, Elementary quadratic chaotic flows with no equilibria. Phys. Lett. A **377**, 699–702 (2013)
82. J.H. Jeans, *Astronomy and Cosmogony* (Cambridge University Press, Cambridge, 1928)
83. T. Kapitula, K. Promislow, *Spectral and Dynamical Stability of Nonlinear Waves*. Appl. Math. Sciences, vol. 185 (Springer, New York, 2013)
84. J.L. Kaplan, J.A. Yorke, *Chaotic Behaviour of Multidimensional Difference Equations*. Lecture Notes in Mathematics, vol. 730 (Springer, New York, 1979)
85. J.B. Keller, S. Kogelman, Asymptotic solutions of initial value problems for nonlinear partial differential equations. SIAM J. Appl. Math. **18**, 748–758 (1970)
86. J. Kevorkian, J.D. Cole, *Perturbation Methods in Applied Mathematics*. Appl. Math. Sci., vol. 34 (Springer, New York, 1981)
87. O.N. Kirillov, F. Verhulst, Paradoxes of dissipation-induced destabilization or who opened Whitney's umbrella? ZAMM. Z. Angew. Math. Mech. **90**(6), 462–488 (2010)
88. O.N. Kirillov, *Nonconservative Stability Problems of Modern Physics*. De Gruyter Studies in Mathematical Physics, vol. 14 (De Gruyter, Berlin, 2013)
89. O.N. Kirillov, F. Verhulst, *From rotating fluid masses and Ziegler's paradox to Pontryagin- and Krein spaces and bifurcation theory*, in Novel Mathematics Inspired by Industrial Challenges, ed. by M. Günther, W. Schilders. Mathematics in Industry, vol. 38. The European Consortium for Mathematics in Industry (Springer, New York, 2022), pp. 201–243
90. M.S. Krol, On a Galerkin-averaging method for weakly non-linear wave equations. Math. Methods Appl. Sci. **11**, 649–664 (1989)
91. M.S. Krol, On the averaging method in nearly time-periodic advection-diffusion problems. SIAM J. Appl. Math. **51**, 1622–1637 (1991)
92. C. Kuehn, *Multiple Time Scale Dynamics*. Appl Math. Sciences, vol. 191 (Springer, New York, 2015)
93. J. Kurzweil, Van der Pol perturbation of the equation for a vibrating string. Czech. Math. J. **17**, 558–608 (1967)
94. Yu.A. Kuznetsov, *Elements of Applied Bifurcation Theory*, rev. ed. (Springer, New York, 2004)
95. H. Lamb, On kinetic stability. Proc. Roy. Soc. A **80**, 168–177 (1908)

96. J.S.W. Lamb, J.A.G. Roberts, Time-reversal symmetry in dynamical systems: A survey. Physica D **112**, 1–39 (1998)
97. J. Laskar, Frequency analysis of a dynamical system. Cel. Mech. Dyn. Astron. **56**, 191–196 (1993)
98. B.V. Lidskii, E.I. Shulman, Periodic solutions of the equation $u_{tt} - u_{xx} = u^3$. Funct. Anal. Appl. **22**, 332–333 (1988)
99. J. Llibre, M. Messias, A.C. Reinol, Global dynamics and bifurcation of periodic orbits in a modified Nosé-Hoover oscillator. J. Dynam. Control Syst. **27**, 491–506 (2021)
100. A.A. Mailybaev, A.P. Seyranian, Instability of a general rotating system with small axial asymmetry and damping. J. Sound Vibr. **332**, 346–360 (2013)
101. L. Martinet, P. Magnenat, F. Verhulst, On the number of isolating integrals in resonant systems with 3 degrees of freedom. Celestial Mech. **25**, 93–99 (1981)
102. Matcont, A. Dhooge, W. Govaerts, Yu. Kuznetsov, W. Mestrom, A. Riet, B. Sautois, Matcont Continuation toolboxes in matlab (2022). Numerical continuation and bifurcation program, available at http://www.matcont.ugent.be
103. R.M. May, Simple mathematical models with very complicated dynamics. Nature **261**, 459–466 (1976)
104. M. Messias, A.C. Reinol, On the formation of hidden chaotic attractors and nested invariant tori in the Sprott A system. Nonlinear Dyn. **88**, 807–821 (2017)
105. M. Messias, A.C. Reinol, On the existence of periodic orbits and KAM tori in the Sprott A system: a special case of the Nosé-Hoover oscillator. Nonlinear Dyn. **92**, 1287–1297 (2018)
106. Y.A. Mitropolsky, G. Khoma, M. Gromyak, *Asymptotic Methods for investigating Quasiwave Equations of Hyperbolic Type* (Kluwer, New York, 1997)
107. E. Ott, Strange attractors and chaotic motions of dynamical systems. Rev. Mod. Phys. **53**(4), 655–671 (1981). Included in [80]
108. H. Pals, The Galerkin-averaging method for the Klein-Gordon equation in two space dimensions. Nonlinear Anal. **27**, 841–856 (1996)
109. H-O. Peitgen, H. Jürgens, D. Saupe, *Chaos and Fractals, New frontiers of Science* (Springer, New York, 1992)
110. H. Poincaré, *Les Méthodes Nouvelles de la Mécanique Céleste*, 3 vols (Gauthier-Villars, Paris, 1892, 1893, 1899)
111. H. Poincaré, Sur l'équilibre d'une masse fluide animée d'un movement de rotation. Acta Math. **7**, 259–380 (1885)
112. M.H. Protter, H.F. Weinberger, *Maximum Principles in Differential Equations* (Prentice-Hall Inc., Upper Saddle River, NJ, 1967)
113. R.H. Rand, W.I. Newman, B.C. Denardo, A.I. Newman, Dynamics of a nonlinear parametrically-excited partial differential equation. Proc. Design Eng. Techn. Conferences **3**, 57–68 (1995). ASME, DE-84-1 (see also Newman et al., Chaos **9**, 242–253 (1999))
114. B. Rink, F. Verhulst, Near-integrability of periodic FPU-chains. Physica A **285**, 467–482 (2000)
115. B. Rink, Symmetry and resonance in periodic FPU-chains. Comm. Math. Phys. **218**, 665–685 (2001)
116. J.A.G. Roberts, G.R.W. Quispel, Chaos and time-reversal symmetry, order and chaos in reversible dynamical systems. Phys. Rep. **21**, 63–177 (1992)
117. M. Roseau, *Vibrations nonlinéaires et théorie de la stabilité* (Springer, Berlin, Heidelberg, 1966)
118. D. Ruelle, F. Takens, On the nature of turbulence. Comm. Math. Phys. **20**, 167–192 (1971)
119. M. Ruijgrok, A. Tondl, F. Verhulst, Resonance in a rigid rotor with elastic support. ZAMM Z. angew. Math. Mech. **73**, 255–263 (1993)
120. J.A. Sanders, Are higher order resonances really interesting? Celestial Mech. **16**, 421–440 (1977)
121. J.A. Sanders, F. Verhulsts, Approximations of higher order resonances with an application to Contopoulos' model problem. Asymptotic Anal. LNM **711**, 209–228 (1979)

122. J.A. Sanders, R. Cushman, Limit cycles in the Josephson equation. SIAM J. Math. Anal. **17**, 495–511 (1986)
123. J.A. Sanders, F. Verhulst, J. Murdock, *Averaging Methods in Nonlinear Dynamical Systems*, rev. ed. (Springer, New York, 2007)
124. A.L. Shtaras, The averaging method for weakly nonlinear operator equations. Math. USSSR Sbornik **62**, 223–242 (1989)
125. J.C. Sprott, Some simple chaotic flows. Phys. Rev. E **50**, R647–R650 (1994)
126. J.J, Stoker, *Nonlinear Vibrations in Mechanical and Electrical Systems*, edition (1950) repr. (Wiley Classics Library Edition, New York, 1992)
127. A.C.J. Stroucken, Verhulst, The Galerkin-averaging method for nonlinear, undamped continous systems. Math. Meth. Appl. Sci. **9**, 520–549 (1987)
128. W. Thomson, P.G. Tait, *Treatise on Natural Philosophy, Vol. I, Part I*, New Edition (Cambridge Univ. Press, Cambridge 1879), pp. 387–391
129. A. Tondl, *On the Interaction Between Self-Excited and Parametric Vibrations* (National Research Institute for Machine Design, Bechovice, Monographs and Memoranda 25, Prague, 1978)
130. A. Tondl, M. Ruijgrok, F. Verhulst, R. Nabergoj, *Autoparametric Resonance in Mechanical Systems* (Cambridge University Press, Cambridge, 2000), 196 p.
131. J.M. Tuwankotta, F. Verhulst, Symmetry and resonance in Hamiltonian systems. SIAM J. Appl. Math. **61**, 1369–1385 (2000)
132. F. Verhulst, J.M. Tuwankotta, *A parametrically excited nonlinear wave equation*, in *Nonlinear Dynamics of Discrete and Continuous Systems*, ed. by A.K. Abramian et al. Advanced Structural Materials, vol. 139 (2020)
133. B. Van den Broek, F. Verhulst, Averaging techniques and the oscillator flywheel problem. Nieuw Arch. Wiskunde 4th Ser. **5**, 185–106 (1987)
134. B. Van den Broek, *Studies in nonlinear resonance, appiications of averaging*. PhD thesis, University of Utrecht (1988)
135. E. Van der Aa, First-order resonances in three-degrees-of-freedom systems. Celestial Mech. **31**, 163–191 (1983)
136. E. Van de Aa, F. Verhulst, Asymptotic integrability and periodic solutions of a Hamiltonian system in 1 : 2 : 2 resonance. SIAM J. Math. Anal. **15**, 890–911 (1984)
137. E. Van der Aa, M.S. Krol, *A weakly nonlinear wave equation with many resonances*. PhD thesis M.S. Krol, University of Utrecht (1990)
138. A.H.P. Van der Burgh, On the asymptotic solutions of the differential equations of the elastic pendulum. J. de Mécanique (4), 507–520 (1968)
139. A.H.P. Van der Burgh, *Studies in the asymptotic theory of nonlinear resonance*. PhD thesis, Technical University Delft (1974)
140. A.H.P. Van der Burgh, On the asymptotic approximations of the solutions of a system of two non-linearly coupled harmonic oscillators. J. Sound Vibr. **49**, 93–103 (1976)
141. W. van der Kallen, F. Verhulst, Explorations for alternating FPU chains with large mass. Int. J. Non-Linear Mech. **137**, 103809 (2021)
142. W.T. Van Horssen, Asymptotics for a class of semilinear hyperbolic equations with an application to a problem with a quadratic nonlinearity. Nonlinear Anal. TMA **19**, 501–530 (1992)
143. F. Verhulst, Asymptotic expansions in the perturbed two-body problem with applications to systems with variable mass. Celestial Mech. **11**, 95–129 (1975)
144. F. Verhulst, Discrete symmetric dynamical systems at the main resonances with applications to axi-symmetric galaxies. Phil. Trans. Roy. Soc. Lond. **290**, 435–465 (1979)
145. F. Verhulst, On averaging methods for partial differential equations, in *SPT98-Symmetry and Perturbation Theory II*, ed. by A. Degasperis, G. Gaeta (World Scientific, Singapore, 1999), pp. 79–95
146. F. Verhulst, *Nonlinear Ordinary Differential Equations and Dynamical Systems*, 2nd rev. ed. (Springer, New York, 2000)

147. F. Verhulst, Invariant manifolds in dissipative dynamical systems. Acta Applicandae Mathematicae **87**, 229–244 (2005)
148. F. Verhulst, *Methods and Applications of Singular Perturbations, Boundary Layers and Timescale Dynamics*. Texts in Applied Mathematics, vol. 50 (Springer, New York, 2005)
149. F. Verhulst, Brouwer's rotating vessel I: stabilization. Z. Angew. Math. Phys. **63**, 727–736 (2012). https://doi.org/10.1007/s00033-011-0177-5
150. F. Verhulst, *Henri Poincaré, Impatient Genius* (Springer, New York, 2012)
151. F. Verhulst, Integrability and Non-integrability of hamiltonian normal forms. Acta Applicandae Mathematicae **137**, 253–272 (2015)
152. F. Verhulst, Profits and pitfalls of timescales in asymptotics. SIAM Rev. **57**, 255–274 (2015)
153. F. Verhulst, Near-integrability and recurrence in FPU cell-chains. Int. J. Bifurcat. Chaos **26**(14), (2016). https://doi.org/10.1142/S0218127416502308
154. F. Verhulst, Interaction of low and higher order Hamiltonian resonances. Int. J. Bifurcat. Chaos **28**(08), (2018). https://doi.org/10.1142/S0218127418500979
155. F. Verhulst, Recurrence and resonance in the cubic Klein-Gordon equation. Acta Appl Math. **162**, 145–164 (2019). https://doi.org/10.1007/s10440-019-00238-4
156. F. Verhulst, Evolution to mirror-symmetry in rotating systems. Symmetry (MDPI) **13**, 1189–1208 (2021)
157. F. Verhulst, High-low frequency interaction in alternating FPU $\alpha$-chains. Int. J. Non-Linear Mech. **131**, (2021). https://doi.org/10.1016/j.ijnonlinmec.2021.103686
158. F. Verhulst, *A Toolbox of Averaging Theorems, Ordinary and Partial Differential Equations*. Surveys and Tutorials in the Applied Mathematical Sciences, vol. 12 (Springer, New York, 2023)
159. F. Verhulst, T. Bakri, Nonintegrability of 3 DoF Hamiltonian resonant systems. Int. J. Bifurcat. Chaos **34**, (2024). https://doi.org/10.1142/S0218127424501682
160. D. Viswanath, The fractal property of the Lorenz attractor. Physica D **190**, 115–128 (2004)
161. G.B. Whitham, *Linear and Nonlinear Waves* (Wiley, New York, 1974)
162. V.A. Yakubovich, V.M. Starzhinskii, *Linear Differential Equations with Periodic Coefficients*, vols 1 and 2 (Wiley, New York, 1975)
163. G.M. Zaslavsky, *The Physics of Chaos in Hamiltonian Systems*, 2nd ext. ed. (Imperial College Press, London, 2015)

# Index




F. Verhulst, *Averaging for Nonlinear Dynamics with Applications and Numerical Bifurcations*, Applied Mathematical Sciences 223,
https://doi.org/10.1007/978-3-032-12745-7

MIX
Papier aus verantwortungsvollen Quellen
Paper from responsible sources
FSC® C105338

If you have any concerns about our products,
you can contact us on
**ProductSafety@springernature.com**

In case Publisher is established outside the EU,
the EU authorized representative is:
**Springer Nature Customer Service Center GmbH**
**Europaplatz 3, 69115 Heidelberg, Germany**

Printed by Libri Plureos GmbH
in Hamburg, Germany